BIOCOMPATIBILITY
Interactions of Biological and Implantable Materials

Volume 1

Polymers

BIOCOMPATIBILITY

Interactions of Biological and Implantable Materials

By
Frederick Silver and Charles Doillon

Volume 1
Polymers

VCH

Frederick H. Silver and Charles Doillon
Biomaterials Center
University of Medicine and Dentistry of
New Jersey
Robert Wood Johnson Medical School
675 Hoes Lane
Piscataway, NJ 08854-5635

Library of Congress Cataloging-in-Publication Data

Silver, Frederick H., 1949-
Biocompatibility : interactions of biological and implantable materials / by Frederick Silver and Charles Doillon.
p. cm.
Bibliography: p.
Includes index.
Contents: v. 1. Polymers.
ISBN 0-89573-317-X (v. 1)
1. Biomedical materials. 2. Biocompatibility.
I. Doillon, Charles. II. Title.
R857.M3S55 1989
617'.9—dc20 89-14828
CIP

British Library Cataloguing in Publication Data

Silver, Frederick H. 1949-
Biocompatibility.
Volume 1. Polymers
1. Medicine. Implantable materials
I. Title II. Doillon, Charles
610'.28

ISBN 0-89573-317-X

Printed in the United States of America.
ISBN 0-89573-317-X VCH Publishers
ISBN 3-527-26861-8 VCH Verlagsgesellschaft

Printing History:
10 9 8 7 6 5 4 3 2 1

Published jointly by:

VCH Publishers, Inc.
220 East 23rd Street
Suite 909
New York, New York 10010

VCH Verlagsgesellschaft mbH
P.O. Box 10 11 61
D-6940 Weinheim
Federal Republic of Germany

VCH Publishers (UK) Ltd.
8 Wellington Court
Cambridge CB1 1HW
United Kingdom

Table of Contents

Part II. Biomaterial End-Use Applications

Chapter 8. Wound Dressings and Artificial Skin

Chapter 9. Replacement of Skeletal Soft Tissue Structures

Preface

Although the use of animal skin and other biologically derived materials dates back centuries, the field of biomaterials has become a scientific discipline only during the last three decades. Most of our understanding of how materials interact with cells and tissues was gained by trial-and-error attempts to replace body tissues using "off-the-shelf" materials. This book attempts to characterize interactions between materials and tissues in terms of basic chemical, physical, and biological concepts. Volume 1 of this series covers biological response mechanisms as well as interactions with polymeric implant materials. Structure and properties of metals, ceramics, and composites used in implants are discussed in Volume 2 in addition to end-use applications of these materials.

The challenge for the student of biomaterials science lies in piecing together the available information to provide practical and timely solutions to important medical problems. As our understanding of basic biological science and pathobiology expands we will be better able to implement improved prototype designs. Although many of the biomaterials in use today are relatively simple by design, they have made a significant impact on delivery of health care. It is important for the student in this field to recognize that simple effective designs can be implemented as part of health care maintainence more easily than complex ones.

It cannot be emphasized enough that the future of biomaterials as a science is intimately tied to our ability to understand the basic discoveries in chemistry, biology, and medicine that at are made daily. However, because of the prolonged development time required for design and testing of new medical devices it is essential that ideas that derive from basic science concepts be focused into practical end-use applications as early as possible. The ability to recognize practical applications of basic science is a lifelong endeavor. The goal of this book is to attempt to provide the knowledge needed to begin this task.

The authors would like to acknowledge Y. Pedro Kato and Michael Marks, graduate students in biomedical engineering, for their assistance in preparing and proof reading the manuscript. We wish to thank Ms. Maria Ayash for spending countless hours typing drafts as well as David Christiansen who organized the preparation of the camera ready copy. Finally, we are grateful to the numerous graduate students at University of Medicine and Dentistry and Rutgers University enrolled in courses entitled "Introduction to Biomaterials", "Biopolymers", and "Artificial Implantable Materials" whose comments were useful in converting class notes into the final manuscript.

Part I - Biomaterials Science

Chapter 1. An Overview

1.1. Introduction

The field of biomaterials has advanced rapidly in the twentieth century mainly as a result of pioneering attempts to replace body tissue with natural and man-made materials (Figures 1.1 and1.2). For instance the use of amnion and placenta, two biological tissues, for wound dressings dates back to the early 1900s (see Thompson and Park,1984, and MacMillan, 1984 for reviews). Tubes of Dacron and Teflon, two man-made plastics, have been used since the 1950s to replace large-diameter blood vessels such as the aorta (see Myer and Parsonnet, 1969, for a review of early work). Medical advances that have prolonged the average life span have generated more need for materials that can be used as tissue and organ replacements. It is likely that during our lifetime we will see the routine clinical use of host cell-seeded artificial organs for such applications as artificial blood vessels, skin, tendon, and ligament, cornea, esophagus, gum, bone, cartilage, pancreas, nerves, and other tissues and organs.

The biomaterials scientist of the 1990s needs to be familiar with the chemistry, physics, biology, and engineering concepts required to design and evaluate materials to replace tissues and organs. The interdisciplinary nature of this field will attract students skilled at assembling information from different scientific disciplines and piecing it together to find the best and most practical engineering solution (Figure 1.3).

The approach used in this text will initially emphasize the basic concepts of chemistry, biology, and medicine necessary to understand the structure and function of body tissues and organs as well as to introduce such pathobiological processes as response to injury that are inherent to mammals.

Chapter 2 of this volume will review the structure of a few of the many biological tissues and organs that have been the subject of interest to researchers in the biomaterials community. These tissues will include skin, tendon,

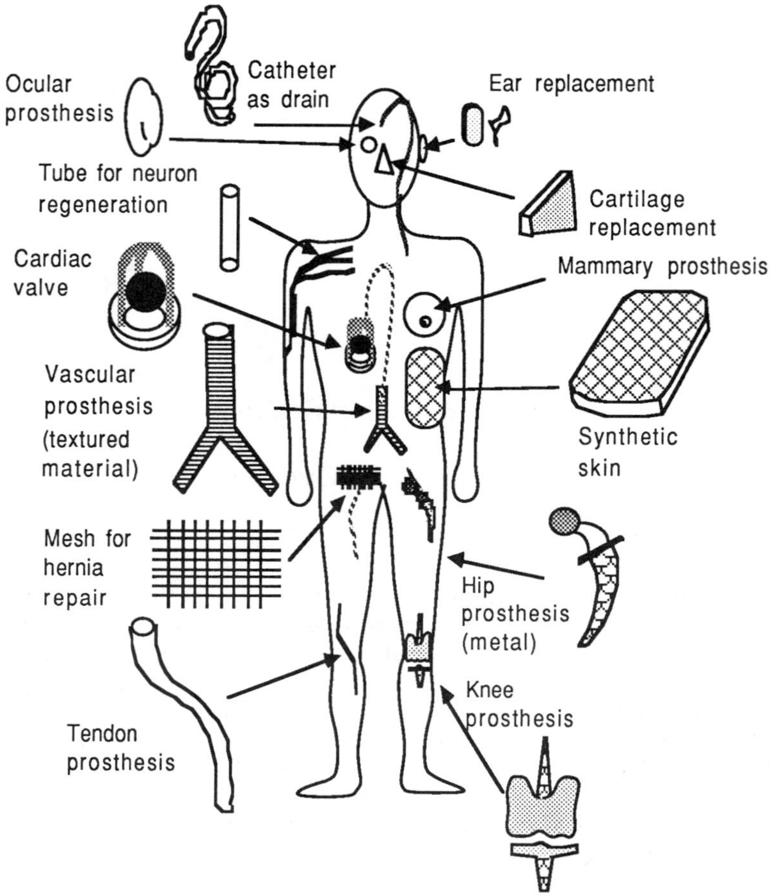

Figure 1.1. Types of tissues replaced with medical devices. These tissues include, eye, ear, nasal, breast, skin, nerve, heart valve, vessel, abdominal wall, tendon, hip joint, and knee joint.

cartilage, bone, and cardiovascular tissues. We will not attempt to cover every tissue and organ of potential interest to the biomaterials scientist. However, we will try to outline some of the properties that require consideration in designing replacements for tissues and organs (Figure 1.4).

Of primary concern to the biomaterials scientist is the biological response that a material induces in either tissues, blood, or other body fluids. Biological responses include changes in cell structure, state of metabolism, and degree of

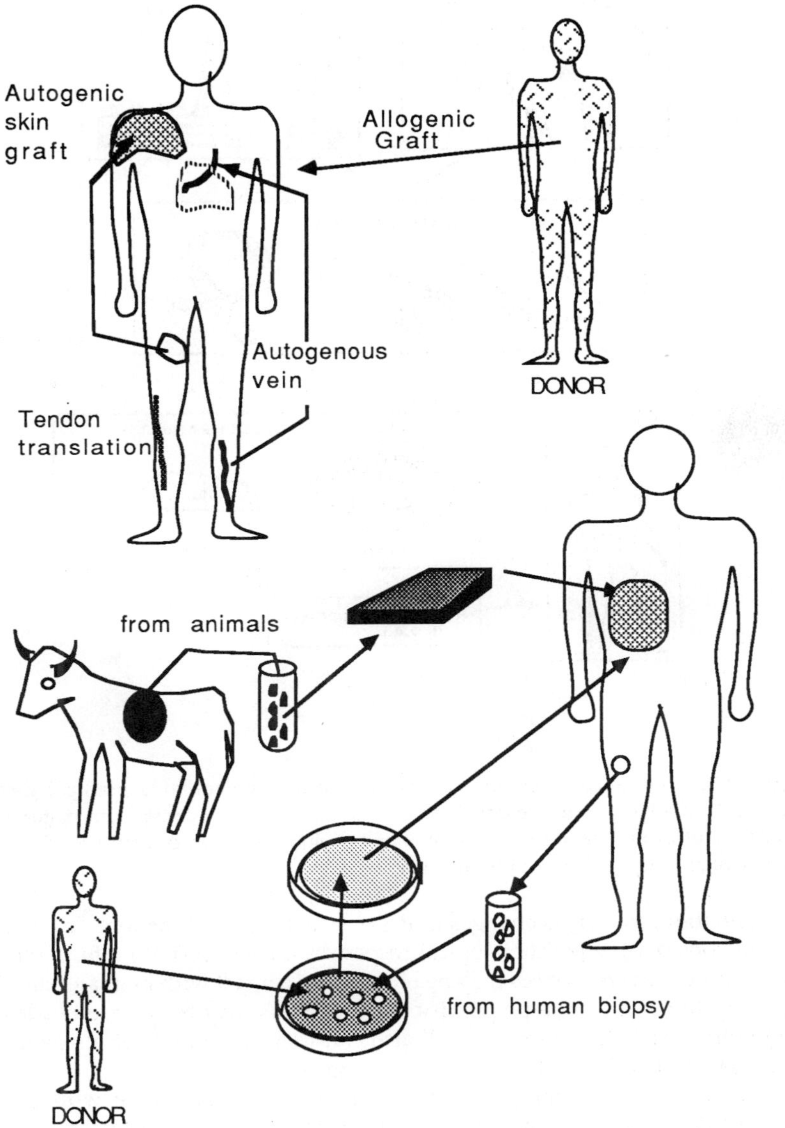

Figure 1.2. Sources of biological materials used to fabricate medical devices. These sources include tissue grafts between different areas of the body (autografts), grafts from other humans (allografts), tissues from animals, and cells expanded in culture.

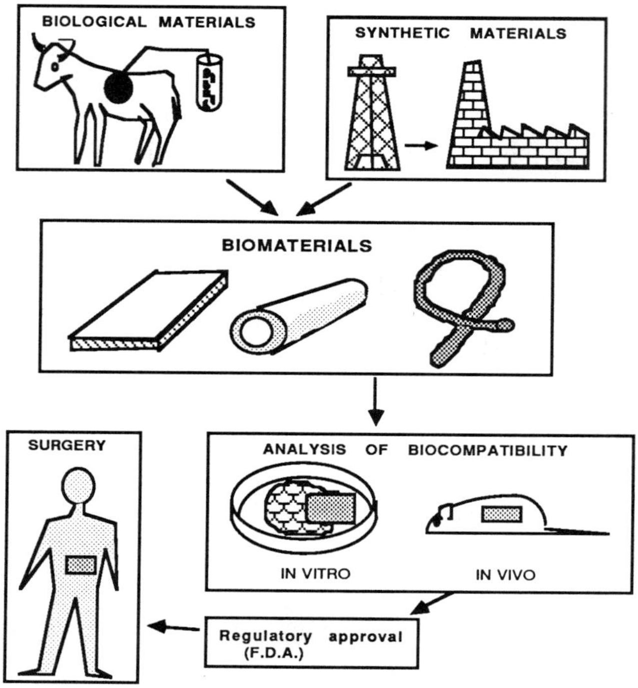

Figure 1.3. Sequence of events involved in development of a medical device. The events involved in device development include: choice of constituent materials, fabrication of prototype designs, analysis of biocompatibility in cell culture and in animal models, regulatory review, and human testing.

differentiation, as well as changes in tissue structure and function. In Chapter 3 we will consider the pathobiological responses associated with injury and discuss how these responses may be induced by contact with biomaterials. It is necessary to consider the possibility that materials may be released or leached from biomaterials, therefore we will also consider how biomaterials degrade in the presence of body fluids and enzymes (Figure 1.5).

One of the most important biological responses is that of wound healing since most biomaterials and devices are used in clinical situations in which a wound must be made to install the material or device. In other situations, the presence of a disease state such as mechanical loss of skin automatically triggers the repair process. In Chapter 4 we will focus on natural biological processes necessary to bring about the filling in of defects in tissues and organs. In addition, we will discuss the changes in wound-healing processes that are elicited by the presence of a material or a device (Figure 1.6).

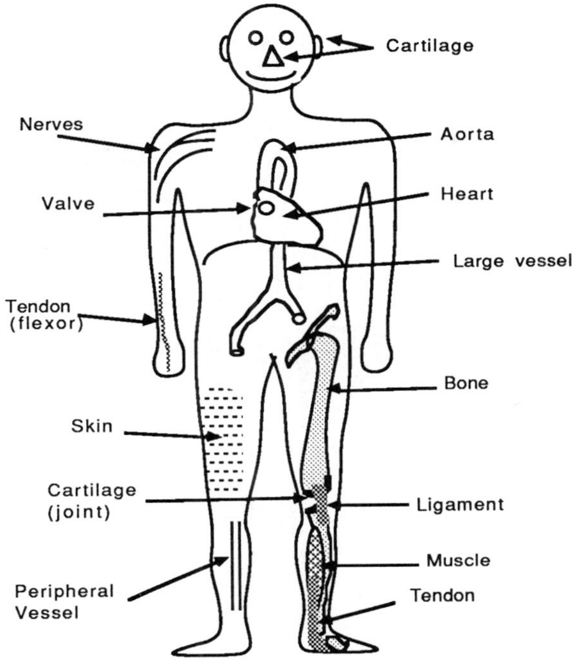

Figure 1.4. Design of replacement tissues. Engineering criteria used in the design include physiology of tissue, mechanical properties, biocompatibility, and end-use functioning of a medical device.

Once we have dealt with the properties of some tissues and organs and the pathobiological responses to injury, we will examine the types of polymers that have been used in medical devices. Chapter 5 deals with the chemistry and physics of potential tissue-replacement materials as well as their biodegradation products (Figure 1.7). Chapter 6 covers the types of tests used to characterize materials to be used in devices. Since the devices that we will discuss in this book contain polymers, we will focus on methods for characterizing polymeric materials (Figure 1.8). Chapter 7 introduces methods for evaluating the biocompatibility of materials using laboratory tests (*in vitro* testing) and animal tests (*in vivo* testing). Several examples are given of the results of biocompatibility testing and how test results are used to modify the design of devices (Figure 1.3).

Part II of this text deals with the engineering aspects of biomaterials design. A brief review of the evolution of prototype designs will be given for tissues, including skin, tendons, and blood vessels, and for drug delivery systems (Figure 1.1). Chapter 8 will introduce the concepts that have evolved over the last two decades in artificial skin design. These include advances in both coverings or

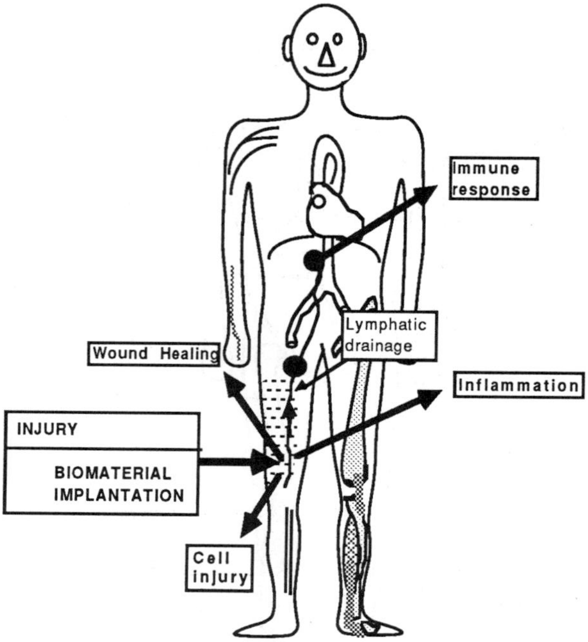

Figure 1.5. Biological response mechanisms. Response mechanisms that are used to evaluate the efficacy of a medical device include cell injury, inflammation, wound healing, drainage of the surrounding tissue, and the immune response to an implant.

dressings as well as cell-seeded replacements for the upper (epidermis) and lower (dermis) layers of skin.

The fields of orthopaedic and facial biomaterials are quite dynamic. In Chapter 9, we will review some of the exciting work in tendon, ligament, and cartilage replacement using polymers.

The field of cardiovascular materials is another prominent area of biomaterials. Chapter 10 reviews the recent progress in developing artificial blood vessels and materials to line the pumping chamber of artificial hearts and assist devices. Clearly the recent successes in clinical use of the artificial heart have been in part attributable to advances in blood compatible materials.

During the last decade, increased interest has arisen in the area of controlled drug delivery and sustained release of biologically active factors such as growth factors, anti—inflammatory agents, and chemotherapeutic agents. Chapter 11 deals with the types of polymers and devices that have been developed for transdermal, intramuscular, and intradermal release of drugs and other factors.

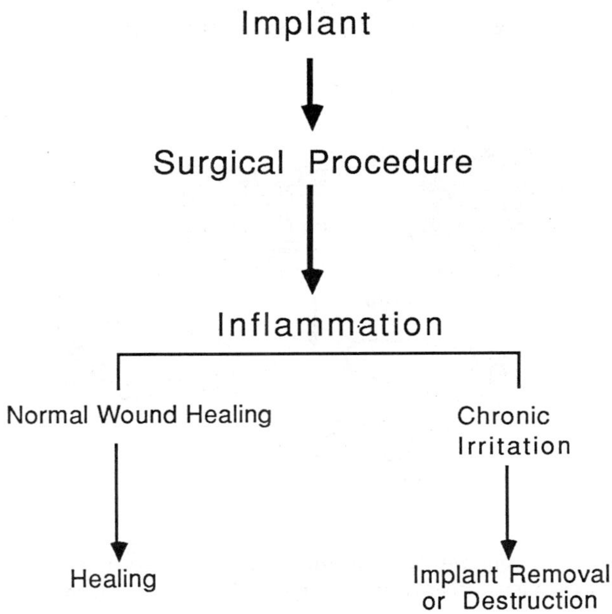

Figure 1.6. Implant wound healing interaction. Diagram showing the relationship between wound healing, acute and chronic inflammation, and implant functioning.

1.2. Market Size

The field of biomaterials has considerable commercial implications. As shown in Table 1.1 the number of procedures per year involving blood dialysis has been estimated to be over 6 million compared to 50,000-100,000 procedures involving vascular replacement, 100,000 procedures involving breast procedures; over 200,000 procedures involving hips, knees, and joints, and 150,000 procedures for coronary bypass.

The estimated 1986 U.S. markets for these products is about 2.58 billion dollars. Table 1.2 lists by type some market estimates for procedures involving biomaterials. Dialysis, orthopaedics and cardiac pacers each have a U. S. market of 600 million dollars a year. U.S. markets for special surgery and cardiovascular and wound care products are 145-240 million dollars each. Worldwide market figures are approximately twice those of the United States. These numbers suggest that the world wide biomaterials market is large and in the neighborhood of 5 billion dollars per year. The continued growth of these markets will depend on continued advancement in the area of biocompatibility.

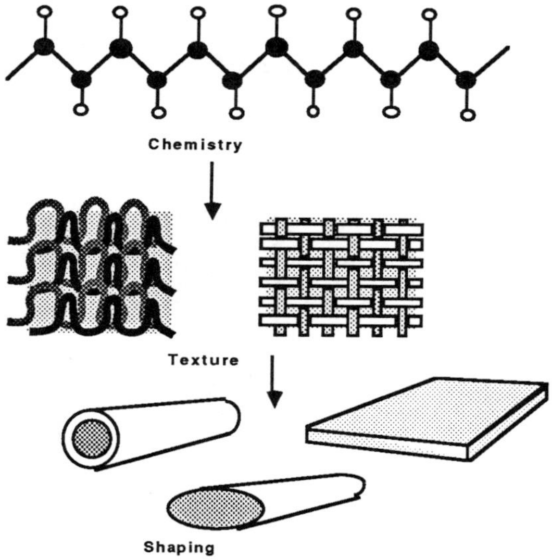

Figure 1.7. Steps in implant fabrication. Diagram illustrating the relationship between polymer chemistry, fiber structure, and implant physical form.

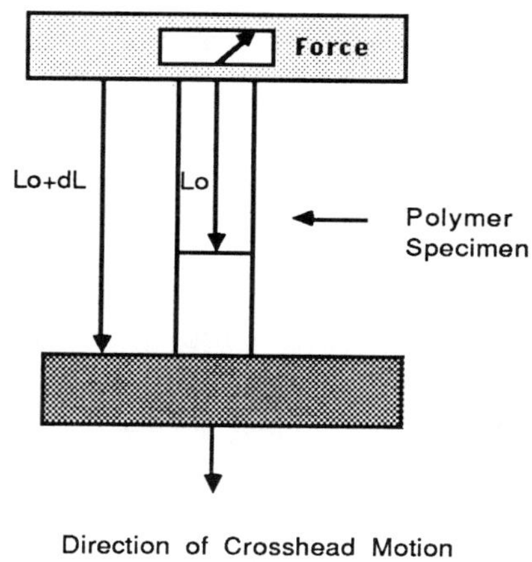

Figure 1.8. Mechanical properties of an implant. Diagram demonstrating mechanical testing of a polymer specimen in uniaxial tension. A specimen is clamped to a moving rigid frame (crosshead) that is lowered at a constant rate stretching the specimen. The upper end of the specimen is clamped into grips attached to a load cell which determines the resultant force.

Table 1.1. Biomaterials Applications in the U.S.

Application	Number/year
Cardiovascular implants	
Heart valves	30,000
Pacemakers	100,000
Vascular grafts	50,000 - 150,000
Dental implants	10,000 - 20,000
Plastic and reconstructive implants	
Breast prostheses	100,000
Cosmetic	78,000
After mastectomy	22,000
Nose, chin, other prostheses	10,000
Penile prostheses	5,000
Opthalmic implants	
Lenses	250,000
Retinal surgery	25,000 - 50,000
Other prostheses	7,000 - 8,000
Orthopedic prostheses	
Hips	110,000
Knees	65,000
Shoulder, finger joints, other areas	50,000
Neurosurgical implants	
Ventricular shunts	21,500
Extracorporeal devices	
Oxygenators	196,000
For coronary bypass procedures	150,000
Renal dialyzers	6,200,000

From Seltzer (1982)

Table 1.2. Economic Impact of U.S. Biomaterials Market

	Sales (in millions of dollars)			Growth (%/year)	
	1976	1981	1986	1976 - 1981	1981 -1986
Dialysis products	160	505	645a	25	5
Orthopedics	220	480	640a	17	6
Cardiac pacers	150	375	660a	20	12
Special surgery	60	120	240a	15	15
Cardiovascular	40	95	145a	19	9
Anesthesia/respiratory	45	90	115a	15	5
Neuromuscular stimulation	0	40	80a	NA	15
Extracorporeal therapy	0	16	50a	NA	25
Drug delivery	0	2	10a	NA	?
Wound care products			210b		

[a] From Ketteringham et. al. (1982).
[b] From Health Industries Handbook (1984).

1.3. Terms and Definitions

Before we begin discussing the structure of different tissues, their mechanical properties, and design criteria for biomaterials, it is first necessary to introduce a few terms that will be used throughout this text.

1.3.1. Components of Tissues

All tissues are composed of (1) fibrous or structural components that give shape to tissues, (2) cells that metabolize nutrients to produce energy and conduct a variety of chemical reactions, (3) cell attachment factors that link cells to structural material, and (4) interfibrillar material that acts as a filler between the fibrous materials as well as a reservoir for water and ions (Table 1.3).

Collagen fibers are the major tissue structural component and are made up of types I, II, and III collagen; however, there are at least nine other types of collagen found in tissues; some of these form fibers (Table 1.4). Cellular attachment factors include fibronectin, laminin, and chondronectin and are secreted by fibroblasts and other types of cells listed in Table 1.5

Proteoglycans are found on the surface of cells, in the interfibrillar matrix, and in tissue fluids such as that found in joint spaces. Proteoglycans are composed of a protein core to which glycosaminoglycan (GAG) side chains are attached. The glycosaminoglycan side chains are polymers of an amino sugar and a hexuronic acid. Table 1.6 lists the different types of GAGs and their tissue location. Hyaluronic acid and heparin are the only GAGs that are not covalently bound to a core protein. Throughout this text we will refer to cells

Table 1.3. Components of Tissues

	Composition	Location
Structural Components		
Collagen fibers	Protein composed of a triple helix	Extracellular banded fiber in all tissues
Elastic fibers	Protein core of amorphous elastin and crystalline component of microfibrillar protein	Extracellular nonbanded fibers found in skin and vessel walls
Inorganic minerals	Hydroxyapatite	Extracellular crystals surrounding collagen fibers (bone, tooth)
Cellular components	Nucleic acids, proteins, lipids and carbohydrates	Line internal and external surfaces of body, line inner surfaces of blood vessels and glands, intermix with fibrous and nonfibrous components of tissues
Cell attachment factors	Glycoproteins	Mediate attachment of cells that line surfaces with fibrous and nonfibrous components
Interfibrillar matrix	Proteoglycans, water, ions, and polysacchrides	Found in the spaces between fibers outside the cell

and cellular components. Table 1.7 lists the cellular components that will be discussed in terms of biocompatibility of implants. These include cell membrane, cytoplasm, and organelles. The organelles include endoplasmic reticulum, mitochondria, ribosomes, Golgi apparatus, lysosomes, and nucleus. The plasma membrane maintains the ionic and chemical gradients between intracellular and extracellular environments. The membrane carries specific markers that are sites of recognition and receptors for hormones and other molecules. The cytoplasm is liquid like and contains enzymes required for generation of energy (ATP) in the absence of oxygen. It also contains the machinery (ribosomal large and small subunits) for protein synthesis. Ribosomes are particles found either in the cytoplasm or bound to membranes of the endoplasmic reticulum. In the presence of mRNA (messenger RNA) and the large and small ribosomal subunits protein synthesis can occur.

The endoplasmic reticulum is a system of membrane limited channels (40-70 nm in diameter) that in association with ribosomes serve as a site of protein synthesis. Proteins synthesized in the endoplasmic reticulum in the presence of

Table 1.4. Location and Tissue Source of Different Collagens

Collagen type	Tissue or Organ	Location
I	Tendon, skin, bone, and fascia	Thick extracellular fibrils and fibers
II	Cartilage	Thin fibrils around cartilage cells
III	Cardiovascular tissue	Intermediate size extracellular fibrils
IV	Basement membranes	Network forming component
VI	Tendon, skin, and cardiovascular tissue	Pericellular matrix around cells
VII	Skin	Anchoring fibrils
VIII	Cardiovascular tissue	Around endothelial cells
IX	Cartilage	Extracellular matrix
X	Mineralizing cartilage	Extracellular matrix
XI	Cartilage	Extracellular matrix
XII	Tendon	

From Martin et al, (1985) and Dublet and Van Der Rest (1987).

ribosomes are targeted for transport extracellularly. These proteins are transported through the endoplasmic reticulum into the Golgi apparatus where they are packaged into vesicles that are later discharged outside of the cell. At this point sugars or lipids are combined with proteins to produce glycoproteins and lipoproteins, respectively. The information required to synthesize proteins as well as that required to regulate cell division is found within the genetic material (DNA) contained in the nucleus. Finally, lysosomes are membrane-lined vesicles containing hydrolytic enzymes that breakdown material intra- and extracellularly. These enzymes are important in determining the rate of degradation of implants as well as the degree of biocompatibility.

Table 1.5. Cells Found in Connective Tissue

Cell	Location	Cell attachment factors
Chondrocytes	Cartilage	Chondronectin, anchorin
Epithelial	Cover exposed body surface; skin	Laminin, epibolin
Endothelial	Line inside of blood vessels	Laminin
Smooth muscle	Found in vessel walls	Fibronectin
Fibroblasts	Found in skin, tendon, and cardiovcascular tissue	Fibronectin, vitronectin
Skeletal muscle	Found connected to tendon and ligaments	Fibronectin (?)
Nerve	Central and peripheral nervous system	Laminin

1.3.2. Components of Implants

Implant materials include polymers (also termed macromolecules), metals, ceramics, and composites of two or more of these materials. In this book we will focus on polymers that are used as implant materials for replacement of tissue. Polymers are materials that are synthesized by the addition of units termed monomers. In the simplest case the monomers are added together via a free radical mechanism such as that diagrammed in Figure 1.9A for polymerization of ethylene. Polymers are also formed using another type of polymerization process termed condensation polymerization. Condensation polymerization involves a chemical reaction between two monomers that releases a by-product such as water. For example, protein synthesis involves the condensation of amino acids to form a polypeptide chain as diagrammed in Figure 1.9B. Table 5.1 lists some of the repeat units found in synthetic and natural polymers used as biomaterials.

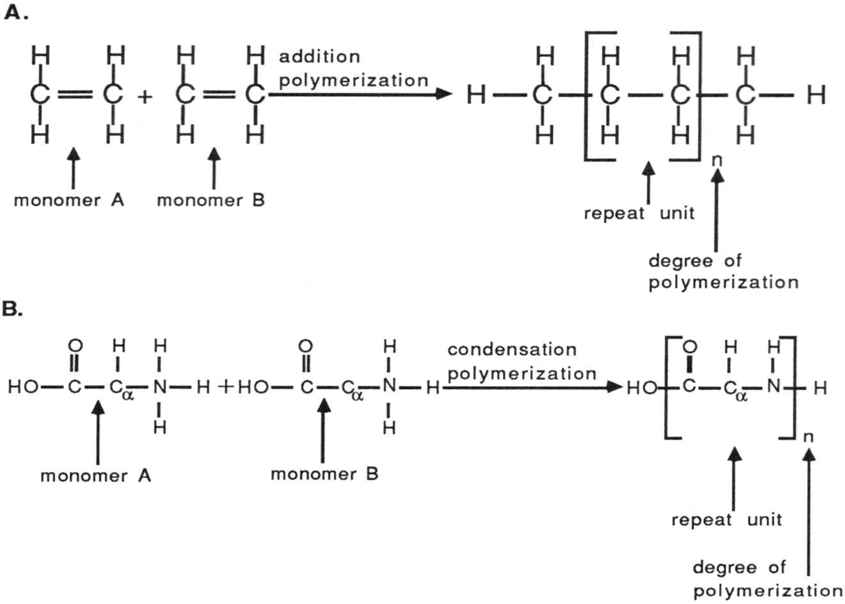

Figure 1.9. Polymerization processes. Diagram showing formation of polymer chains by addition and condensation polymerization. Formation of an addition polymer (A) from a monomeric unit of ethylene. The repeat unit of the polymer chain is shown in brackets and n is the degree of polymerization. Condensation polymerization (B) of a peptide chain from two amino acids of glycine results in the release of a mole of water as a by-product.

1.3.3. Physical Properties of Polymers Used as Biomaterials

Both natural and synthetic polymers exhibit characteristic properties depending on temperature. At high temperatures all polymeric chains exhibit maximum mobility and entropy and minimum bonding and internal energy. The chemical nature of the repeat unit of a polymer chain dictates the conformational freedom of the chain. For polypeptides, conformation maps (termed Ramachandran plots) have been constructed based on the coordinates of atoms in a peptide unit and minimum atomic contact distances. These plots indicate that the conformational freedom (a single conformation is dictated by the rotational states of each polypeptide backbone bond) is dependent on the size of the groups of atoms attached to the backbone. Figure 1.10 a shows a conformational plot for a polypeptide of glycine (R group in Figure 1.10 is a hydrogen). As the area of allowed conformations of the conformational map is decreased (large R groups) the entropy of the polymeric chain is decreased.

Table 1.6. Generalized repeat Units of Glycosaminoglycans

	Disaccharide repeating unit			
	Hexuronic acid	**Hexosamine**	**Sulfate**	**Tissue location**
Hyaluronic acid	D-glucuronic acid	D-glucosamine		Umbilical cord, skin, developing tissues, synovial fluid
Chondroitin 4-sulfate	D-glucuronic acid	D-galactosamine	0-sulfate	Found as mixture in articular cartilage, tendon, skin, cornea
Chondroitin 6-sulfate	D-glucuronic acid	D-galactosamine	0-sulfate	Found as mixture in articular cartilage, tendon, skin, cornea
Dermatan sulfate	L-iduronic acid or D-glucuronic acid	D-galactosamine	0-sulfate	Skin, cartilage, lung, aorta, tendon
Keratan sulfate	D-galactose	D-glucosamine	0-sulfate	Cornea, skeletal tissues
Keratan sulfate	D-glucuronic acid or L-iduronic acid	D-glucosamine	0-sulfate and N-sulfate	Cornea, skeletal tissues
Heparin	D-glucuronic acid or L-iduronic acid	D-glucosamine	0-sulfate and N-sulfate	Inside most cells
Heparin Sulfate				Basement membranes, liver, plasma membranes

Table 1.7. Cell Constituents and Their Functions

Component	Structure	Function
Cell membrane	Lipid bilayer containing peripheral and integral proteins	Maintains ionic balance and chemical gradients carries cell-specific markers and receptors participates in intercellular communication regulates cell division
Cytoplasm	Liquid like	Contains enzymes for generation of ATP in the absence of oxygen as well as protein synthesis
Endoplasmic reticulum	Branching system of membrane limited channels inside of cell	Cellular site of biosynthesis for proteins transported extracellularly
Mitochondria	Cigar shaped organelles	Enzymatic reactions of tricitric acid cycle, respiratory chain, and oxidative phosphorylation occur converting glucose to ATP in the presence of oxygen
Ribosomes	Large and small subunits	Involved in protein synthesis
Golgi apparatus	Membrane-bound tubular system forming stacks	Involved in transport and packing of proteins for extracellular discharge
Lysosomes	Membrane-covered vesicles 0.1-0.8 μm in diameter	Contain hydrolytic enzymes for digestion of both intracellular and extracellular materials
Nucleus	Membrane-limited region within cell	Contains genetic material (chromosomes), DNA that directs protein synthesis and cell division

From Sheeler and Bianchi (1980).

Table 1.8. Definitions of Mechanical Terminology Used in Text

Terms	Definitions
Strain	Change in length/initial length
Extensibility	Strain at failure
Stress	Force/cross-sectional area
Stiffness or modulus	Change in stress/change in strain
Toughness	Strain energy absorption: area beneath stress strain curve
Elasticity	Refers to material that recovers its orginal dimensions when unloaded

Table 1.9. Mechanical Properties of Tissues

Tissue	Failure stress (MPa)	Failure strain	Maximum stiffness (MPa)
Tendon	> 100.0	0.10	500.0
Skin	> 4.0	0.50	> 6.0
Aorta	> 0.5	0.50	> 1.0

From Silver (1987).

Polymeric chains with bulky residues therefore have less entropy and undergo less backbone reorientation.

Backbone reorientation associated with a particular chemistry of the repeat unit depends on the thermal energy (temperature) of the chain. As pointed out above, at high temperatures all polymer chains are mobile, undergoing rapid backbone reorientations. As the temperature is lowered the number of polymer chain conformations decreases and the chains begin to be held together, which lowers the internal energy. If the polymer chain is flexible and does not have large or bulky residues it can form a regular crystalline structure. If the chains are not regular enough to form a crystal on lowering the temperature, the chains form a structureless rigid state termed a glass (Figure 1.11). Typical melting temperatures and glass transition temperatures of polymers are given in Table 5.4. All polymers above the melting and glass transition temperatures are soft (have low stiffnesses) while below these temperatures polymers have much higher stiffnesses.

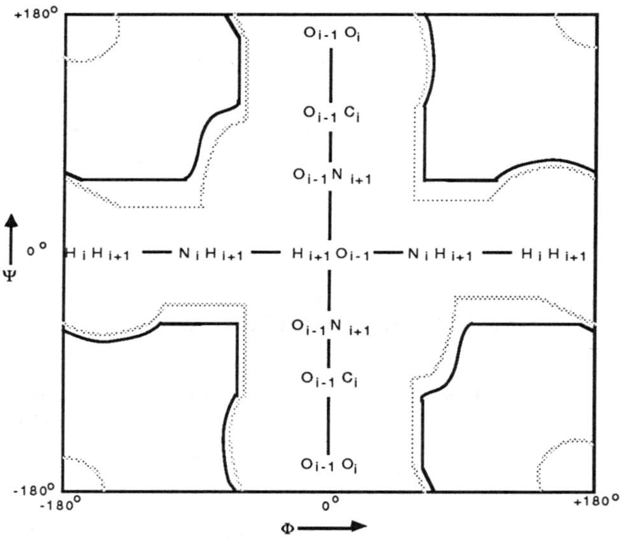

Figure 1.10. Polymer chain conformational plot. Plot of allowable conformations for a peptide of glycine including fully allowed (_____) and partially allowed (.........) based on interatomic distances found in crystal structures. Dihedral angles (Φ,Ψ) in center of plot are unallowed because of contacts between backbone atoms on the i th, i th + 1, and i th - 1 units (see Silver, 1987 p. 15 for additional details).

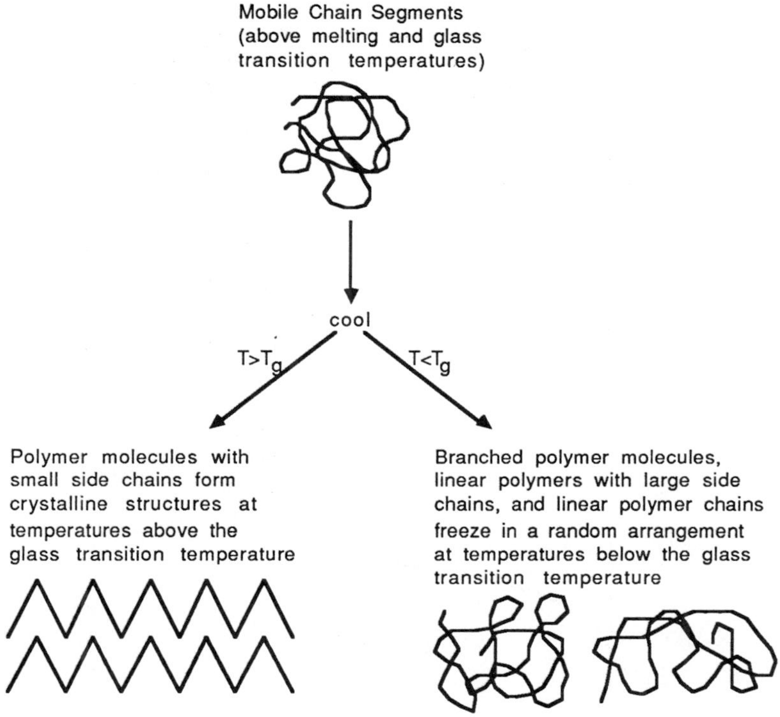

Figure 1.11. Formation polymer structures. When polymer chains are rapidly cooled below the glass transition temperature (T_g) they are frozen into a random state termed an amorphous glass. In comparison, if they are cooled below the melting temperature but above the glass transition temperature, they form crystals.

1.3.4. Mechanical Properties of Polymeric Biomaterials

All polymeric materials exhibit a phenomenon termed viscoelasticity when the temperature is above the glass transition and melting temperatures. This phenomenon is illustrated in Figure 1.12 and involves the time-dependent increase in strain of a polymer that is subjected to a fixed load. Viscoelasticity of polymers arises because this type of material has properties characteristic of a viscous liquid as well as a solid. The net result is that unlike solids the force per unit area (stress) is not uniquely related to the strain (change in length divided by the initial length). Thus, in order to characterize the state of stress in a polymer, one needs to know the state of strain and when the stress was applied. Even after defining these parameters, it is not always likely that a unique relationship exists between stress and strain for a polymer.

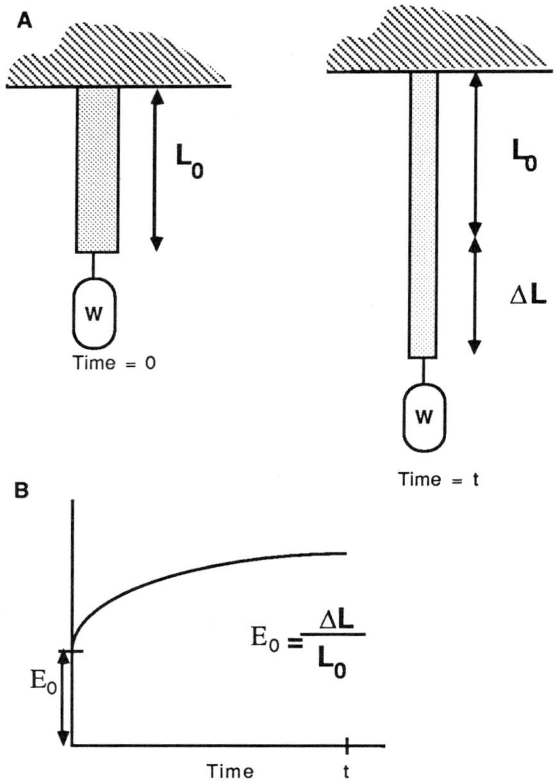

Figure 1.12. Viscoelastic behavior of a polymer.
When a fixed weight, W, is placed on the specimen (see A) the strain, $\Delta L/Lo$ initially adopts a value of Eo but then increases with time (see B). Time dependence of the strain that arises from a fixed load is termed creep and is a mechanical characteristic of viscoelastic behavior.

Table 1.8 is a list of mechanical terminology used throughout this text. Besides the terms stress and strain we will refer to extensibility, stiffness, toughness, and elasticity. Extensibility refers to the strain required to cause a material to fail while elasticity is a property of materials such as rubber that can be loaded and unloaded without a change in the original length. The stress required to result in a unit strain is termed the stiffness while the toughness refers to how much energy is required to cause a material to fail.

In many applications the mechanical properties as well as the biocompatibility of a material are important design criteria. Table 1.9 lists some of the mechanical properties of tissues. The properties listed were

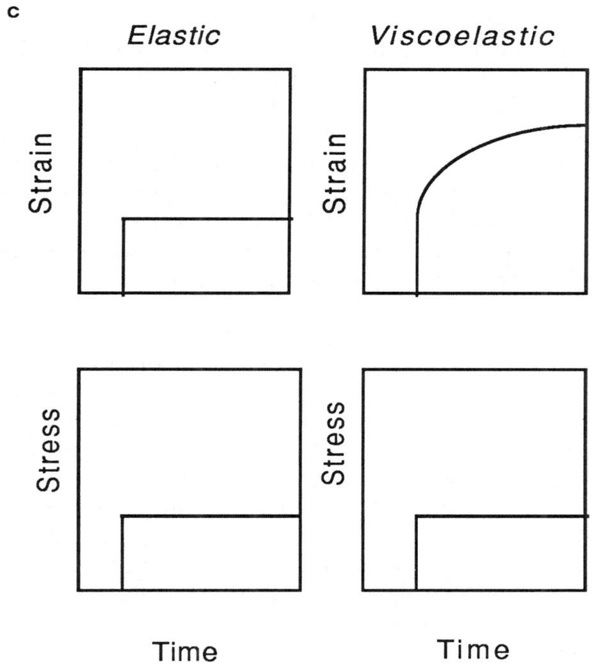

Figure 1.12. (continued)
A comparison between elastic and viscoelastic behavior is shown in C. Elastic behavior is characterized by time-independent strain that results from a fixed load while viscoelastic strains are time dependent.

measured shortly after a load was applied and therefore the time of observation is approximately zero.

1.4. Recent Books on Biomaterials

There are numerous books that have been published recently on the topic of biomaterials. These books are excellent reference texts for the student, researcher, or clinician who is interested in getting references to specific papers or techniques used in the field. Below we will give a brief description of some of these texts.

Fundamental Aspects of Biocompatibility; Biocompatibility of Clinical Implant Materials; Biocompatibility in Clinical Practice; Techniques of Biocompatibility Testing; Blood Compatibility; and Biocompatibility of Tissue

Analogs are sets of books that make up the CRC Series in Biocompatibility (1981-1984) edited by David F. Williams. These texts contain excellent in-depth reviews of different areas of the field. Other books of the same series are: Biocompatibility of Orthopaedic Implants; Blood Compatibility; Biocompatibility of Dental Materials; Systemic Aspects of Biocompatibility, and the most recent books are entitled: Techniques in Biocompatibility Testing and Blood Compatibility. Macromolecular Biomaterials edited by Garth W.Hastings and Paul Ducheyne (1984) is a book in the CRC Series on Structure Property Relationship of Biomaterials. This book is a series of chapters written by different authors that review several important areas in the field of biomaterials.

Contemporary Biomaterials edited by John W. Boretos and Murray Eden (1984) is an extensive compilation by experts of several subfields of biomaterials. These fields include the status of biomaterials, properties and performance of biomaterials, tissue response to biomaterials, clinical experience with biomaterials, and new biomaterials technology. Handbook of Biomaterials Evaluation edited by Andreas F. von Recum (1986) is as its title indicates, a handbook that has assembled information concerning the many subspecialties within the field. Topics covered include the types of materials used and their characterization, toxicological evaluations, *in vitro* and animal models for assessing biocompatibility, histology of soft and hard tissues, regulatory aspects of device development, and clinical trials.

Implants by Wilfred Lynch (1982) is a book that contains many excellent diagrams and pictures illustrating prototypes of devices that are used as skeletal and joint prostheses, cardiovascular implants, urology devices, and implants for the visual, auditory, and neurological systems.

Biomaterials, Science and Engineering by Joon Bu Park (1984) is a basic text covering materials used in implants, tissue response to implants, soft and hard tissue replacements. Unlike most of the other books on biomaterials it can be used as a basic course text.

These are several of the many excellent books on biomaterials. The goal of this text is to organize and present the basic concepts that have been developed in other texts and reference books in a manner that underscores the application of basic chemistry, physics, and biology to the evaluation of biocompatibility of biomaterials used as implants.

References

Boretos, J.W. and Eden, M. (editors) Contemporary Biomaterials, Noyes Publications, Park Ridge, N.J. 1984.

Dublet, B. and Van Der Rest, M. J. Biol. Chem. 262,17724,1987.

Hastings, G.W. and Ducheyne, P. (editors) Macromolecular Biomaterials, CRC Press, Boca Raton, Florida, 1984

Ketteringham, J.M., Cavanaugh, L.M., Charest, M.C., Couvillon, L.A. and Fanucci, M.H. Commercial Opportunities In The Health Care Industry, Arthur D. Little Impact Study, Nov. 1982, p 6.

Lynch, W. Implants, Van Nostrand Reinhold Company, New York, 1982.

MacMillan, B.G. Burn Wound Coverings, Vol. I, edited by D.L. Wise, CRC Press, Boca Raton, Florida, p. 115,1984.

Martin, G.R., Timpl, R., Muller, R.K. and Kuhn, K. Trends Biochem. Sci. 2,285,1985.

Myer, G.H. and Parsonnet, V. In: Engineering In The Heart And Blood Vessels, Wiley-Interscience, Chapter 11,1969.

Park, J.B. Biomaterials, Science and Engineering, Plenum Press, New York, 1984.

Schulz, G.E. and Schirmer, R.H. Principles of Protein Structure Springer—Verlag, New York, Chapter 2,1979

Seltzer, R.J. Chem. & Engineering News. Nov. 15,1982.

Sheeler, P. and Bianchi, D.E. Cell Biology: Structure, Biochemistry And Function. John WileyAnd Sons, New York, Chapters 15, 16, 18, 19, 20, 21, 1980.

Silver, F.H. Biological Materials: Structure Mechanical Properties and Modeling of Soft Tissues, NYU Press, New York, p. 224,1987.

Thomson, P.D. and Park, D.H. Burn Wound Coverings, Vol. I, edited by D.L. Wise, CRC Press, Boca Raton Florida, p. 48, 1984.

von Recum, A.F. (editor) Handbook of Biomaterials Evaluation, Macmillan Publishing Company, New York, 1986.

Williams, D.F. (editor) Fundamental Aspects of Biocompatibility, Volumes I and II, CRC Press, Boca Raton, Florida, 1981.

Williams, D.F. (editor) Techniques of Biocompatibility Testing, Volumes I and II, CRC Press, Boca Raton, Florida, 1986.

Williams, D.F. (editor) Blood Compatibility, Volumes I and II, CRC Press, Boca Raton, Florida,1987.

Williams, D.F. (editor) Biocompatibility of Clinical Materials, Volumes I and II, CRC Press, Boca Raton, Florida, 1981.

Williams, D.F. (editor) Biocompatibility of Tissue Analogs, Volumes I and II, CRC Press, Boca Raton, Florida, 1984.

Chapter 2

Organ and Tissue Structure

2.1. Introduction

The biomaterials engineer is often involved in the design of devices that are used to reinforce or replace body tissues such as ligaments, tendons, and blood vessels. Design of prosthetic devices is accomplished best by first understanding the structure and function of each tissue to be replaced. In this chapter, we will briefly cover the structure of organs and tissues that are commonly dealt with by biomaterials scientists and engineers.

The human body is a complex system of structural and functional units. These units must operate in unison in order that complex operations can be performed efficiently. These complex operations include control of growth (endocrine system), defense against infectious agents (immunological system), locomotion, food processing and digestion, respiration, reproduction, hearing, and vision (Table 2.1). Locomotion requires the transmission of force from muscle to bone via soft and hard tissue components. Food processing involves chewing, food transport, food digestion, and adsorption. Treatment, purification and excretion of waste involves the intestines, liver, kidneys, and lungs. Respiration involves oxygen and carbon dioxide transport into and out of the lungs and distribution via the cardiovascular system throughout the body. Visual and auditory data are obtained by collecting images or sound respectively, and by signal processing through neuronal influx into the brain. Understanding the structure and function of these systems is a challenge for the biomaterials researcher.

Below we will deal with each of the systems individually and discuss the clinical need for tissue replacements. However, before we deal with each organ system it is first necessary to introduce connective tissue components that are common to all organ systems and that provide structure and shape.

2.2. Connective Tissue Components

All organs contain connective tissue networks that separate different functional units as well as provide integrity. Connective tissue is composed of a

Table 2.1. Structural and Functional Units Found in Organ Systems

Locomotion:	Force generated by muscle → Transmission of force by tendon → Movement of joint
Food processing and waste treatment:	Mastication → Swallowing → Digestion → Adsorption → Excretion
Respiration:	O_2 Air → Nostrils → Nasal Cavity → Pharynx → Larynx → Trachea ↓ Bronchi → Bronchioles → Aveoli CO_2 out
Visual information:	Light → Cornea → Aqueous humor → Lens → Vitreous humor ↓ Retina → Optic Nerve → CNS
Auditory information:	Sound → Tympanic membrane → Ossicles → Auditory nerve → CNS

fixed number of cell types (Table 2.2) that synthesize components, including cell attachment factors, fibrous components, and interfibrillar matrix (Table 2.3). Cellular components are responsible for synthesizing the building blocks that are required to provide structure and stability to functional units. Fibrous components form the structural network that physically maintains tissue shape and prevents mechanical disruption. The three-dimensional organization of fibrous components vary from organ to organ, however, the basic components, collagen and elastic fibers, are found in most organs or organ systems.

Table 2.2. Cells Found in Connective Tissues

Cell	Location
Chondrocytes	Cartilage
Epithelial	Cover exposed body surface; skin
Endothelial	Line inside of blood vessels
Smooth muscle	Found in vessel walls
Fibroblasts	Found in skin, tendon and cardiovascular tissue

2.2.1. Collagen

There are at least twelve types of collagen (Tables 1.4 and 2.4). On a mass basis, types I, II, and III predominate in mammalian tissues. These types form structural fibers in skin, tendon, bone, cartilage, and cardiovascular tissue. Type IV collagen is found in the form of a sheetlike mat termed the basement membrane that forms a support for epithelial cells. Other collagen types (V through XII) are found in minor amounts in cartilage, cardiovascular tissue, and in the matrix around cells (see Timpl *et al.*, 1981; Dublet and van der Rest, 1987).

Type I, II, and III collagens are triple-helical molecules that are composed of over 333 repeat triplets of the form Gly-*X*-*Y* where *X* and *Y* are commonly the amino acids proline and hydroxyproline. In tissues these rodlike molecules align in a modified quarter stagger array with a hole between the ends of molecules that form a linear array Figure 2.1). After staining with heavy metals, the fibers have a repeat pattern with a spacing of 64-67 nm that is a characteristic finger print of collagen (Figure 2.2).

Self assembly of types I, II, and III collagens occurs in neutral solution via a nucleation and growth process (Figure 2.3) that involves a series of linear and lateral growth steps, while in connective tissues collagen fibers are believed to form from discontinuous collagen fibril bundles (Figure 2.4) that later grow in length (McBride *et al.*, 1985, 1988; Silver, 1987). Collagen fibers limit connective tissue deformation and prevent premature mechanical failure.

Table 2.3. Location and Composition of Connective Tissue Components

Component	Composition	Building blocks (repeat unit)	Location
Collagen	Protein	Amino acids	Found in banded extracellular fibrils, basement membranes and thin fibrils extracellularly
Elastin	Protein	Amino acids	Structureless component of elastic fibers
Microfibrillar protein	Protein	Amino acids	Fibrous core of elastic fibers
Proteoglycans	Protein core and large side chains (glycosaminoglycans)	Amino acids and disaccharides	Associated with collagen fibers
Hyaluronic Acid	Polysaccharide	Disaccharides	Found between collagen fibers
Fibronectin	Glycoprotein	Amino acids and monosaccharides	Attachment (factor between collagen fibers and fibroblasts)
Laminin	Glycoprotein	Amino acids and disaccharides	Attachment factor, component of basement membrane
Amelogenin	Glycoprotein	Amino acids and monosaccharides	Involved in odontogenesis
Osteonectin	Glycoprotein	Amino acids and monosaccharides	Binding glycoprotein in bone
Vitronectin	Glycoprotein	Amino Acids and monoisaccharides	Focal contact between cell/ECM (in cell culture)
Anchorin	Glycoprotein	Amino acids and monosaccharides	In myoblast (laminin binding glycoprotein in chondrocyte membrane (collagen and binding glycoprotein)
Chondronectin	Glycoprotein	Amino acids and monosaccharides	Binding glycoproteins in cartilage
Entactin	Sulfated glycoprotein	Amino acids and monosaccharides	Assist the connection to the cell surface (basement membrane)
Thrombospondin	Glycoproteins	Amino acids and monosaccharides	Epidermo-dermal junction cell attachment to ECM
Epibolin	Glycoproteins	Amino acids and monosaccharides	Epithelial cell attachment

Table 2.4 Location and Tissue Source Of Different Collagens

Collagen Type	Tissue or Organ	Location
I	Tendon, skin, bone and fascia	Thick extracellular fibrils and fibers
II	Cartilage	Thin fibrils around cartilage cells
III	Cardiovascular tissue	Intermediate size extracellular fibrils
IV	Basement membranes	Network forming component
V	Tendon, skin and cardiovascular tissue	Pericellular matrix around cells
VI	Cardiovascular tissue, placenta, uterus, liver, kidney, skin, ligament and cornea	Extracellular matrix
VII	Skin	Anchoring fibrils
VIII	Cardiovascular tissue	Around endothelial cells
IX	Cartilage	Extracellular matrix
X	Cartilage	Extracellular matrix
XI	Cartilage	Extracellular matrix
XII	Embyonic chick tissues (tendon)	Extracellular matrix

2.2.2. Elastic Fibers

Elastic fibers are composed of an amorphous core of the protein elastin and an outer layer of microfibrillar protein (Figure 2.4). The ability of many tissues including lung, skin, blood vessels and some ligaments to recover their original dimensions after a load is applied and removed is a result of the elastic-fiber network. Microfibrillar protein is found in elastic tissue as small fibrils about 14 nm in diameter (Fanning *et al.*, 1981). Biosynthetic studies suggest that the precursor of elastin is tropoelastin, a 72,000 M r polypeptide (Rosenbloom, 1984).

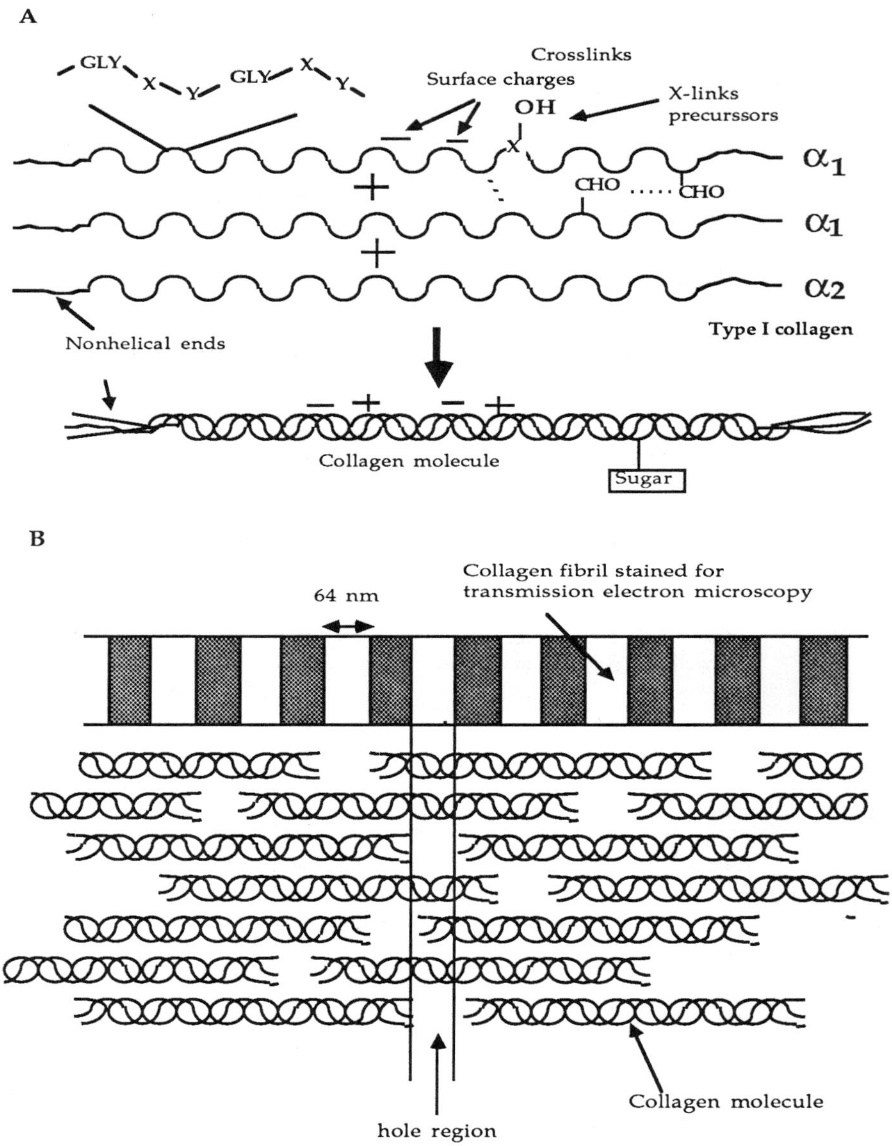

Figure 2.1. Collagen molecule. (A). Type I collagen molecule consists of the assembly of 3 chains (2 α1(I) and 1 α2 (I)) which are crosslinked together and form a triple helical region and two nonhelical regions at both ends. The sequence of amino acids (GLY-X-Y) is almost unique to this molecule as well as the high concentration of hydroxylysine and hydroxproline (GLY= glycine). Carbohydrate is also presented bound to collagen molecule. (B). Diagram illustrating lateral arrangement of collagen molecules in a fibril showing the 64nm repeat pattern seen when stain penetrates the hole region.

C

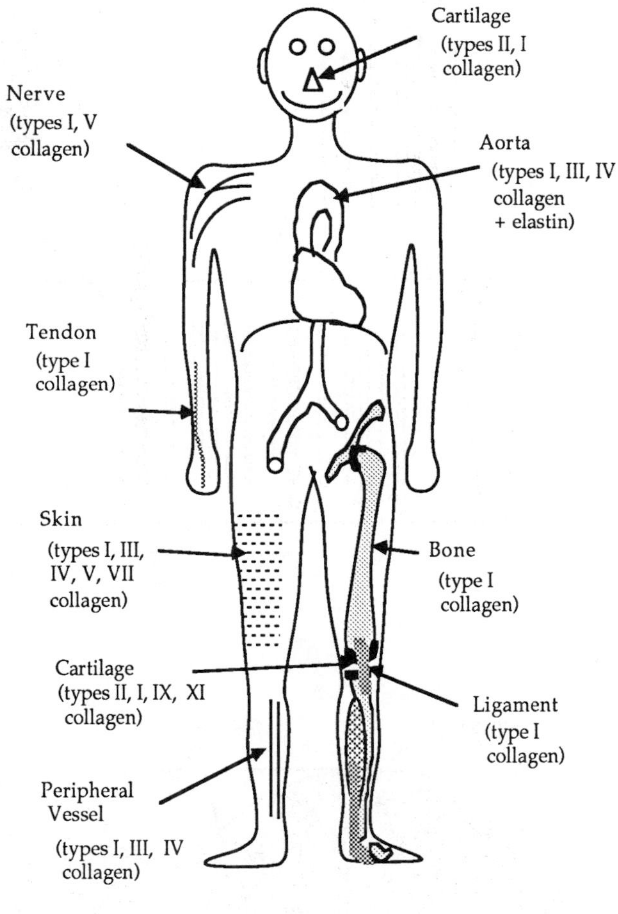

Figure 2.1 (cont.) (C). Location of collagen types in the body in nerve, tendon, aorta, bone, cartilage and ligament.

2.2.3. Proteoglycans, Glycosaminoglycans, and Glycoproteins

Material found between collagen fibers in tissues is termed interfibrillar matrix and is composed of hyaluronic acid (HA), proteoglycans (PGs), and link protein (Table 2.3). Hyaluronic acid is an unbranched polysaccharide that is found in tissues in molecular weights that range from less than 100,000 to several millions. It is believed not to be covalently associated with protein and under physiological conditions; hyaluronate has large domains of freely coiling structures, but there are elements of stiffness in the molecule (Phelps, 1984).

Proteoglycans are high molecular weight macromolecules consisting of a

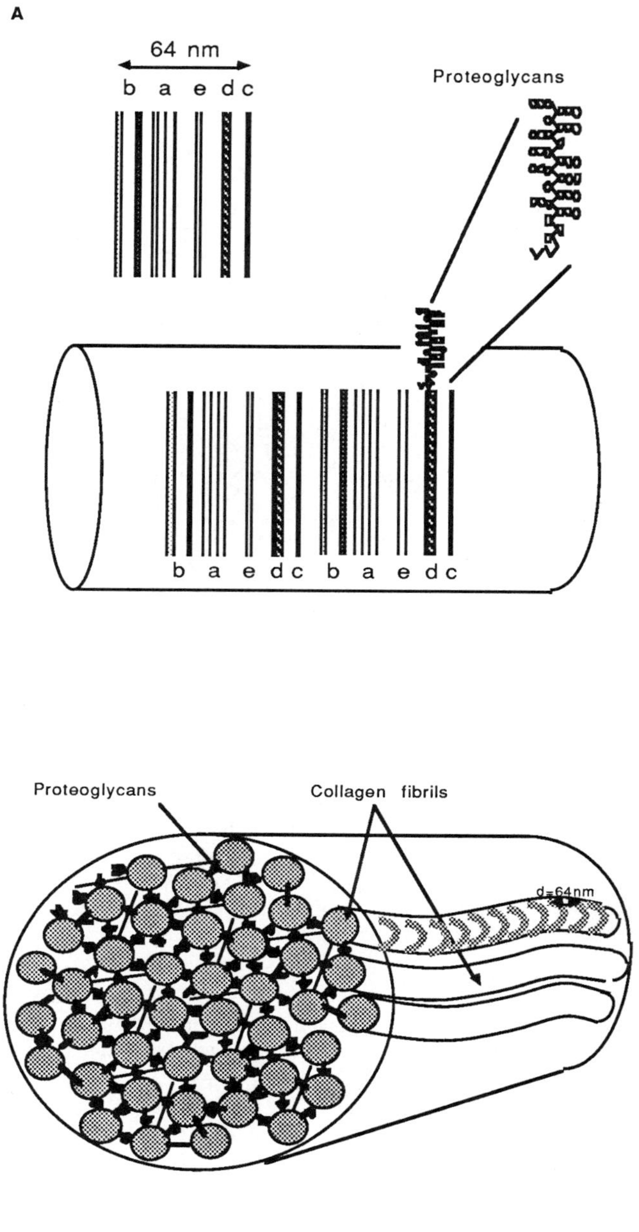

Figure 2.2. Structure of collagen fibril. (A). The D periodic pattern consists of several bands that are observed when a stain binds directly to the charged residues on collagen. The characteristic D period consists of 5 bands (11 sub-bands) that are repeated every 64 nm along the fibril axis. Dermatan sulfate proteoglycan is seen attached to the d band. (B). Section of a collagen fibril bundle showing the interactions between collagen fibrils and proteoglycans in the interior of the bundle. Collagen fibrils are crimped in this arrangement.

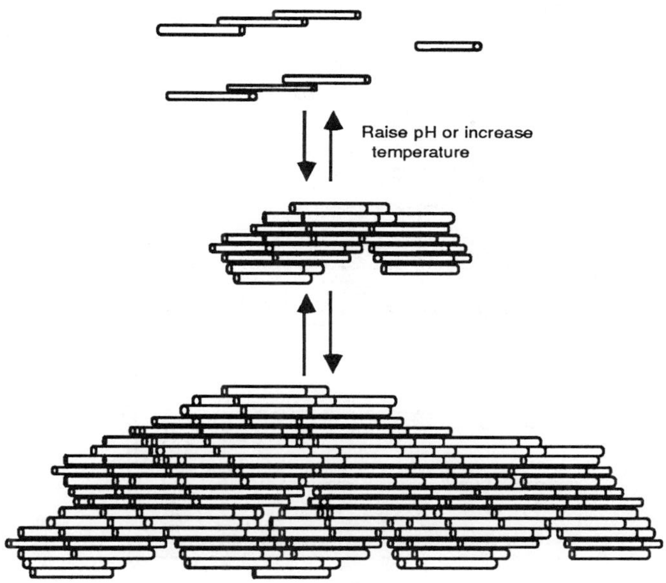

Figure 2.3. Fibrillogenesis of collagen. Schematic presentation of prenucleation and nucleation steps of collagen molecules that lead to formation of collagen fibrils and collagen fibril bundles in vitro.

protein core with a molecular weight ranging from a few thousand to 250,000 to which glycosaminoglycan side chains are attached (Figure 2.5). Proteoglycan monomers from cartilage have molecular weights between 1 x 10^6 and 3 x 10^6 (Rosenberg *et al.*, 1979; Hascall and Hascall 1981). Recent data suggests that connective tissues contain two types of PGs, i.e. small and large proteoglycans. Small PGs from tendon have an average molecular weight of about 37,000 for the dermatan sulfate chain and a molecular weight of 48,000 for the core protein. Large PGs from tendon are composed of chondroitin-sulfate chains with average molecular weight of 17,000 and a core protein molecular weight of 200,000 (Vogel and Heinegard, 1985). Large proteoglycans from tendon, aorta, and cartilage aggregate in the presence of hyaluronic acid and link proteins to form proteoglycan aggregates (Rosenberg *et al.*, 1979; Poole, 1986). However, larger, nonaggregating proteoglycans in skin and muscle, as well as small, nonaggregating proteoglycans from bone and cartilage, do not form high molecular weight complexes (Poole, 1986). Clearly there are a variety of proteoglycans with properties that are specific to each tissue.

Proteoglycan monomers have a structure similar to a bottle brush when totally extended. The backbone of the bottle brush is composed of a protein core while the arms or side chains of the brush are composed of glycosaminoglycans. Glycosaminoglycan side chains found on proteoglycans are listed in Table 2.5. The side chains are composed of a repeat unit containing a sugar acid derivative (hexuronic acid) and an amino sugar derivative (hexosamine). Proteoglycan monomers that aggregate bind to a linear polysaccharide of hyaluronic acid and form proteoglycan aggregates.

A

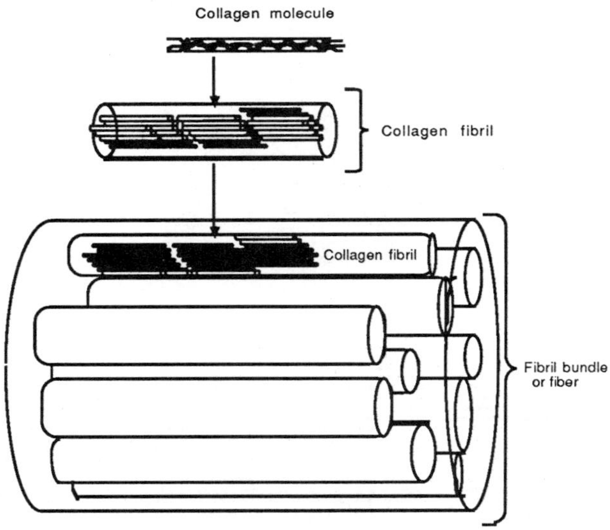

Figure 2.4. Collagen and elastic fiber structure. (A). Schematic diagram showing formation of collagen fibril bundle (also termed collagen fiber) by end-to-end addition of collagen fibrils in developing tendon.

Cartilage contains two link proteins of molecular weights 46,000 and 51,000. Recent evidence suggests that the 51.000-M_r link protein increases proteoglycan aggregate size while the 46,000-M_r link protein may be required to effectively stabilize the proteoglycan aggregate against dissociation (Rosenberg and Buckwalter, 1986). Electrostatic interactions between the negative side chains of glycosaminoglycans of proteoglycan aggregates and positive amino-acid residues on collagen fibrils bind the interfibrillar matrix to collagen fibers (Gelman and Blackwell, 1973). Specific binding of dermatan sulfate proteoglycan to collagen fibrils has been observed in a variety of tissues by Scott and co-workers (Scott and Orford, 1981).

Heparan sulfate proteoglycan is a specialized proteoglycan that is found associated with and embedded in all cell membranes. This proteoglycan glues the cell surface to the interfibrillar matrix and collagen fibers via glycoprotein attachment factors and is found in a variety of tissues (Poole, 1986). Other glycoproteins such as fibronectin, chondronectin, osteonectin, laminin, and vitronectin (Tables 1.5 and 2.3) mediate adhesion between cells and the collagenous fibers that form the tissue scaffold.

2.3. Classification of Connective Tissue

Connective tissue can be classified as loose or dense based on its appearance under the light microscope (Ham and Cormack, 1979). In addition, the term "reticular connective tissue" or "reticulin" refers to a fibrous structure composed of

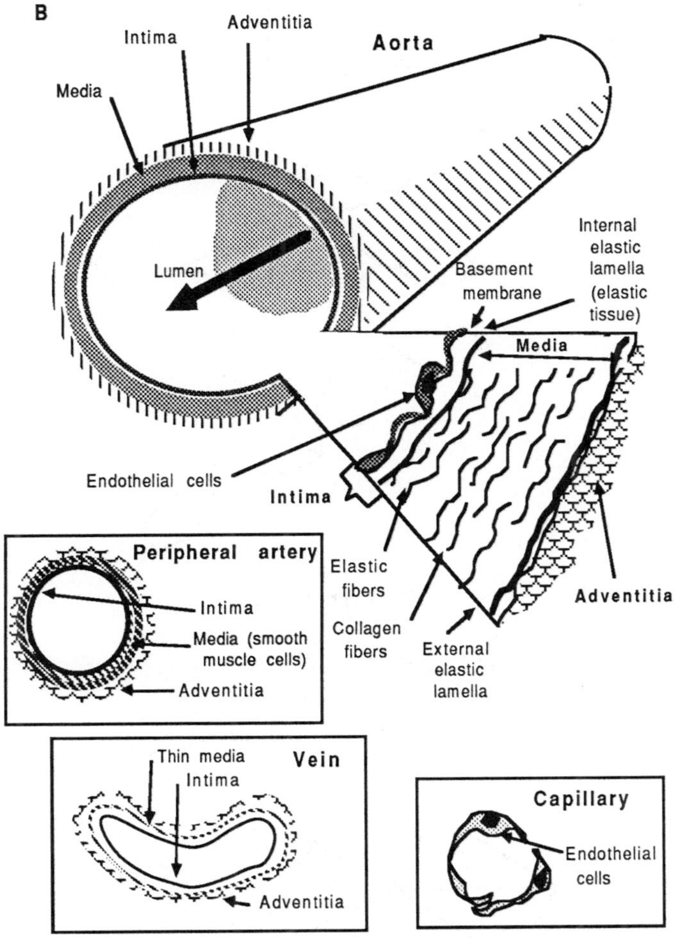

Figure 2.4. (cont.) (B). Elastic fibers are mostly composed of amorphous elastin surrounded by microfibrillar protein.

type III collagen and fibronectin and a noncollagenous glycoprotein (Unsworth *et al.*, 1982). This material is seen in many connective tissues but is not included in the categories loose or dense connective tissues.

The terms "regular" and "irregular" refer to the organization of fibrous components within the connective tissue and the terms "dense" and "loose" refer to the tightness of the collagen fibers. When fibrous components are aligned in a single direction such as in tendon and ligament the tissue is termed regular. Ligament and tendon are examples of dense regular connective tissue. The dermis of skin is an example of an organ containing dense irregular connective tissue (dermis) since the collagen fibers are found in more than one direction within the plane of skin.

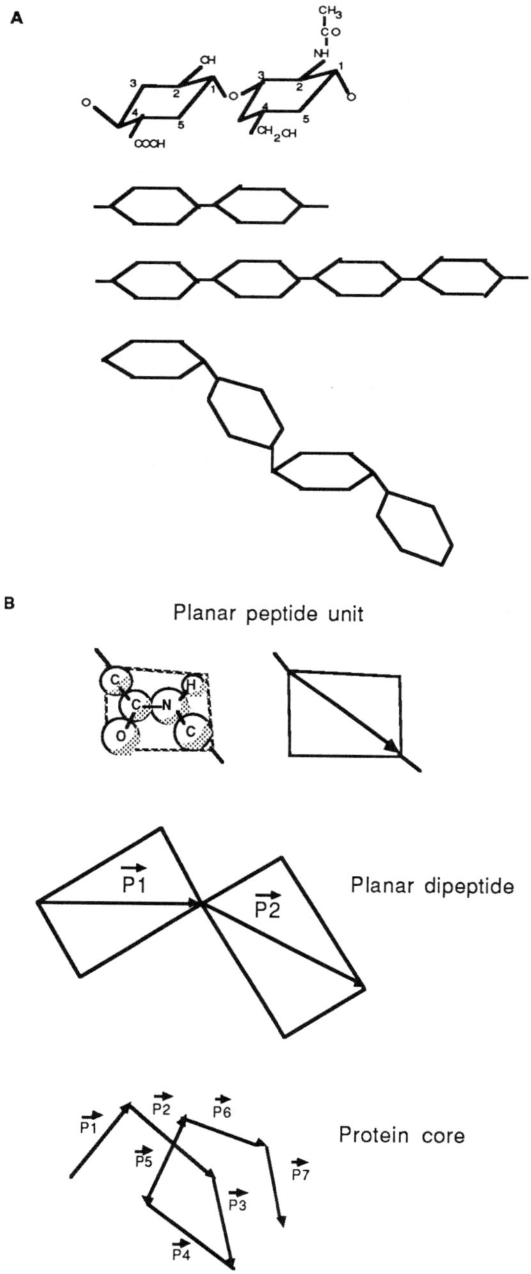

Figure 2.5. Proteoglycan structure. (A). Diagram of the sugars composing the molecule of hyaluronic acid in stereochemical boat structure and planar projected structures. Similar structures are observed for other glycosaminoglycans. (B). Diagram of protein core structure showing peptide unit, dipeptide and vectors representing core protein.

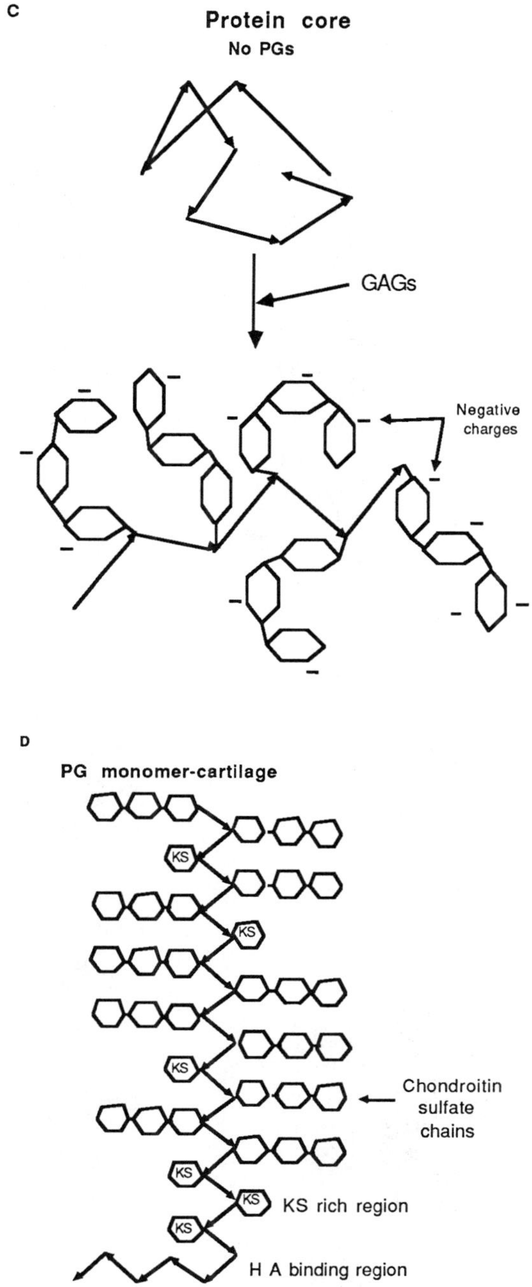

Figure 2.5. (cont.) (C). Addition of sulfated, negative charged glycosaminoglycans to protein core results in expansion of the molecule. (D). Planar projection of proteoglycan monomer containing keratan sulfate (KS) and chondroitin sulfate chains and the hyaluronic acid binding region (HA).

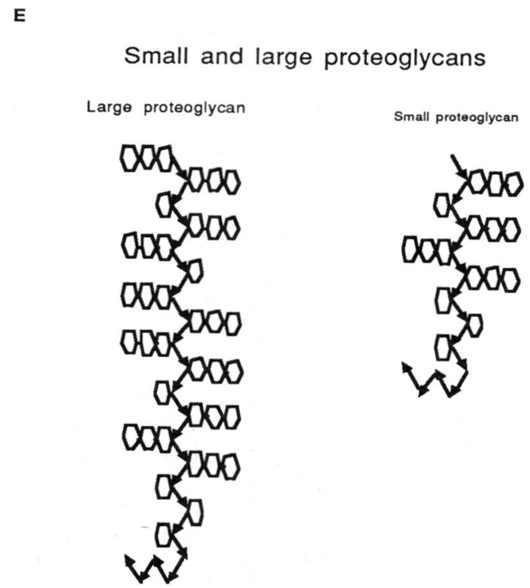

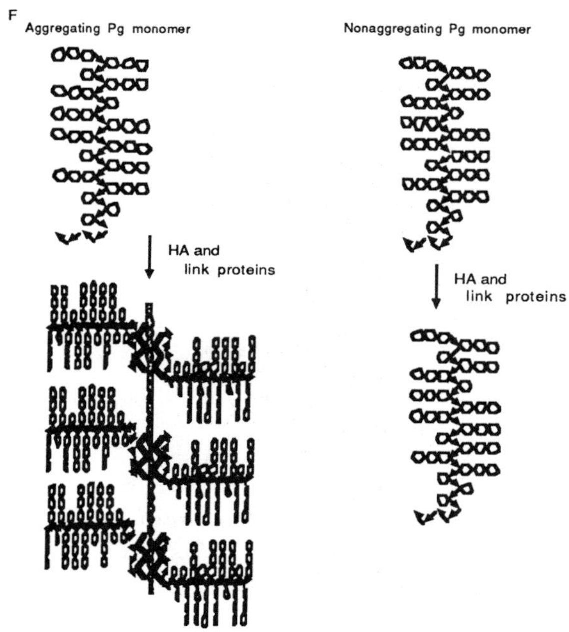

Figure 2.5. (cont.) (E). Proteoglycans are found in small and large molecular sizes. (F). Proteoglycans exhibit varying aggregating properties.

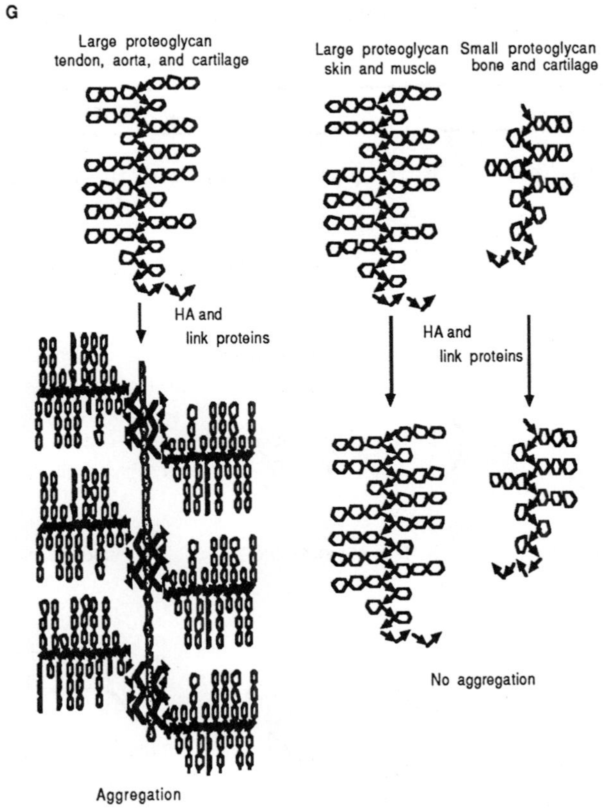

Figure 2.5. (cont.) (G). Aggregation properties of proteoglycans depend on the tissue source and size.

2.3.1. Loose Connective Tissue

Loose connective tissue is characterized by an abundance of cells with thin undulating collagen fibers (Figure 2.6). It is found as the external sheath of many organs. Microscopically it is composed of cellular components with inflammatory cells (macrophages, plasma cells, and mast cells) and fibroblasts.

2.3.2. Dense Connective Tissue

Dense connective tissue is characterized by a high content of wide, aligned collagen fibers (Figure 2.7). Some tissues such as skin contain mixtures of loose (papillary layer of the dermis) and dense connective tissue (reticular layer of the dermis).

Dense connective tissue is composed of the same components as loose connective tissue except that there are fewer inflammatory cells and more fibroblasts. Collagen fibers and elastic fibers are concentrated within this tissue and

Table 2.5. Repeat Units Found in Glycosaminoglycans

Glycosaminoglycans	Repeat disaccharide
Hyaluronic acid	β- D - glucuronic acid + β- D - N - acetyl glucosamine
Chondroitin sulfate	β- D - glucuronic acid + β- D - N - acetyl galactosamine
Dermatan sulfate	β- D - glucuronic acid or β- L - iduronic acid + β- D - N - acetyl glucosamine
Heparan sulfate, Heparin	β- D - glucuronic acid or β- L - iduronic acid + α- D - N - acetyl glucosamine
Keratan sulfate	β- D - glucosamine + β- D - glucosamine

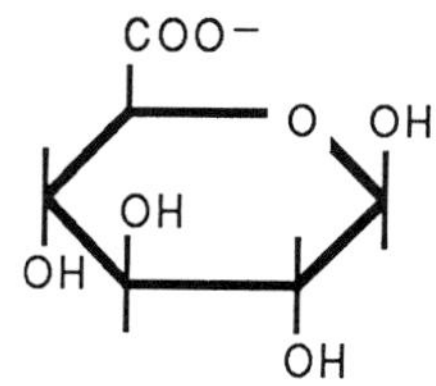

β- D- glucuronic acid

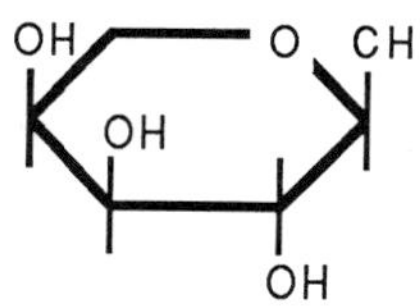

β- D- galactose

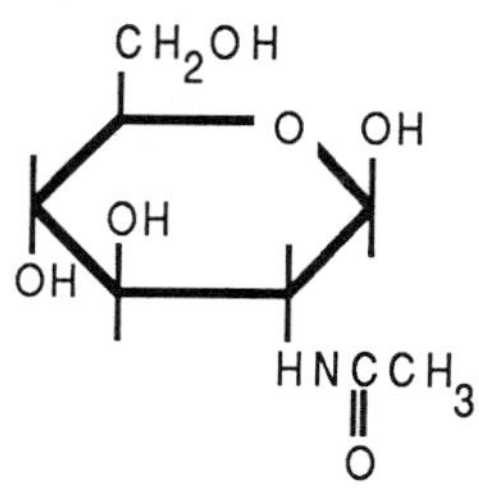

β- D- N-acetyl glucosamine

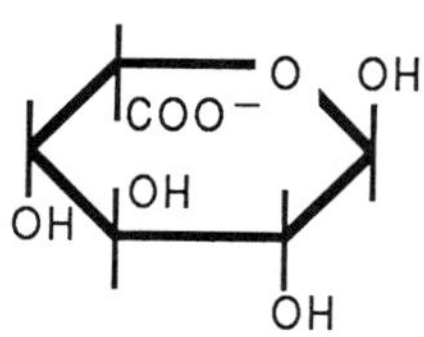

β-L- iduronic acid

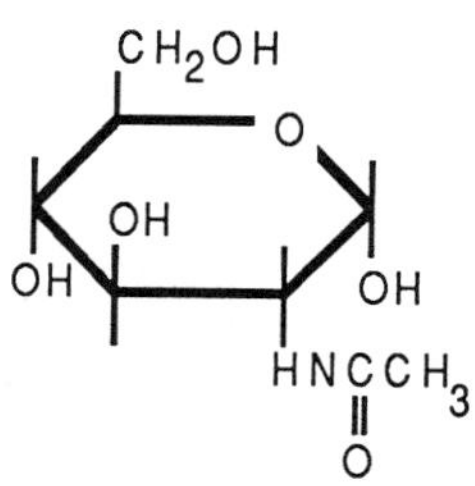

α-D- N- acetyl glucosamine

proteoglycans and glycoproteins are not as apparent as in loose connective tissue. This type of connective tissue is found in tendons, ligaments, aponeurosis (fascia), bone, and dermis. However, in the case of bone, dense connective tissue is mineralized.

2.3.3. Basement Membrane

Basement membranes are extracellular sheets that separate connective tissue from cells that line internal and external body surfaces. Ultrastructural observations reveal a fine, membranelike structure that is the support for epithelial and epidermal cells. Basement membrane (Figure 2.7B) consists of three layers: lamina lucida, lamina densa, and sublamina densa. The lamina lucida which is rich in laminin (a glycoprotein), is directly attached to the cell surface, and is linked via proteoglycan to the second layer (lamina densa). Lamina densa is composed of proteoglycan (heparan sulfate proteoglycan) and type IV collagen, which may have a reticular pattern (Stanley *et al.*, 1982). Sublamina densa is composed of type VII collagen in the form of anchoring fibrils that connects the basement membrane to collagen fibrils in connective tissue below the basement membrane. Other macromolecules such as fibronectin, entactin, and nidogen are also present in small amounts and their role is not well understood.

Ultrastructural examination has demonstrated that basement membrane is formed from interconnected networks of 3- to 8- nm wide cords concentrated in the lamina densa which join to the cell membrane by extensions through the lamina lucida (Inoue *et al.*, 1983; Laurie *et al.*, 1983). Type IV collagen contains two genetically distinct polypeptide chains, $\alpha 1(IV)$ and $\alpha 2(IV)$; each has about 1700 amino acids (Timpl and Martin, 1982). The chain composition of this collagen is heterogeneous and may vary between different basement membranes (Kefalides *et al.*, 1985). Sequence similarities have been reported for at least parts of the $\alpha 1(IV)$ and α 2(IV) chains (Kurkinen *et al.*, 1987).

Type IV collagen molecules make up about 30-40% of the dry weight of basement membranes. These molecules appear by electron microscopy as flexible rods 400 nm long with a 17- nm globular head at the carboxyl terminus (Timpl *et al.*, 1981) and form tetramers by antiparallel association of the short amino terminal regions (Duncan et al., 1983). Aggregation of the carboxyl-terminal globular ends of tetramers leads to a loosely packed three-dimensional structure that has been modeled by several investigators (Timpl *et al.*, 1981; Yurchenco *et al.*, 1986). Light microscopic and x-ray diffraction studies support the presence of an ordered structure at the fibrillar and tissue levels (Junqueira *et al.*, 1983; Barnard *et al.*, 1987).

2.4. Geometry of Connective Tissues

Connective tissue is synthesized by a variety of different cell types in the body. Although the cell types involved in synthesis vary, the geometrical forms of the biosynthetic products can be seen repeatedly throughout the body. These forms include cables, sheets and tubes (Tables 2.6- 2.9).

Cablelike geometries are found in nerves, tendons (Figure 2.8), and ligaments. These tissues are involved with the transmission of electrical signals (nerves) and contractile forces (tendons and ligaments). Alignment of structural units is required in tendon and ligament primarily because the transmission of force is most efficiently achieved by use of the most direct path, a straight line. For this reason force transmission units (the collagen fibers) in these tissues are aligned along the axis of the tendon. Tendon and ligament also contain connective tissue

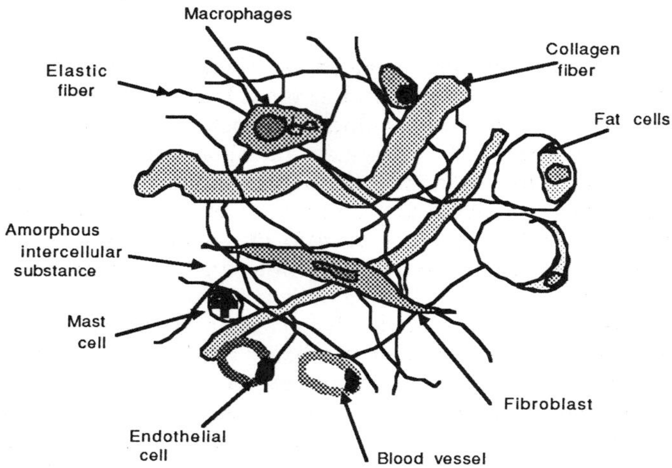

Figure 2.6. Loose connective tissue. Diagram representing cells and extracellular matrix that can be present in a loose connective tissue. Cells such as mast cells and macrophages can be activated during an inflammatory reaction (see Chapter 4).

tubes that serve as protective wrappers.

Connective tissue tubes are found in nerve, tendon, bone, blood vessels, spinal cord, reproductive system, and intestine (Figures 2.7-2.9 for some examples). Connective tissue tubes, in general, have three layers that reinforce the tube wall and separate the contents of the tube lumen from the material in the external environment of the tube. For instance, blood vessel walls are composed of three layers, the intima, media, and adventitia. The inner layer maintains the fluidity of the blood while the media and adventitia structurally reinforce the vessel wall preventing "blow-outs". Muscle cells are found in one of the layers (wall of uterus, intestines, and blood vessels) to further reinforce the tube and to regulate the tube diameter.

Connective tissue sheets such as the dermis of skin (Figure 2.10) and pleural and pericardial serous membranes surrounding the abdominal and thoracic cavities and the heart, respectively, dura mater enveloping the brain and spinal cord, articular cartilage covering long bones, cartilage of the nose and the tympanic membrane in the ear are planar arrangements of connective components. Connective tissue sheets are designed primarily to support cellular sheets (epidermis and epithelium), act as a mechanical and chemical barrier between different organ compartments (dura mater and pericardium) and prevent surface-to-surface adhesion (articular cartilage). Collagen fibers are arranged in several orientations in these sheets, i.e. parallel to the surface of these sheets, biaxially (dura and skin) or even with three principal directions of orientation (pericardium). These arrangements optimize the mechanical strength in the physiological loading directions.

Although connective tissue cables, tubes or sheets found in different organs are not structurally and chemically identical, they have enough similarity to be considered geometrical subclasses of connective tissues.

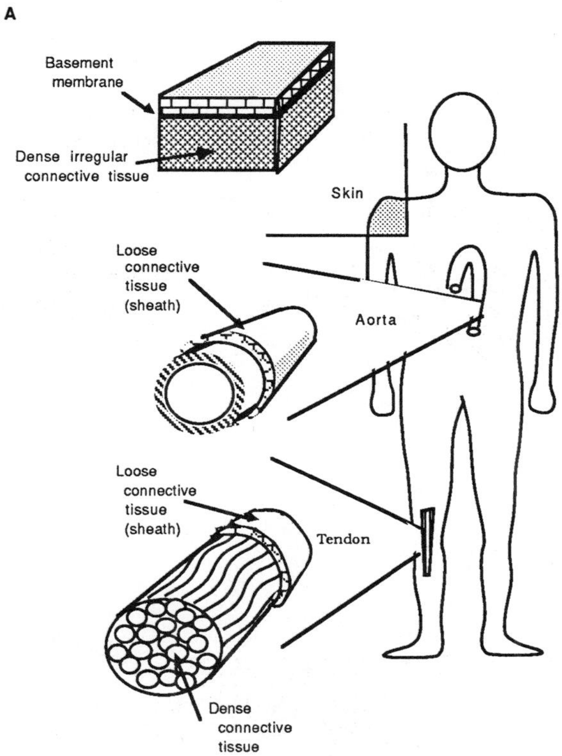

Figure 2.7. Distribution of connective tissue. (A). Diagram illustrating loose and dense connective tissues that are found in skin, aorta, and tendon.

2.5. Tissues Involved in Locomotion

2.5.1. Nerve and Muscle

Nerve impulses conveyed from all parts of the body travel through axons, the long processes of nerve cells and come together in the central nervous system. The peripheral nervous system comprises all nervous tissue outside of the brain and spinal cord and functions to keep all tissues of the body in communication with the central nervous system (brain and spinal cord).

There are three main categories of neurons (Figure 2.11): sensory neurons, which convey impulses from receptors to the central nervous system; motor neurons, which convey impulses from the central nervous system to effector cells; and interneurons which are interposed between sensory and motor neurons (Ross and Reith, 1985). The majority of all neurons are of the interneuron category.

Most neurons share certain structural features, including a cell body, dendrites, and axons. The cell body contains the nucleus and main concentration of organelles of the cell.

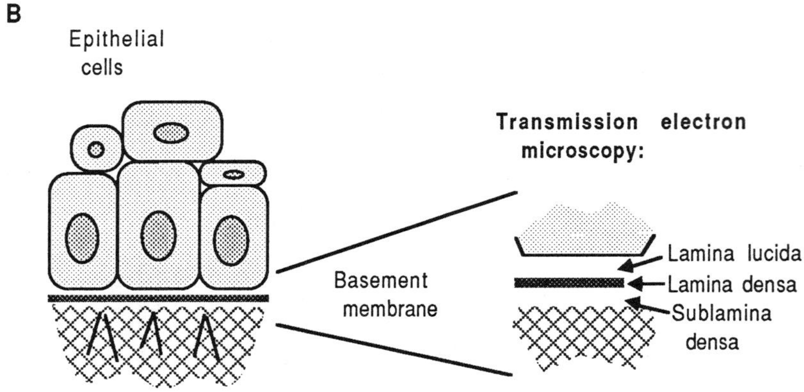

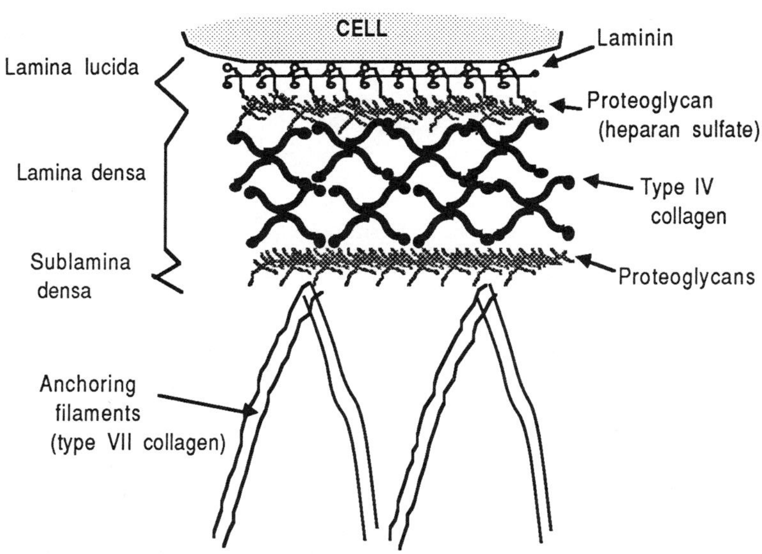

Figure 2.7 (cont.). (B). Basement membrane is composed of lamina lucida, lamina densa, and sublamina densa. Lamina lucida contains laminin and heparan sulfate proteoglycan. Lamina densa containing type IV collagen is attached by proteoglycans (sublamina densa) and type VII collagen anchoring filaments to underlying tissue.

Dendrites are extensions of the cell body and serve as sites of signal input into the cell body. Synapses occur between neuron and striated muscle and are the

Table 2.6. Cordlike Biosynthetic Products

Tissue	Structure
Tendon	Unidirectional bundles of collagen fiber, type I collagen elastin fibers, proteoglycans
Nerve (peripheral)	Schwann cells (myelin sheath), nerve fibers (axons/dendrites), connective tissue (endo, peri, and epineurium)
Long bone	See bone/rigid structures

site of release of neurotransmitters that cause muscle contraction. The axon serves for transmission of impulses as well as for transport of molecules to and from its ends. Axons in peripheral nerves are surrounded by a lipid—rich sheath of myelin, which is externally covered by a sheath composed of a Schwann cell that is wrapped around the axon in a spiral fashion. Movement of a limb such as a leg initiates with an electrical signal transmitted through axons to the appropriate muscles.

The bulk of peripheral nerve is found in the form of nerve fibers and Schwann cells. Groups of axons are wrapped by a loose connective tissue layer called perineurium that serves as a semipermeable barrier and forms a nerve unit termed the fascicle. The axons are surrounded by loose connective tissue termed endoneurium. Groups of fascicles are surrounded by a connective tissue covering termed epineurium that carries blood vessels and provides blood to the nerve. Groups of fascicles form the nerve bundle (Figure 2.9).

The collagen found in peripheral nerve contains fibrils of uniform diameter (Juinquera *et al.*, 1979). Collagen fibrils in the epineurium form a tight cuff around each nerve fiber; however, the rest of the connective tissue is, as discussed above, loosely arranged (Friede and Bischhausen, 1978). During development the collagen-fibril diameter increases in the endoneurium (Parry and Craig, 1984) and in the embryo nerve contains high proportions of type III collagen. Endoneurial collagen has been postulated to protect the nerve fibers against compressive forces (McFarlane *et al.*, 1980) whereas resistance to tensile stretching has been attributed to the epineurium (Haftek, 1970). Recent structural studies give additional details concerning nerve connective tissue structure (Ushiki and Ide, 1986; Salonen *et al.*, 1987).

Stimulated nerve cells conduct electrical impulses via the axon and directly stimulate the sarcomere, the contractile unit of skeletal muscle (Figure 2.12). The sarcomere contains actin and myosin filaments that contract on stimulation. Muscle contraction at the level of the sarcomere results in tension applied to tendons that connect muscle to bone.

Table 2.7. Sheetlike Biosynthetic Products

			Cell types
Skin	Epidermis		Epidermal cells (stratification), hair, glands (secretion) basement-membrane (keratin)
	Cutaneous appendages		
	Dermis	Papillary layer	Dense connective tissue: type I, III and VII collagens, glycoproteins, glycosaminoglycans
		Reticular layer	Dense connective tissue: type I and III collagens, glycoproteins, glycosaminoglycans
Cartilage	hyaline (joint)		Chondrocytes matrix: types II, IX and X collagens, glycosaminoglycans proteoglycans, and glycoproteins
	fibrocartilage (disc)		Collagen (type I collagen); (type II collagen in nucleus)
	elastic cartilage		Elastin fiber
Cardiac valves			Endothelial cells, basement membrane Dense connective tissue rich in elastin fibers (type III collagen) (tendinous cords: dense collagenous connective tissue)
Meninges	pia mater		collagen, elastic fiber network
	arachnoid		Collagen (elastic fibers)
	dura mater		Dense connective tissue (elastic fibers)
	Flat bone (skull)		(See bone)

Table 2.8. Tubelike Biosynthetic Products

Tissue			Inner layer	Middle layer	External layer
			Cell types		
Vessels	Artery	Elastic	Endothelial cells, internal elastic lamina (elastic fibers)	Elastic fibers (collagen: type I & III)	External elastic lamina (elastic fibers)
		Muscular		Smooth muscle (elastin) (type V collagen)	Loose connective tissue (collagen type I)
	Vein		Endothelial cells, internal elastic lamina (elastic fibers)	Collagens (type I & III) (elastic fibers smooth Muscle fibroblasts	External elastic lamina (elastin fibers)
Pulmonary Track		Trachea Bronchi	Cliated cells Basement membrane (lamina propria)	Cartilagenous dense connective tissue (smooth muscle)	Loose connective tissue
Urinary Track		Ureter Bladder Urethra	Epithelium Basement membrane Collagen fibers elastic fibers	Smooth muscle	Loose connective tissue
	Bile Ducts		Epithelium Basement membrane	Dense connective tissue Smooth muscle	Loose connective tissue
	Intestinal track		Epithelium & glands, basement membrane	Smooth muscle (collagen type III)	Loose connective tissue
	Reproductive track Deferents, fallopian tubes		epithelium Basement membrane	Smooth muscle Dense connective tissue	Loose connective tissue

Table 2.9. Rigid Biosynthetic Products

Tissue	Cellular components	Extracellular organic matrix	Extracellular crystal line phase
Bone (compact cortical)	osteoblast, osteocyte, osteoclast	Collagen type I (90%) proteoglycans,[a] fibronectin, sialo-protein, glycoproteins, Gla-protein, phosphoprotein	Calcium phosphate crystals (calcium hydroxyapatite $Ca_{10}(PO_4)_6(OH)_2$
	Ameloblast	Enamel amelogenins (pro teins), enamelins (proteins)	Large crystal (calcium hydroxyapatite)
	Odontoblast	Predentin, dentin collagen (type I) (glycoproteins, Gla proteins, proteoglycans, phosphoproteins, phosphoryns)	Calcium hydroxy-apatite (high phosphate contents)

[a]Chondroitin - 4 - sulfate

2.5.2. Tendon and Ligament

Tendon has a cablelike structure that is composed of roughly circular fascicular units that contain collagen fibrils oriented approximately parallel to the long axis of the tendon (Figure 2.8). The collagen fibers are composed of bundles of collagen fibrils that are buckled into a crimped planar pattern. The uncrimping of collagen fibrils occurs at low strains and is believed to absorb high-strain-rate muscle loads. The highly organized structure of collagen fibers in tendon occurs by the biosynthesis and ordered deposition of connective tissue components by fibroblasts. These cells form long rows that lay down collagen fibers within a cylindrical template that is aligned with extensions of fibroblast membrane (McBride *et al.*, 1985). The collagen fibers are believed to grow in length by end-to-end addition of collagen fibril bundles until they reach a critical length required to reinforce the interfibrillar matrix (McBride *et al.*, 1988).

Patellar tendon (Figure 2.13) is connected to the patellar bone and continues in the form of a ligament until it connects to the tibia, a long bone. Several other ligaments are important in the stability of the knee. These include the collateral and cruciate ligaments (see Muller, 1983, and Figure 2.14). The anterior cruciate ligament is commonly injured in sports; it prevents the femur from sliding forward over the tibia and is essential for locomotion that requires a change in direction.

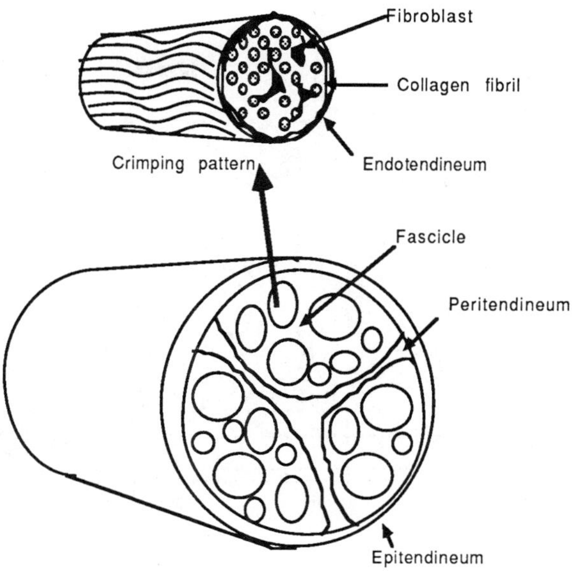

Figure 2.8. Tendon structure. Tendon consists of a cable-like structure containing different connective tissue elements that separate collagen fibril bundles.

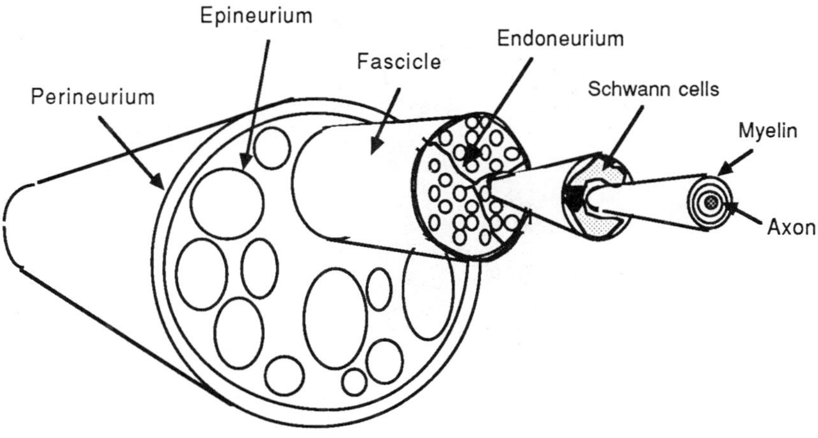

Figure 2.9. Peripheral nerve structure. Diagram of a peripheral nerve showing a cable-like structure containing layers of connective tissue and neural cells (axon).

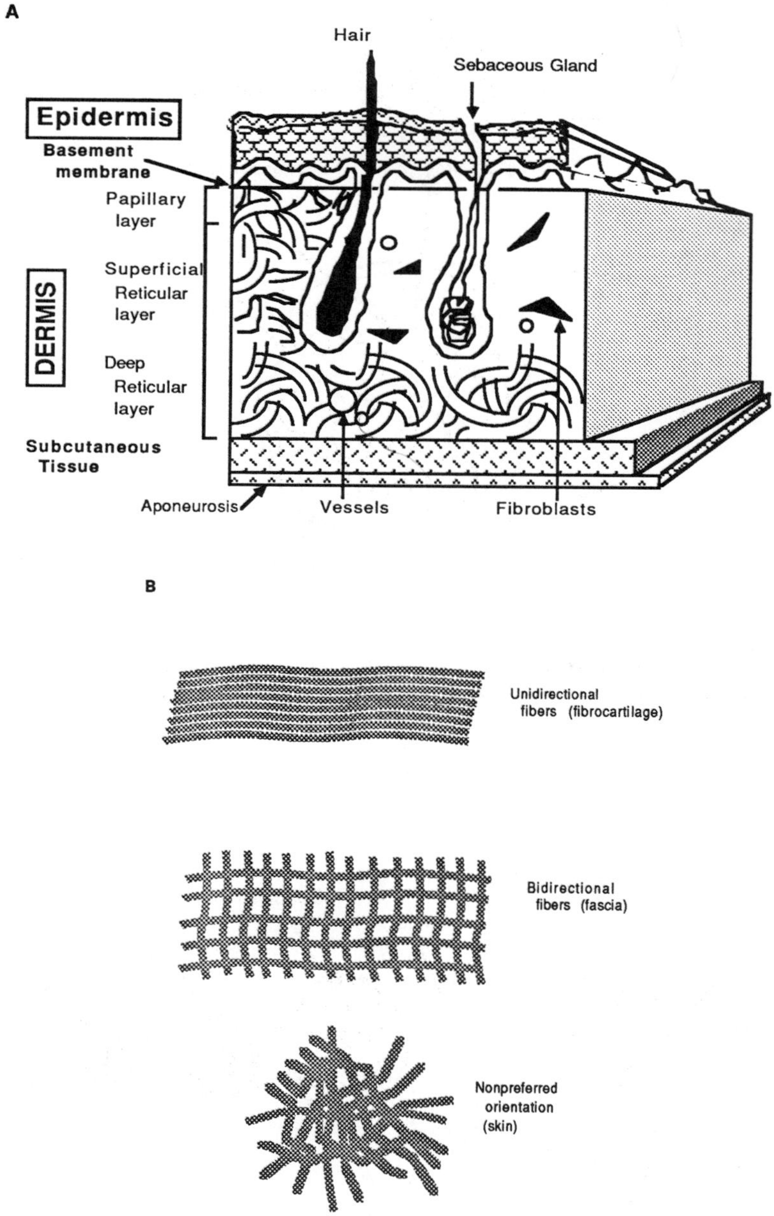

Figure 2.10. Skin structure. (A). The different layers of skin and subcutaneous tissue form a sheetlike structure. Epidermis is composed of several layers of cells while dermis is a dense connective tissue with large collagen fibers present in the deep layer of the reticular dermis. In the subepidermal layer (papillary layer of the dermis) a loose connective tissue is observed. (B). Diagram illustrating the arrangement of collagen fibers in skin as compared to fibro- cartilage and fascia.

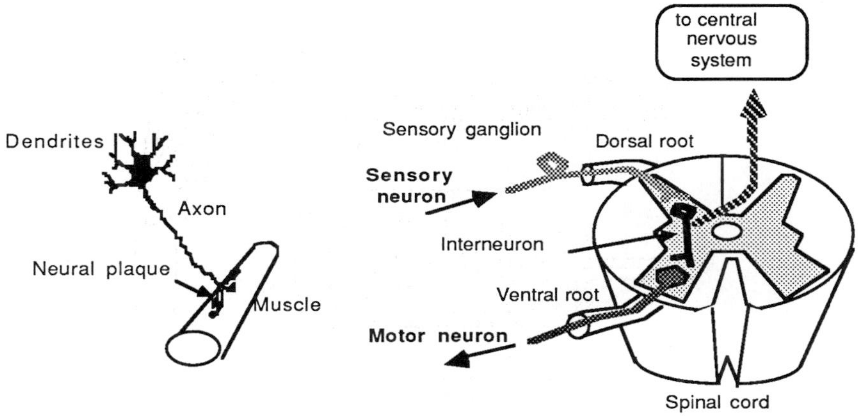

Figure 2.11. Type of nerves. Illustration of connection between sensory neuron, interneuron, and motor neuron. Motor neurons contain dendrites through which impulses are collected and transferred through an axon of muscle.

2.5.3. Bone

Long bones such as the femur and tibia are composed of compact and trabecular bone (Figure 2.15). In compact bone osteocytes, also known as bone cells, are arranged in concentric cylindrical layers within a unit termed osteon (Figure 2.16).

Osteons contain mineralized collagen fibrils in concentrically arranged layers with the orientation of the collagen fibrils alternating between $\pm 45^{o}$ with the long axis of the osteon. Hydroxyapatite is the mineral phase found deposited in the hole region within the quarter-stagger arrangement of collagen fibrils. The mineral phase increases the stiffness and decreases the strain at failure for the collagen fibril network of bone. Osteoclasts are constantly degrading the mineralized collagen matrix that is being laid down by osteoblasts within the osteon. For this reason the osteon can be considered a dynamic unit that is continually being added to and broken down. Maintenance of bone stability and osteoblast activity is associated with continued mechanical stresses that are borne by the osteon.

The center of long bone is composed of open channels of cancellous bone (spongy bone) as diagrammed in Figure 2.15. The mineralized collagen fibers of cancellous bone are woven into a network that is less organized as compared to the non-woven, aligned network found in compact bone. The channels found within cancellous bone are filled with marrow containing precursor blood cells. Compact and spongy bone are wrapped with a connective tissue layer termed periosteum.

Bone contains a variety of noncollagenous components in addition to type I collagen. These include osteonectin, bone proteoglycan, bone sialoprotein, and phosphoproteins (see Termine, 1985 for a review). Noncollagenous bone-specific components are found in abundant quantities at all stages of osteogenesis.

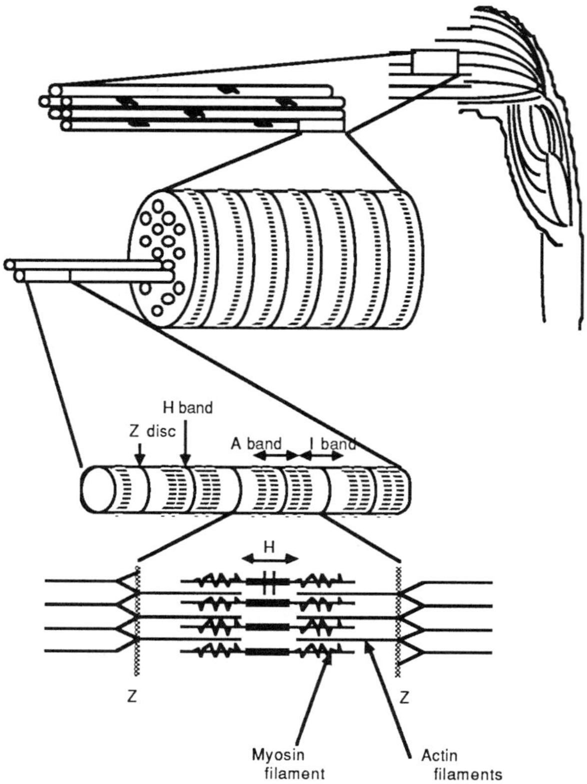

Figure 2.12. Skeletal muscle. Schematic presentation of the organization of a striated muscle (skeletal muscle). The muscle is covered by a dense connective tissue in the form of a sheetlike structure. Skeletal muscle consists of actin and myson filaments that contract after an electrical impulse is received by the muscle cell.

2.5.4. Articular Cartilage

Articular cartilage is a special type of hyaline cartilage (Figure 2.17) composed of at least three layers. The surface layer (superficial layer) consists of aligned collagen fibers with cells dispersed between the fibers. The intermediate layer is composed of a bracing system of collagen fibers oriented at angles of about 45° with respect to the surface and contains cartilage cells termed chondrocytes (Minns and Stevens, 1977). Chondrocytes are observed to be surrounded by almost transparent regions termed lacunae that are rich in proteoglycans. Proteoglycans are rich in negatively charged sugar residues causing the intermediate zone of cartilage to be highly water swollen. When a compressive load is applied to cartilage, water and joint fluid is squeezed out of the intermediate zone into the synovial cavity. The deep zone is composed of collagen fibers, which are oriented perpendicular to the underlying bone, and serves to attach cartilage to the surface of compact bone.

Bone-to-bone contact and force transmission are normally prevented by articular cartilage that lines the joint surfaces. A layer of viscous fluid (synovial

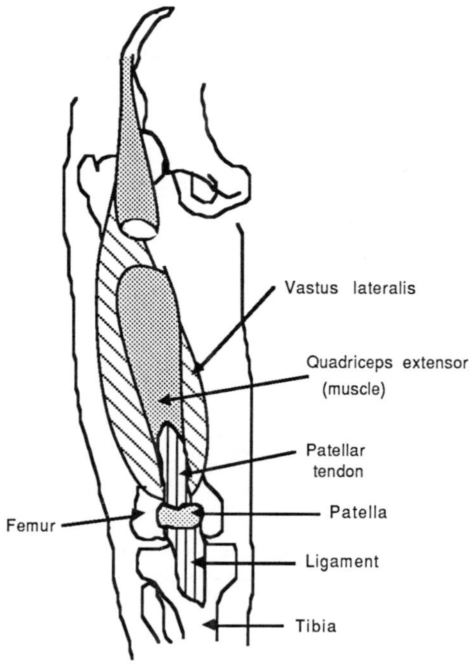

Figure 2.13. Example of tendons, ligaments, bones, and muscles. Anterior view of the thigh and knee showing tendon, muscle, bone, and ligaments of the knee.

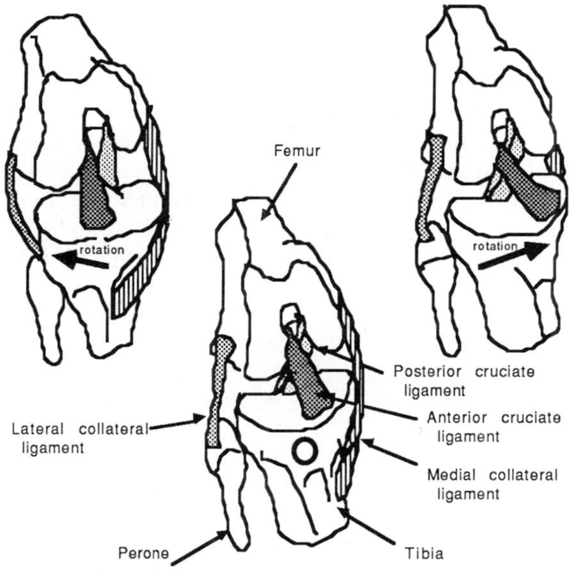

Figure 2.14. Ligaments of the knee. Diagram showing the cruciate and collateral ligaments of the knee which are involved in rotation.

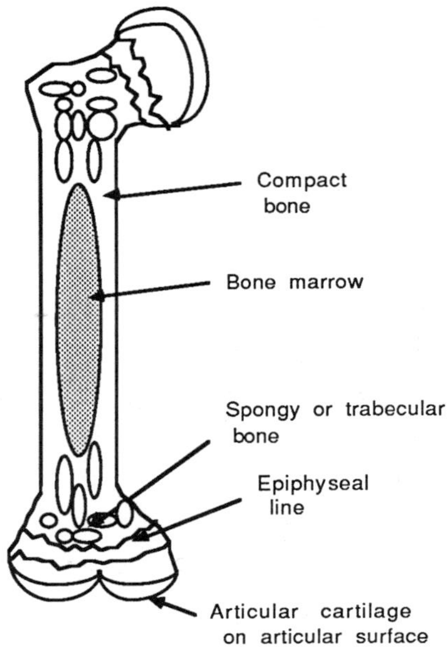

Figure 2.15. Structure of long bone. Long bone consists of a tube like structure with several layers of well-organized connective tissue to support compressive loads. Compact and trabecular bone are found in the ends and center, respectively.

fluid) is present on the surface of cartilage so that cartilage surfaces "hydroplane" across each other without significant energy loss or surface wear. A pathologic process termed osteoarthritis, occurs as a result of cartilage degeneration. This disease is characterized by a loss of proteoglycans from cartilage resulting in erosion of the surface and intermediate zones and eventual joint pain due to bone-to-bone contact. Cartilage under these conditions no longer acts as a shock absorber and does not dissipate mechanical loads applied to the joint.

Results of recent studies suggest that the mechanical properties of bone depend on the mass and continuity of the trabecular lattice (Mosekilde *et al.*, 1987) while those of articular cartilage depend on the load and collagen/proteoglycan ratio (Akizuki *et al.*, 1986; Kiviranta *et al.*, 1987).

2.6. Tissues Involved in Food Processing and Digestion

Food ultimately is the source of energy required for adenosine triphosphate (ATP) production and muscular contraction. In order for water-insoluble food stuffs to be processed for digestion they must be ground into small fragments by the teeth (Figure 2.18).

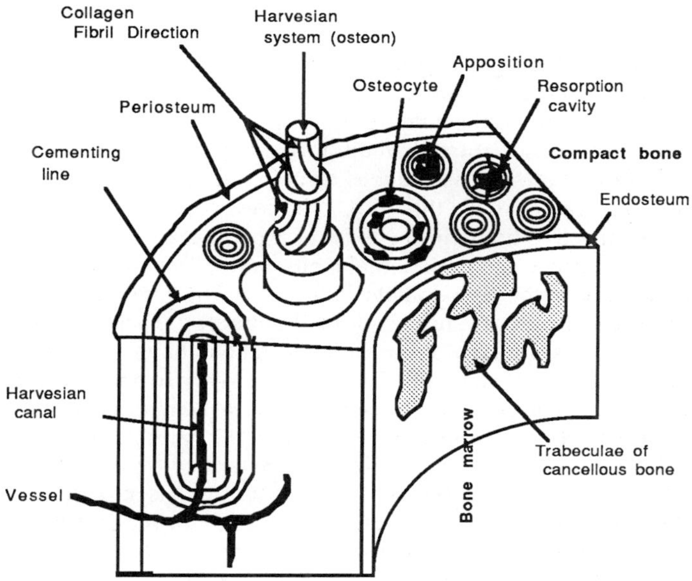

Figure 2.16. Haversian system. Diagram of a bone cross-section showing osteons in compact bone.

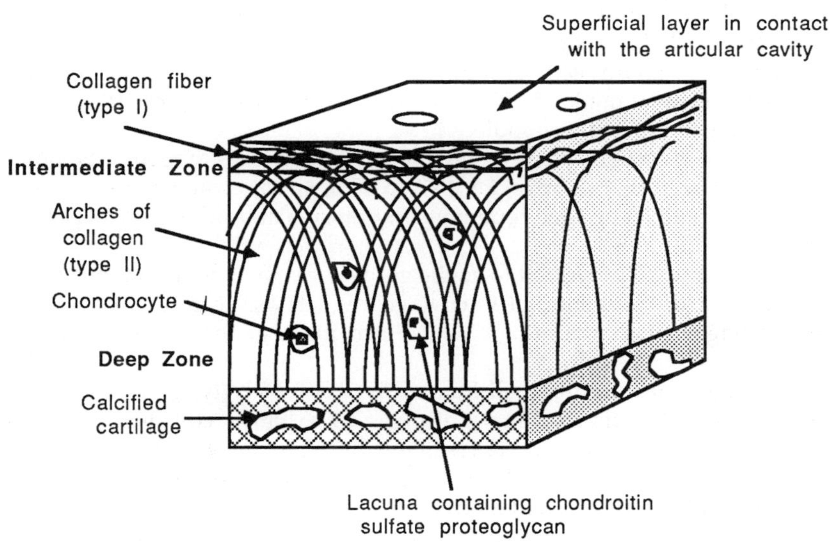

Figure 2.17. Cartilage structure. Diagram showing trilaminar structure of articular cartilage consisting of superficial zone, intermediate zone (arches of collagen), and deep zone that interfaces with the subchondral bone.

2.6.1. Teeth

Teeth are composed of mineralized and non-mineralized connective tissue in the form of enamel and dental pulp that are in contact with the non-mineralized gum and mineralized connective tissue of the jaw bone.

The jaws are bony structures that are formed from the maxillae (upper jaw) and mandible (lower jaw). These bones contain sockets to which the teeth are attached and are covered with the gum. The lower part of the bone socket is covered with connective tissue in the form of a periosteum. The socket is lined with a membrane termed the periodontal membrane. Teeth are embedded in the socket and held close to the periodontal membrane by the bony cementum.

There are four types of adult teeth in the mouth. These are the incisors, canines, premolars and molars. Premolars and molars have more than one root while the others have only one root. Types of teeth vary in shape and number of roots, but all teeth are composed of the same structural components.

The central portion of the tooth (Figure 2.18) consists of a pulp cavity surrounded by a thick layer of mineralized tissue termed dentin. The surface of the tooth that is exposed to the oral cavity (crown) is covered by a hard substance termed enamel and the surface embedded in the jaw is covered by cementum (root). The tooth is attached to the surrounding alveolar bone by a collagenous periodontal ligament (Ross and Reith, 1985).

Enamel is a cell—free extracellular tissue and contains about 96% mineral in a mature tooth. Enamel proteins have been found to represent at least two major classes of polypeptides: (1) amelogenins and (2) enamelins (Slavkin, 1985).

Dentin composes the bulk of the tooth substance and is approximately 70% mineral (hydroxyapatite) and 30% organic components. The bulk of the organic phase is collagen and the remainder consists of phosphoproteins and proteoglycans, and glycoproteins similar to those found in bone (Termine, 1985).

Dental pulp contains cells including odontoblasts, fibroblasts, macrophages, and migrating cells from blood. Blood vessels and nerves are present in the root canal and leave the pulp as a bundle.

Cementum is a mineralized tissue that covers the roots of teeth. Its mineral content is similar to that of bone, i.e. 45% to 50%, and its water content is about 50% to 55%. The main organic component is collagen and proteoglycans are also present.

Gingiva commonly referred to as the gum covers the underlying bone and is firmly attached to the bone. It is covered by a layer of epithelial cells and contains collagen fibers that attach to the cementum of the tooth and alveolar bone.

The collagen of fully matured human teeth is mostly type I collagen while the pulp tissue of the diseased tooth displays a mixture of collagen types I and III (Karjalainen and Soderling, 1984). Results of other studies indicate that antibodies to types I and III collagen react with fibroblasts of the dental pulp (Karjalainen, *et al.*, 1986). Collagen turnover is considered to be faster in periodontal ligament and gingiva compared to non oral tissues such as skin (Imberman *et al.*, 1986; Narayanan and Page, 1983; Simpson, 1983).

Gathecole *et al.* (1987) found that the axial period in collagen from periodontal ligament and gingiva resembles the value found for dermal collagen (65 nm) as opposed to that for tendon collagen (67 nm). The chewing surface that grinds food is made up of a hard material termed enamel, while the rigid portion

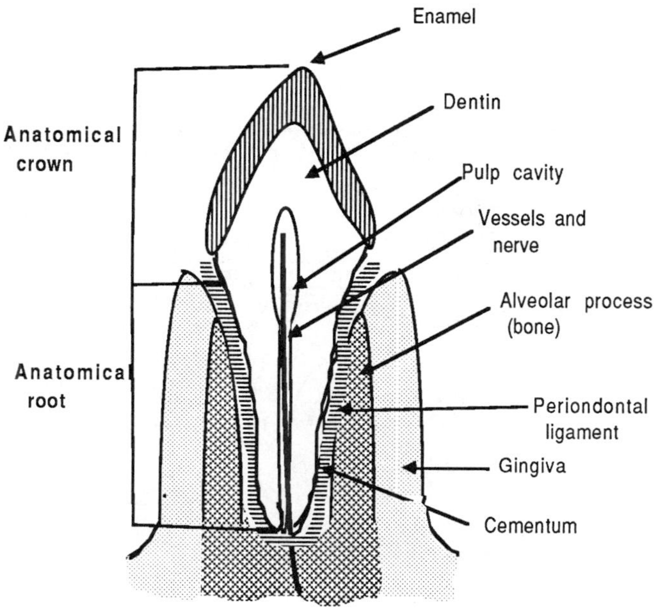

Figure 2.18. Tooth structure. Diagram of a tooth (sectioned) showing the relationship between tooth and mandible (periodontal ligament).

of the tooth below the enamel is composed of dentin.

2.6.2. Gut and Intestine

Food is chewed by the teeth and then moved backwards into the throat by the tongue. During swallowing, food is pushed through a muscular opening (pharynx) and into the esophagus (Figure 2.19). Swallowing requires lubricating action that is provided by saliva which is produced by the submandibular, parotids and other salivary glands of the mouth. The esophagus is a muscular tube behind the trachea (wind pipe) that is 23- 25 cm in length and connects the pharynx to the stomach. Muscular squeezing action (peristalsis) causes food movement from the pharynx to the stomach. The entry of food into the stomach is controlled by a ring of muscle termed the cardiac sphincter.

The stomach is a muscular J-shaped bag as diagrammed in Figure 2.20. Cells in the stomach release pepsinogen, a precursor to the proteolytic enzyme pepsin, hydrochloric acid, and a lubricating mucus that prevents digestion of the stomach wall. The stomach is divided into three regions, i.e. cardiac, pyloric, and body. These three regions of the stomach have wall structures that include mucosa, submucosa, muscularis externa, and serosa.

The mucosa is the inner surface of the stomach that is folded and contains numerous grooves. The inner surface contains gastric pits and gastric glands that are lined by epithelial cells that secrete mucous which protects the stomach lining from damage due to acid and pepsin.

The mucosa is a membrane consisting of elongated (squamous) epithelial

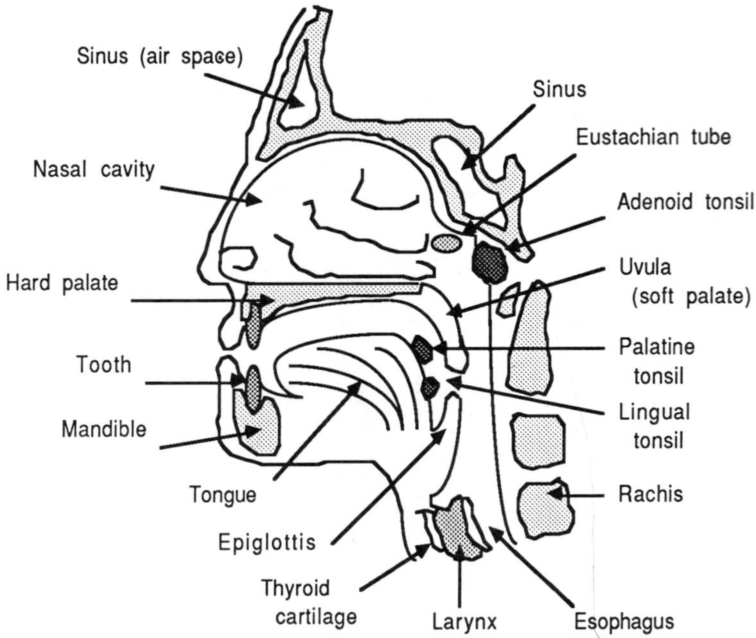

Figure 2.19. Oral and nasal cavities. Schematic presentation of a section through the cavities of the nose and throat showing the superior respiratory (larynx) and digestive (esophagus) tracks.

cells and a small amount of underlying connective tissue. Submucosa is composed of dense, irregular connective tissue and contains blood and lymphatic vessels and nerves. Muscularis externa of the stomach consists of three layers of smooth muscle; an outer longitudinal layer, a circular middle layer, and an inner oblique layer. The serosa is a sheath containing loose connective tissue.

The enzyme- and acid- hydrolyzed food then passes into the small intestine, a tube 3.25 m long. Absorption of digested products occurs through the epithelial wall of the small intestine into the blood via the capillary walls with the aid of secreted enzymes and mucus. The wall of the intestine contains layers that are similar to that of stomach. The submucosa in the small intestine has been shown to have a crossply arrangement of collagen fibers that are parallel to the surface of the wall and make an angle of between 30° and 55° to the longitudinal axis. The submucosa is thought to offer resistance to excessive distension of the wall.

Absorbed products are transported through the intestine and accumulate in the lumen of blood vessels for transport throughout the body. Undigested products then move to the large intestine (colon), a tube 1.5 m long and 6.5 cm in diameter. The remaining food is excreted at the end of the large intestine (rectum) through the anus that is composed of several rings of muscles (anal sphincter).

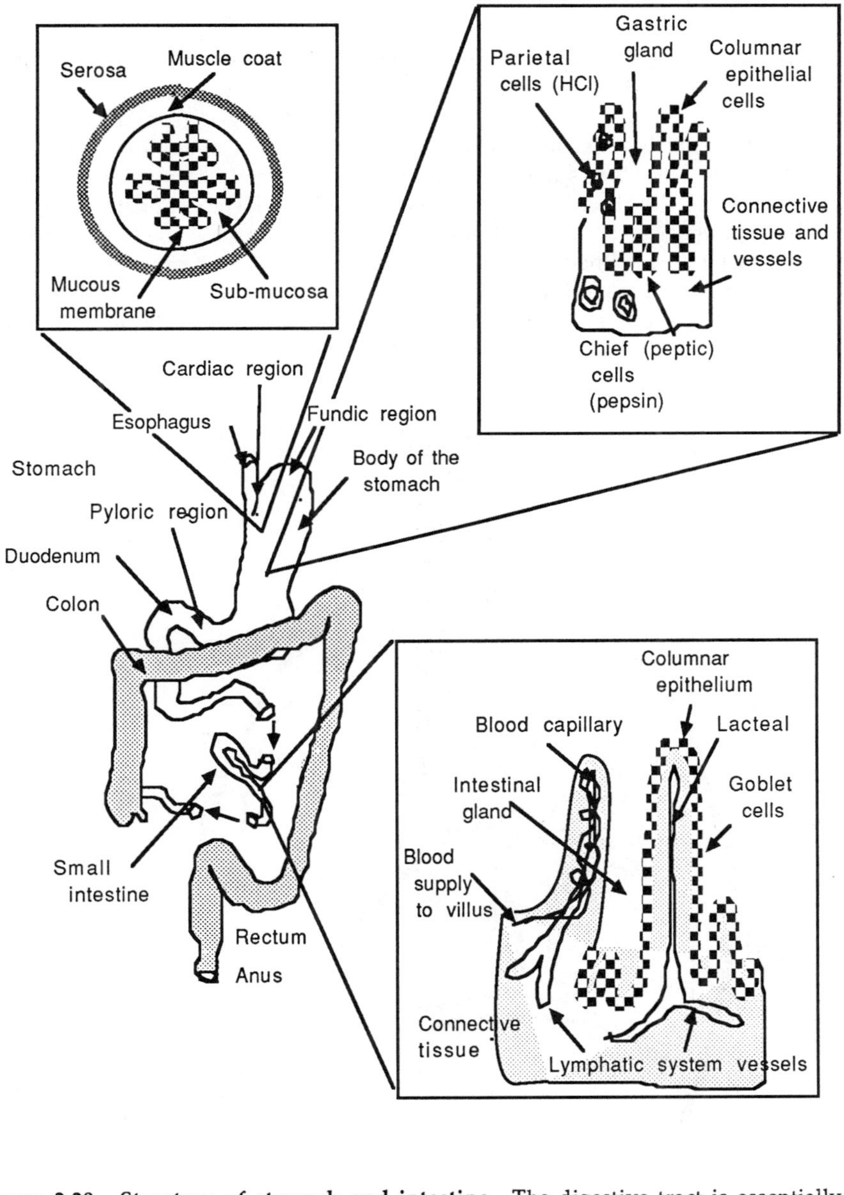

Figure 2.20. Structure of stomach and intestine. The digestive tract is essentially a tubelike structure containing 3 layers: (1) A serosa consisting of an external connective tissue sheet surrounding the intestine. This sheet allows sliding of the gut in the peritoneal cavity. (2) A muscular coat (muscularis mucosa) consisting of smooth muscle cells surrounded by a small amount of connective tissue. (3) A submucosa consisting of loose connective tissue containing blood supply and supporting the mucosa. Mucosa consists of cells of different biological properties which vary depending on the location, i.e., stomach versus intestine. Diagram illustrates the structure of the regions where HCl is released into the stomach (upper right hand corner) and nutrients are absorbed in the intestine (lower right hand corner).

2.7. Respiration

The transport and storage of oxygen and excretion of CO_2 is another physiologic process that is required for energy generation (ATP) that is used for locomotion and other bodily functions. Oxygen is obtained from air that enters the nose through the nostrils and passes through the nasal cavities, where it is filtered by the nasal hairs and warmed and humidified. Air passes from the nasal cavities to the throat (pharynx) and voice box (larynx) and then through the windpipe termed the trachea (Figure 2.21). The trachea is 10 cm long and 2.5 cm in diameter and is lined on the inner surface by a ciliated mucous membrane (epithelial cells). It is reinforced along its length by loops of cartilage and covered on the outside by a layer of elastic connective tissue. On the inside, the middle layer is also lined by smooth muscle.

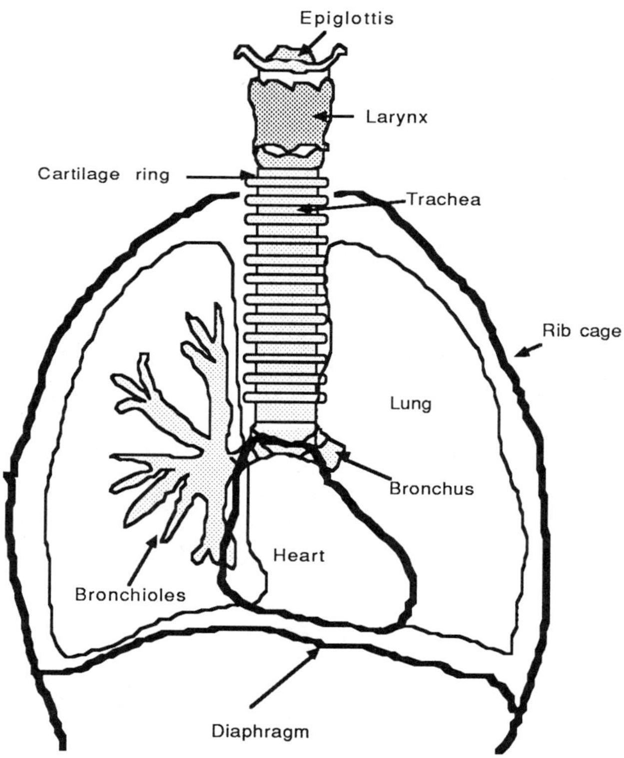

Figure 2.21. Respiratory apparatus. The respiratory tract is composed of different segments with tubelike structures such as larynx, trachea, bronchus, and bronchioles. The trachea consists of alternate cartilage rings which give its biomechanical properties.

At the base, the trachea branches into two divisions termed bronchi. The bronchi are tubes that are reinforced with C-shaped cartilage hoops. The inner wall is lined by a ciliated mucous membrane. The bronchi divide twice into secondary bronchi connecting to the two lobes of the left lung and the three lobes of the right lung. Smaller branches of secondary bronchi are the bronchioles which connect the different segments of a lobe. The bronchioles in turn branch into terminal bronchioles that do not have cartilage supporting the walls but are lined by cuboidal epithelium and have smooth muscle in their wall. Terminal bronchioles connect into alveoli, air sacs lined with epithelium through which oxygen is transported to the pulmonary capillaries and veins and CO_2 is exchanged (Figure 2.22).

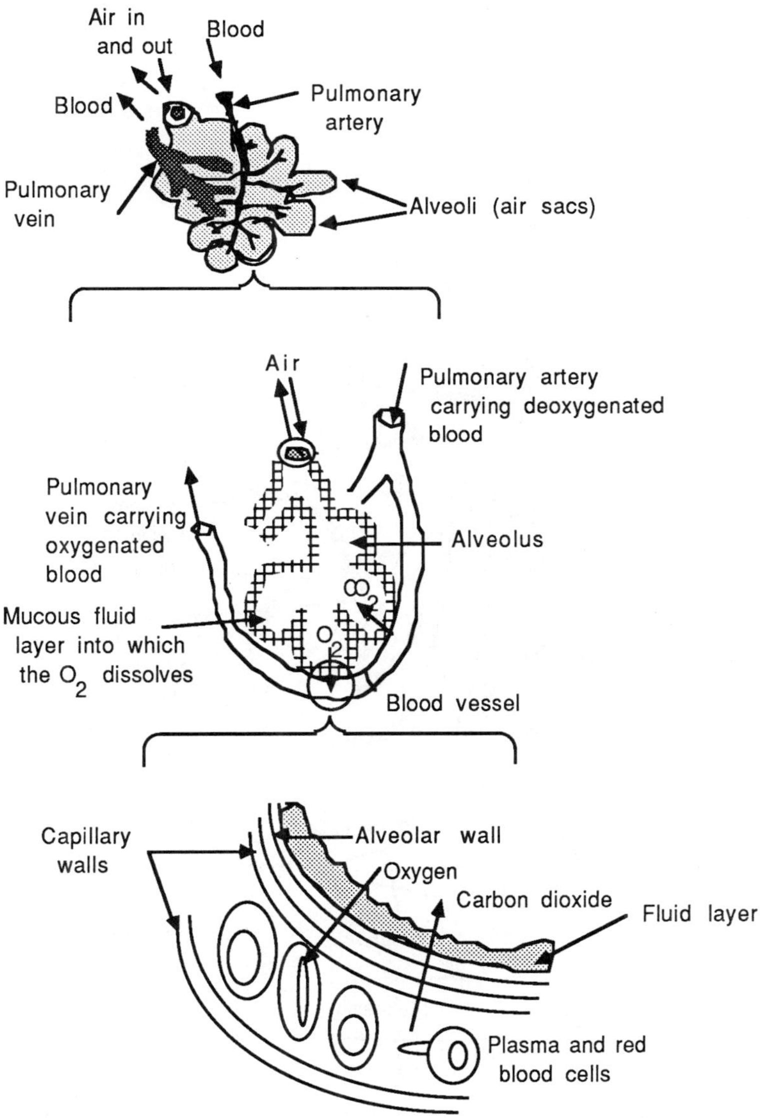

Figure 2.22. Alveoli and gas exchange. Alveoli consist of sacks where air enters and leaves via the bronchioli. Vessels surround each alveoli and oxygen and carbon dioxide rapidly exchange within capillaries in the alveolar wall.

2.8. Cardiovascular System

The heart and blood vessels are the major components of the cardiovascular system. Blood is pumped by the heart from the vena cava through the right atrium to the pulmonary artery, right atrio-ventricular valve, right ventricle and then through the bicuspid valve (Figure 2.23). At the same time, the heart pumps blood from the pulmonary veins to the aorta through the left atrium, left atrio—ventricular valve, left ventricle and finally through the aortic valve (tricuspid valve).

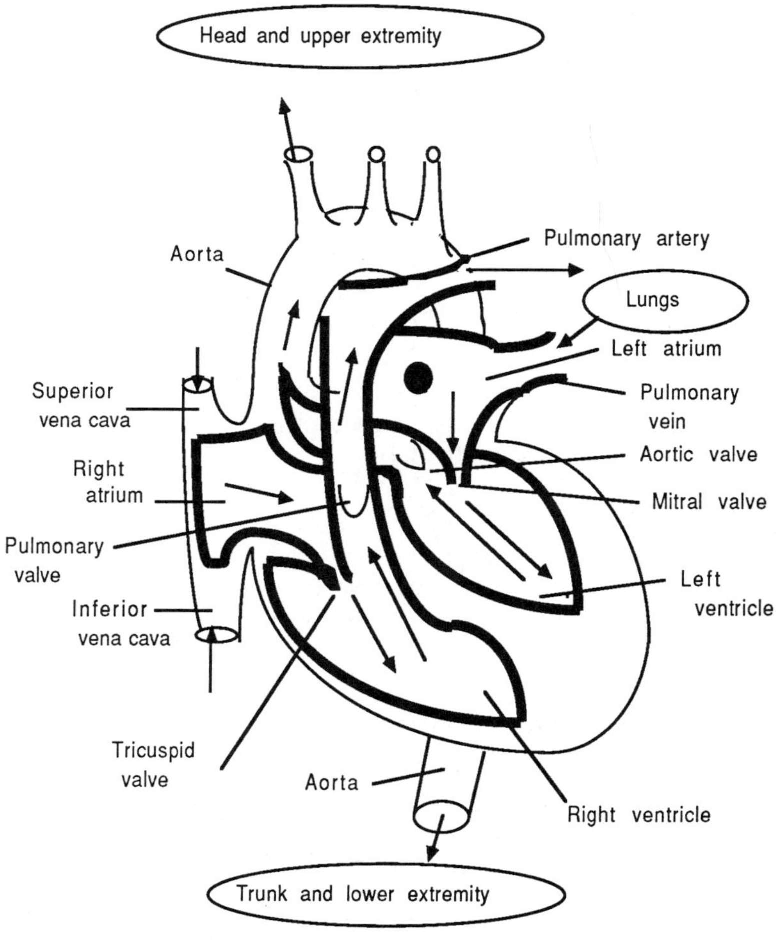

Figure 2.23. Heart and cardiovascular system. Diagram showing the flow of blood from the left ventricle to the aorta and to the peripheral circulation. Blood is returned to the right atrium via the vena cava.

2.8.1. Heart

The heart is composed of three layers, the pericardium, epicardium, and endocardium. The pericardium is a sheet of connective tissue composed of collagen fibers which are oriented at angles of about 60°, 180°, and 300° with respect to the vertical direction, and connective tissue cells. This connective tissue layer prevents overdilation of the heart.

Epicardium is rich in cardiac muscle cells and muscle fibers while the endocardium is a thin layer of loose connective tissue that is lined with endothelium, elongated cells that maintain the fluidity of blood.

Atrio-ventricular cardiac valves are composed of three leaflets that are opened and closed by flexible cables termed cordae tendineae. The organization of the leaflets is very similar to tendon since under polarized light (which is used to study the orientation of fibrous materials), a crimped aligned pattern is seen in both leaflets and tendons. A high content of aligned crimped collagen fibers is found in connective tissues that are cycled continuously during physiological functioning. Pathological conditions result in poor valve leaflet closure and leakage of blood during cardiac contraction and refilling. "Floppy" cardiac valves lead to the inability of the heart to pump sufficient blood to peripheral tissues.

2.8.2. Arteries and Veins

The aorta is an inverted U-shaped tube (Figure 2.24) that acts as an auxiliary pump and is the main conduit of oxygenated blood to the entire body. Branches of the aorta to various tissues of the body exist. When the aortic valve is open the aortic root expands in the radial and circumferential directions accommodating 50% of the blood pumped by the left ventricle. When the aortic valve is closed the retractive force generated by the aortic root serves to maintain elevated blood pressure. The aorta is composed of three layers as is the case for all blood vessels (Figure 2.24B). These layers are: (1) the intima, a layer of loose connective tissue that is lined by endothelial cells; (2) the media, a series of concentric rings of elastic lamellar units containing elastin, collagen fibrils and smooth muscle cells, and (3) the adventitia, a dense connective tissue layer that anchors the vessel wall to surrounding tissues. Aortic narrowing, wall dilatation (aneurysm) formation, and wall separation are pathological processes associated with atherosclerosis, normal aging, and connective tissue disorders.

Arteries that circulate blood to heart muscle cells are termed the coronary arteries. Maintenance of blood flow through these vessels is essential to the contraction of the left and right ventricles.

The aortic root is rich in elastic tissue, although significant levels of collagen are present. However, in the portion of the aorta in the abdominal region, the composition of the aorta is rich in collagen although elastic tissue is still present. Due to this composition difference the aortic root stiffness is lower than the stiffness in the abdominal region. The aorta has several branches that occur in the thoracic and abdominal region. These arteries have wall structures similar to that of the aorta although the thickness of the media varies from vessel to vessel. The aorta is an elastic artery with a well-developed media while the media of most other arteries is not as well-developed.

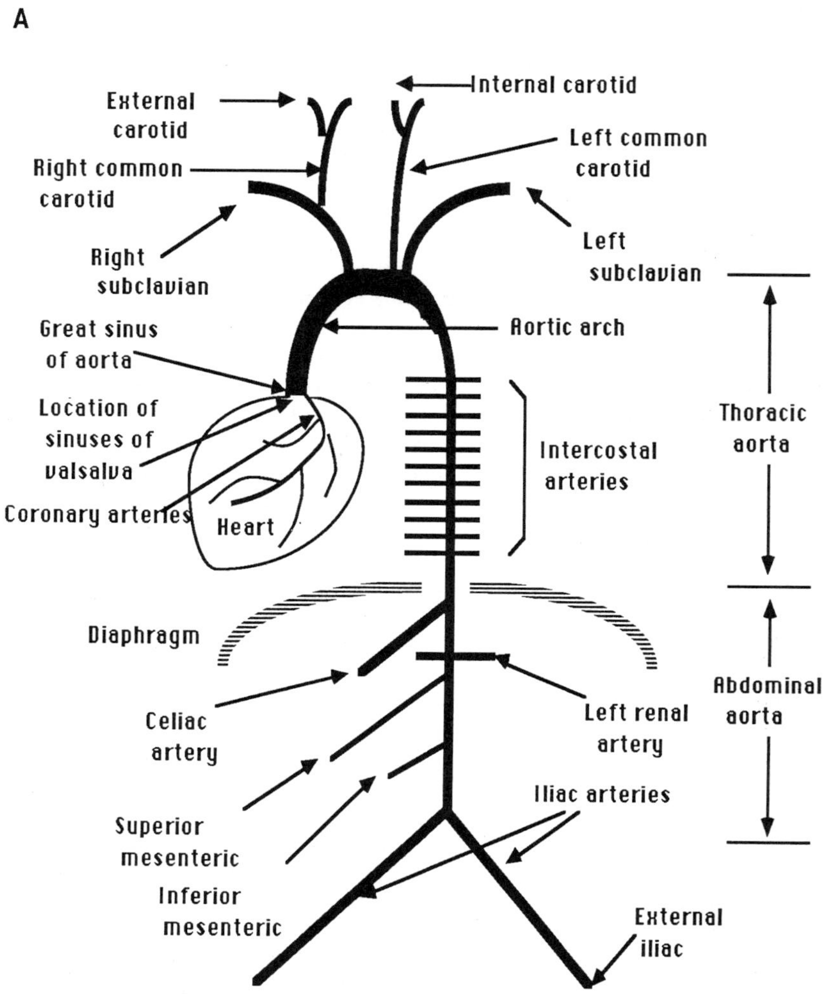

Figure 2.24. Peripheral vasculature. (A). Diagram of arterial systems showing the branches of the aorta.

Arteries that feed peripheral tissues and organs are observed to branch into arterioles and finally capillaries. Peripheral arterie, like the aorta, are composed of three layers. The media of these vessels is rich in smooth muscle cells. Capillaries contain an endothelium and some connective tissue but are only wide enough for a single red blood cell to flow through. Oxygen is released to tissues and carbon dioxide is absorbed in capillaries.

Blood is returned to the heart through the venous system consisting of venules and veins. The major veins returning to the heart are the superior and inferior vena cava. Veins have a wall structure similar to arterial vessels except the media is thinner and vessel walls have a lower modulus.

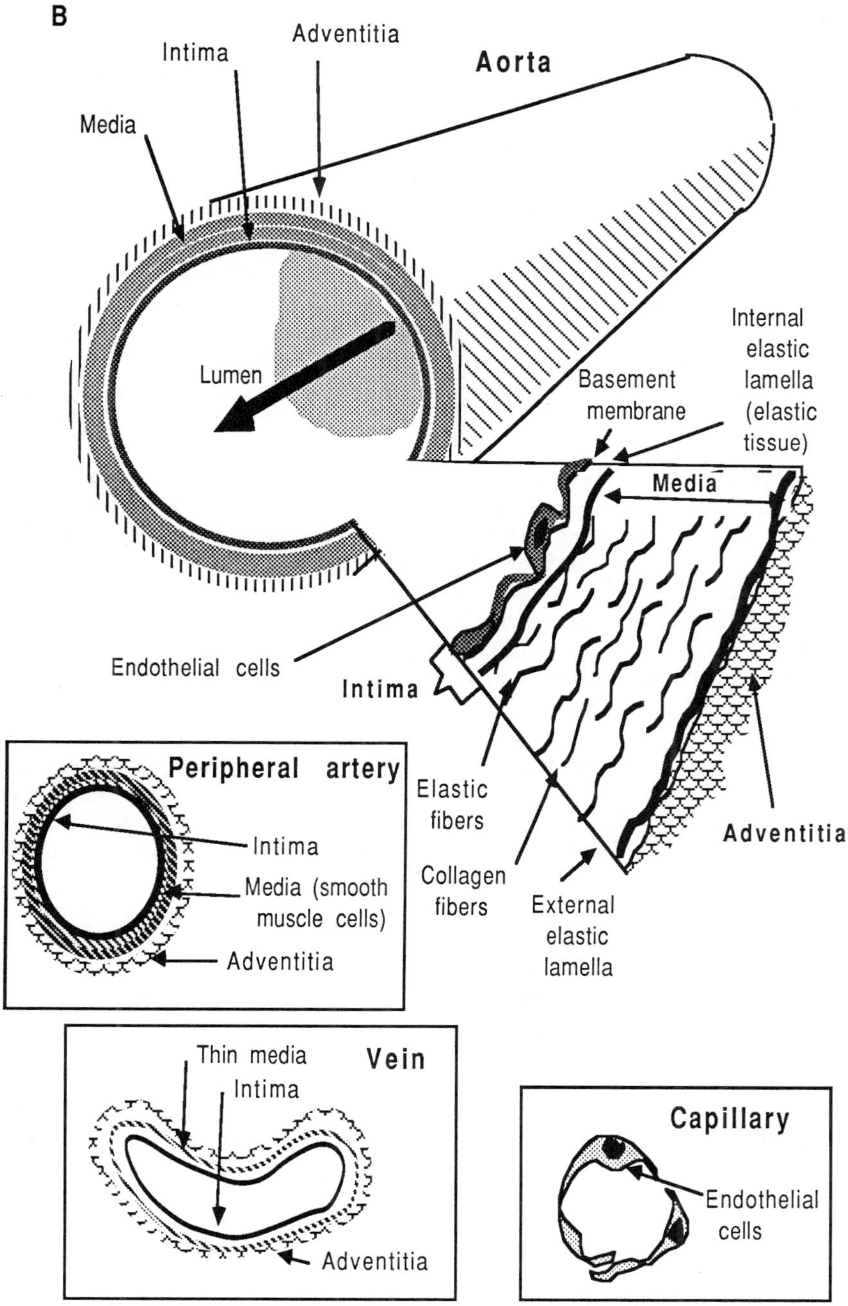

Figure 2.24. (cont.) (B). Cross-sectional view of aorta, veins, and capillaries showing comparative development of the vessel wall.

2.9. Visual System

The eye is lodged within the bony orbit in the skull and padded by fat. At the rear it is held within the eye socket by extrinsic eye muscles. The optic nerve located at the rear of the eyeball connects directly to the brain. The white of the eye is the outermost layer and is called the sclera (Figure 2.25). At the front of the eyeball the sclera becomes transparent and is termed the cornea. The transparent cornea allows for the transmission of light, which is focused by the lens onto the retina in the back of the eye.

The middle layer of the eyeball, the choroid, is rich in blood vessels and carries much of the blood supply to the innermost layer of the eyeball. At the front of the eye the choroid contains the ciliary body that has smooth muscle and to which the suspensory ligaments attach and position the lens. The iris is made up of smooth muscle fibers that help regulate the amount of light collected by the opening and closing of the pupil, an aperture through which the light passes.

The retina lines the rear portion of the eye. The inner portion off the retina is made up of nerve cells that produce and conduct nerve impulses to the brain via the optic nerve. Nerve cells sensitive to light are termed photoreceptors and consist of two classes, rods and cones. Rods are sensitive to dull light and give vision of movement and shape. Cones are sensitive to bright light; they are receptors of color and sharp outline. Cones are clustered around the central area of the retina (fovea centralis) where the vision is sharpest. The rods are found on the outer edge of the retina. Nerve impulses are created as a result of light impinging on the pigmented epithelium at the back of the retina (Figure 2.25).

The lens is a transparent structure composed of protein which is stretched into an oval shape by the suspensory ligaments. The front (anterior) chamber of the eye between the cornea and lens is filled with aqueous humor which circulates around the lens and exits from the eye to return to the blood. An intraocular pressure of about 24 mmHg is maintained by continuous production and drainage of the fluid. The rear (posterior) chamber of the eye behind the lens is filled with a jellylike material termed vitreous humor.

2.10. Auditory System

The ear consists of three parts: the external, middle, and inner ear (Figure 2.26). Sound waves enter the external auditory canal and vibrate the tympanic membrane (ear drum) in the middle ear. The vibration of the tympanic membrane causes motion of the bones (malleus, anvil, and stapes) behind the membrane. Movement of the stapes against the oval window results in stimulation of the auditory nerve and signal transmission to the brain.

2.11. Blood and Lymph Tissues

Blood contains cells and macromolecules (Figure 2.27) that are important in the functioning of a variety of tissues and organs. Red blood cells (anucleated cells) carry oxygen to peripheral tissues and contain blood group antigens (blood A, B, O) on their cell surface. These antigens must be matched for tissue transplantation (see Chapter 10). Platelets are cells lacking nuclei which are

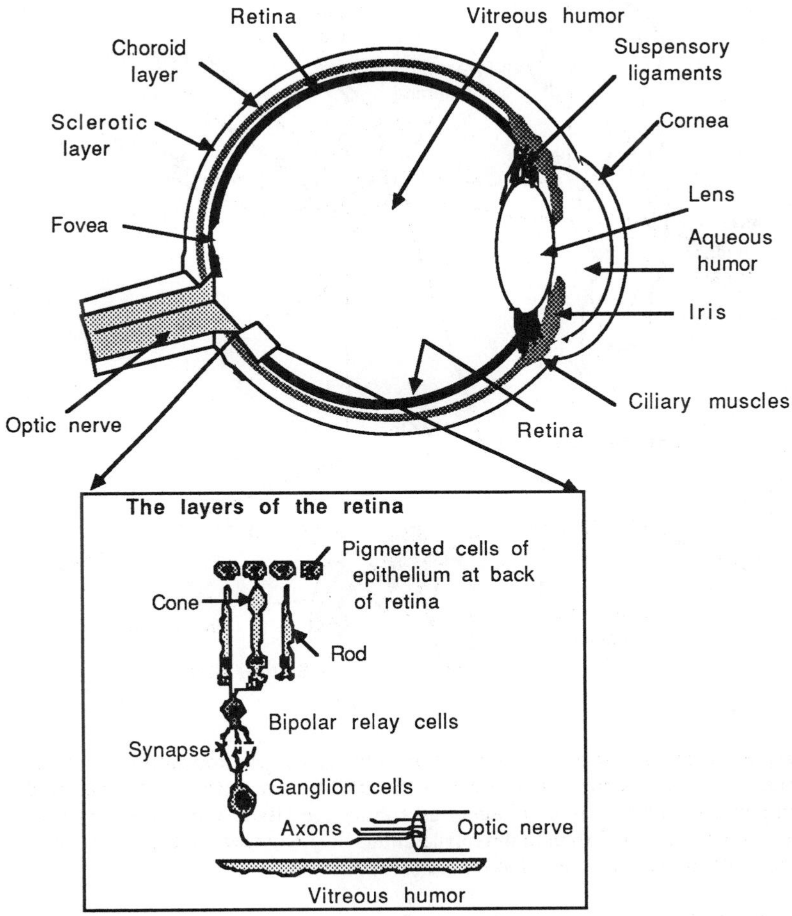

Figure 2.25. Structure of eye. Section of the eye shows several layers mong them a dense connective tissue (sclera) that has a protective role. Retina is a sensory tissue and consists of several layers of neural cells, which converge to form the optic nerve. Cornea is composed of aligned, collagen fibril bundles which gives the transparency. Vitreous humor is rich in hyaluronic acid.

required for blood to clot when a vessel is cut. When platelets contact with exposed collagen fibrils they adhere and aggregate and initiate the formation from soluble blood proteins of a crosslinked network of fibrin.

Blood is also a vehicle for white blood cells that are derived from bone marrow and lymphoid tissue. Neutrophils and monocytes are cells that are capable of phagocytosis that results in engulfment of foreign bacteria and materials. Lymphocytes are cells that are derived from lymphoid tissue (Figure 2.28) such as the lymph nodes, spleen, and thymus, and are involved in immunological responses of the body to foreign cells, tissues, and molecules. Lymphocytes include B-lymphocytes (antibody producing cells) and T-lymphocytes, that are involved in the rejection reaction and cell—mediated

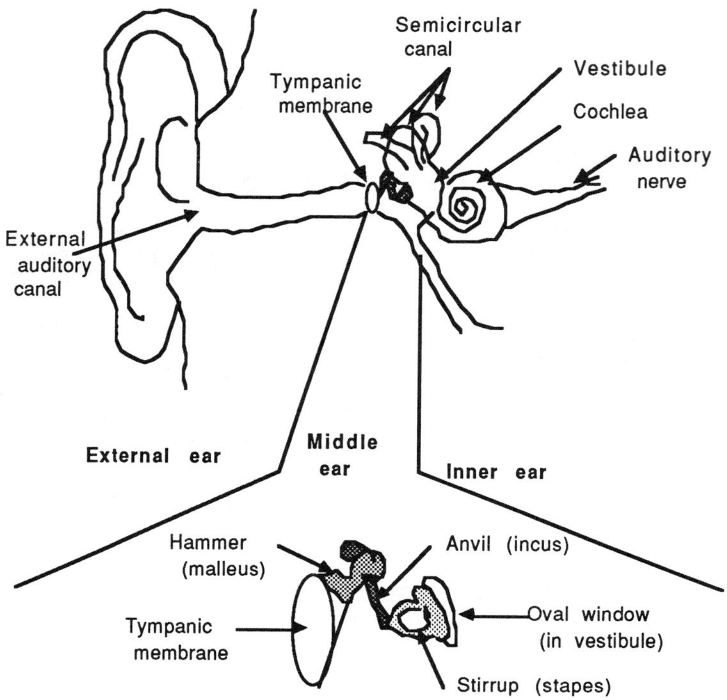

Figure 2.26. Structure of ear. The ear is anatomically separated into external, middle, and inner ear. Fibrocartilage is present in the external ear (auricle). In the middle ear, the tympanic membrane is a collagenous connective tissue that transmits the sound through bony ossicles and then to the cartilaginous oval window. In the inner ear, sound is translated into an electrical signal to the auditory nerve.

immunity. Other blood cells include leukocytes, platelets, and red blood cells.

Blood plasma, the acellular portion, contains a variety of macromolecules including albumin (maintains osmolarity of blood), immunoglobulins (antibodies), globulins involved in lipid, hormone, and metal ion transport, low- and high-density lipoproteins, proteins involved in blood clotting cascade and fibrinolysis, proteins involved in the complement system, and proteins involved in the kinin system (see Chapter 4).

Lymphoid tissues are involved in the defense of the body against infectious agents and materials of natural and synthetic origin such as bacteria and implants. They consist of encapsulated organs such as lymph nodes, spleen, and thymus, which are connected by lymphatic ducts (through which tissue fluid termed lymph flows) and unencapsulated tissue, which is present throughout the body.

Capsulated lymphoid tissue is surrounded by dense connective tissue (capsule) and contains additional connective tissue that acts as a partition that separates the organ into lobes filled with lymphoid cells. Immature lymphocytes in lymphoid tissue are activated on contact with foreign substances (primary immune response) and again on further exposure (secondary immune response).

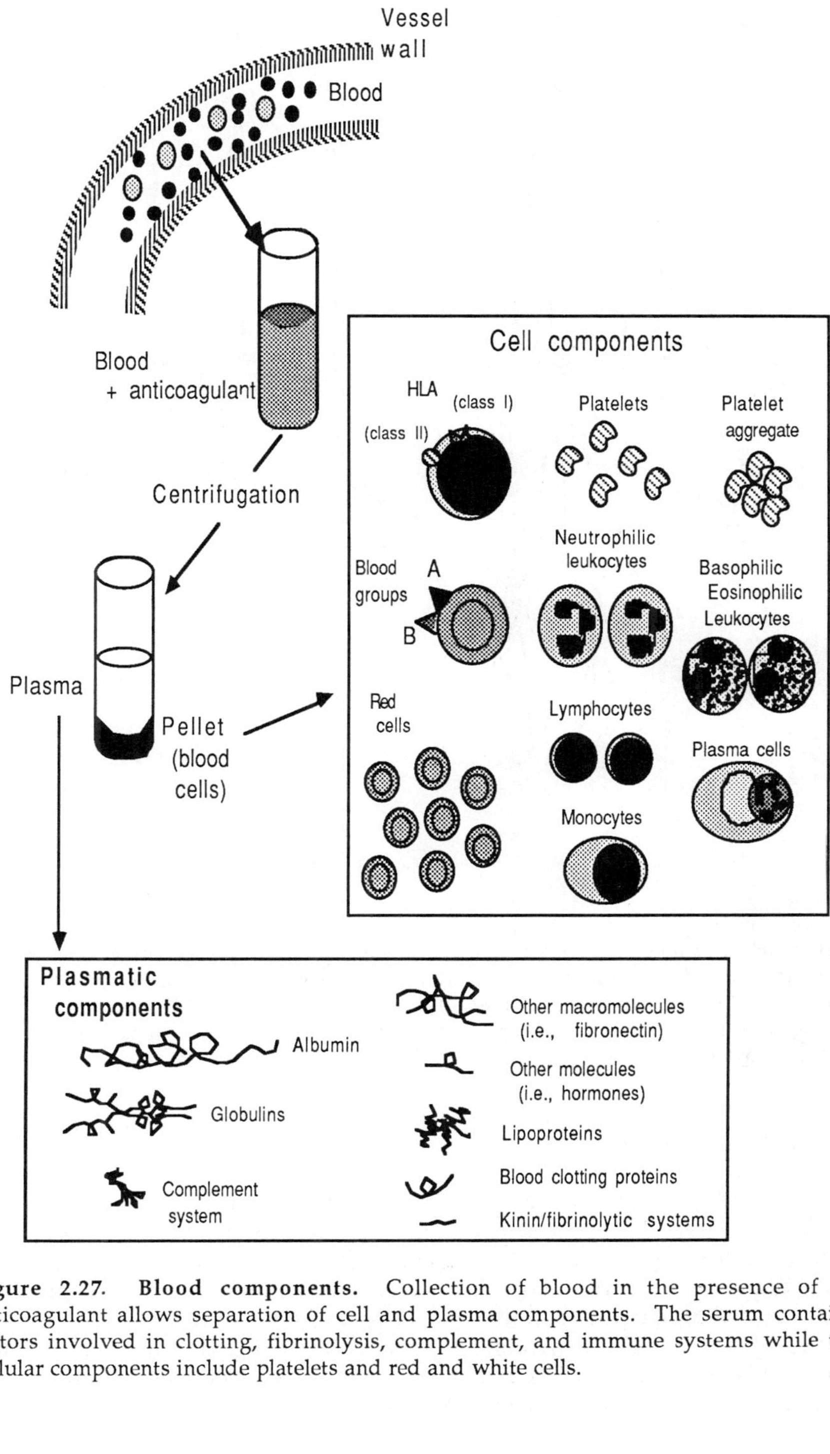

Figure 2.27. Blood components. Collection of blood in the presence of an anticoagulant allows separation of cell and plasma components. The serum contains factors involved in clotting, fibrinolysis, complement, and immune systems while the cellular components include platelets and red and white cells.

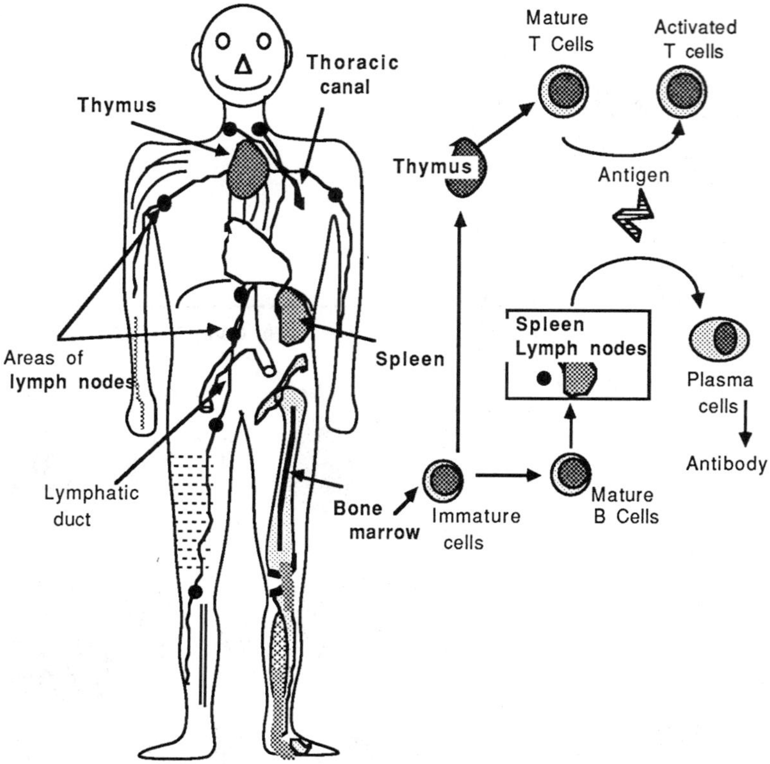

Figure 2.28. Immune system. The immune system is composed of several organs and tissues found in lymph nodes, spleen, and thymus. Stem cells are formed in the bone marrow. B and T cells and monocytes are the major cells involved in immune response. The thymus is a specific organ where T cells become mature; this organ disappears during adulthood. Spleen and lymph nodes contain immune cells such as T cells, B cells, and monocytes.

2.12. Endocrine System

The endocrine system (Figure 2.29) consists of glands dispersed throughout the body surrounded by a capsule and separated into lobes and lobules by connective tissue partitions. These glands, after stimulation, release chemicals termed hormones into the blood. Hormones released into the blood are involved in the control of cell growth in tissues such as the endometrium (see Chapter 11). Endocrine tissues include thyroid, parathyroid, adrenal glands, ovary and testes, hypophysis and pancreas. The release of hormones is also under the influence of the other organs such as the hypothalamus.

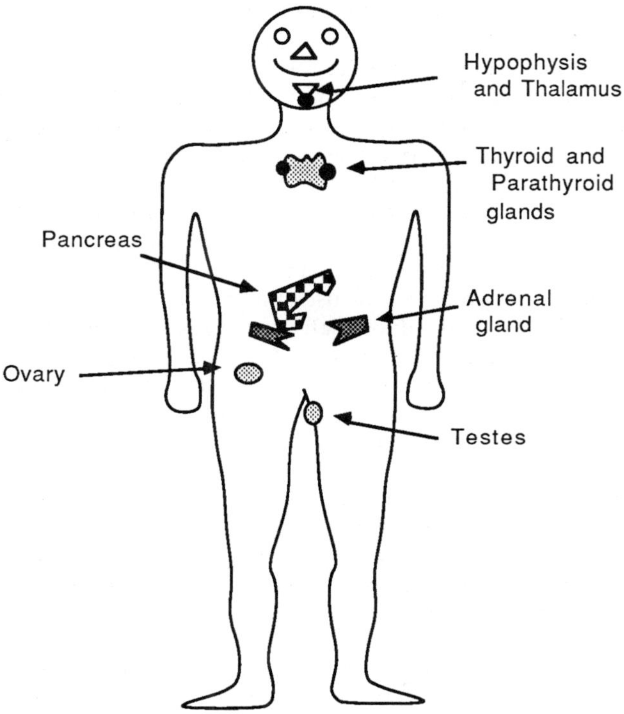

Figure 2.29. Endocrine system. Endocrine glands are situated either in specific organs such as adrenal glands, thyroid, parathyroid, and hypophysis or in organs having another function such as pancreas, hypothalamus, testes, and ovaries.

2.13. Organ and Tissue Replacement: An Overview

Traumatic injury to tissue due to mechanical and/or chemical as well as pathological conditions often requires repair or replacement of tissues and organs. Some of the clinical procedures used to accomplish repair or replacement involve the use of natural tissues from the host (autografts) other humans (allografts) or animals (xenografts) while other procedures involve the use of implants that incorporate polymeric materials.

Tendon/ligament injuries in the hand (flexor tendons) and knee (collateral ligaments) that lead to tendon fraying or rupture are repaired by suturing a ribbon—like supportive stent in the front and back of the torn ends of the tendon/ligament. These stents are made from other tendons/ligaments or from synthetic polymers. Autografts such as fascia lata (thigh), plantaris tendon (foot), or palmaris longus (hand) are often used. Synthetic tendon materials are discussed in detail in Chapter 9.

Injury to bone is repaired using metal wires, screws, nails, plates, and combinations of these fixation and anchoring devices. Metals, ceramics, and polymers are used as materials of construction. Bone grafting using fresh bone from the iliac or tibia is also used to correct nonunion defects. Frozen bone from a bank as well as freeze—dried cadaver bone is also used to stimulate fusion of long and vertebral bones. Joint disfunction is commonly corrected using metallic/plastic prostheses.

Cardiovascular surgery to correct vascular and heart conditions requires use of a variety of biological tissues and implants. Ventricular/atrial septal defects allow leakage of blood between heart chambers and is corrected using a patch made from pericardium. Heart valves often fail to close tightly and are replaced with either porcine or mechanical valves (see Chapter 10).

Large vessels such as the aorta are replaced with woven and non-woven textile tubes while small diameter vessels such as the coronary artery are bypassed using the saphenous vein or mammary artery.

Open heart surgery requires the use of a heart-lung machine that pumps and oxygenates the blood, while kidney dialysis exposes blood to a large surface area of membrane for exchange of urea and other blood-borne elements. In both of these applications blood components are exposed to "foreign surfaces" that may remove or modify any of the components present in blood. Therefore a variety of membrane and pump lining materials have been evaluated to minimize destruction of blood components.

Other tissue reconstructions that are done include replacing a defective segment of the intestine with normal segments of the small or large intestine. Esophagus reconstruction can be achieved using a section of the colon.

Repair of the reproductive apparatus includes use of a tube made out of a synthetic polymer that is inserted as a temporary support for growth of tissue for repair of a duct.

Sheetlike tissues are repaired using a skin graft, skin flap, and muscle—skin flap. Cartilage of the nose or ear can be used for cosmetic reconstruction of the nose. Metallic or plastic mesh wire is used to support a muscular defect in the abdominal wall (hernia).

Specialty surgery to the eye makes use of corneal transplants, plastic intraocular lenses, and viscoelastic gels.

In the ear, interruption of normal hearing due to changes in the properties of tympanic membrane is corrected by replacement with fascia, vessel walls, or a synthetic polymer. Replacement of the bones of the ear is accomplished using bone grafts or prosthetic materials.

2.14. Summary

The structure of a variety of different tissues and organs has been reviewed in this chapter. Although it is beyond the scope of this text to analyze structure-function relationships for tissues and organs in general, it has been possible to review in depth several tissues and organs. The reader is immediately impressed by the structural detail of each organ and the extensive research that has led to our current understanding. A lifetime of work is required to even begin to understand the complexity of the human body.

The biomaterials scientist must be familiar with the structure and function of tissues and organs so that he or she may design the simplest device to restore function.

An attempt has been made to give an overview of some of the applications for biomaterials that exist in medicine. However, because of space limitations it will not be possible to go into the details of all of these applications.

References

Akizuki, S., Mow, V.C., Muller, F., Pita, J.C., Howell, D.S. and Manicourt, D.H. J. Orthopaedic Research 4,379,1986.

Barnard, K., Gathercole, L.J. and Bailey, A.J. FEBS Letters 212,49,1987.

Bloom, W. and Fawcett, D.S. A Textbook of Histology. W.B. Saunders, Philadelphia, Chapters 6, 9, 13 and 23, 1975.

Buckwalter, J.A. Clin. Orthop. Relat. Res.172,207,1983.

Dublet, B., and van der Rest, M. J. Biol. Chem. 262, 17724,1987.

Duncan, K.G., Fessler, L.I., Bachinger, H.P. and Fessler, J.H. J. Biol. Chem. 258,5869,1983.

Fanning, J.C., Yates, N.G. and Cleary, E.G. Micron 12, 339,1981.

Friede, R.L. and Bischhausen, R. J. Neurol. Sci. 38,83,1978.

Gathercole, L.J., Porter, S. and Scully, C. J. Periodontal Res. 22,408,1987.

Gelman, R. A., and Blackwell, J. Conn. Tiss. Res. 2,31,1973.

Haftek, J. J. Bone Joint Surg. 52B,354,1970.

Ham, A. W. and Cormack, D.H. Histology, J.B. Lippincott, Philadelphia, 1979.

Hascall, V.G. and Hascall, G.K. In: Cell Biology of Extracellular Matrix, E.D. Day, Plenum Press, New York, p. 39,1981.

Herbage, D. Ann. Biol. Clin. 44,152,1986.

Imberman, M., Ramamurthy, N., Golub, L. and Schneir, M. J. Periodontal Research 21,396,1986.

Inglis, J.K. A Textbook of Human Biology, Pergamon Press, New York, Chapters 7, 8, and 13, 1986.

Inoue, S., Lebland, C.P. and Laurie, G.W. J. Cell Biol., 97, 1524,1983.

Junqueira, L.C.U., Montes, G.S., and Krisztan, R.M. Cell Tiss. Res. 202,453,1979.

Junqueira, L.C.U., Montes, G.S., Toledo, O.M.S., Bexiga, S.R.R., Gordilho, M.A. and Brentani, R.R. Histochemical J. 15, 785,1983.

Karjalainen, S. and Soderling, E. J. Biol. Bucccale 12,309,1984

Karjalainen, S., Soderling, E., Pelliniemi, L. and Foidart, J.M. Archs. Oral Biol. 31,801,1986.

Kefalides, N.A., Howard, P. and Ohno, N. In: Basement Membranes, edited by S. Mishimi—Shi, Elsevier Science Publishers Co., Amsterdam, p. 73,1985.

Kiviranta, I., Jurvelin, J., Tammi, M., Saamanen, A.—M. and Helminen, M.J. Arthritis Rheumatism 30,801,1987.

Kurkinen, M., Condon, M.R., Blumberg, B., Barlow, D.P. Quinones, S., Saus, J. and Pihlajaniemi, T. J. Biol. Chem. 262, 8496,1987.

Laurie, G.W., Leblond, C.P. and Laurie, G.W. J. Cell Biol. 97,1524,1983.

McBride, D.J., Hahn, R.A., and Silver, F.H. Int. J. Biol. Macromol. 7,71,1985.

McBride, D.J., Trelstad, R.L. and Silver, F.H. Int. J. Biol. Macromol, 10,194, 1988.

McFarlane, K.R., Pollock, M. and Myers, D.B. Acta Neuropathol. 50,217,1980.

Minns, R.J. and Stevens, F.S. J. Anta. 123,437,1977.

Mosekilde, Li, Mosekilde, Le, and Danielson, C.C. Bone 8, 79, 1987.

Muller, W. The Knee, Springer—Verlag, New York, p. 67,1983.

Narayanan, S.A. and Page, R.C. Collagen and Related Research 3,33,1983.

Parry, D.A.D. and Craig, A.S. In: Ultrastructure of The Connective Tissue Matrix. edited by A. Ruggeri and P.M. Motta, Martinus Nijhoff, Boston, p. 34,1984.

Phelps, C.F. In: Molecular Biophysics of the Extracellular Matrix, edited by S. Arnott, D.A. Rees, and E.R. Morris, Humana Press, Clifton, New Jersey, Chapter 2, 1984.

Piez, K.A., and Reddi, A.H., Extracellular Matrix Biochemistry, Elseiver, New York, Chapters 1 and 9,1984 .

Poole, A.R. Biochem. J. 236,1,1986.

Prockop, D.J., Kivinikke, K.I., Tuderman, L., and Guzman, N.A., New Engl. J. Med. 301,13,1979.

Rhodin, J.A.G. Histology, A Text and Atlas, Oxford University Press, New York, 1974.

Rosenberg, L.C. and Buckwalter, J.A. In: Articular Cartilage Biochemistry, edited by K . Kuettner et al., Raven Press, N.Y. p. 39,1986.

Rosenberg, L., Choi, H., Pal, S. and Tang, L. In: Carbohydrate—Protein Interaction, ed. I.J. Goldstein, Vol. 88, ACS Symposium Series, Washington, D.C. p. 186,1979.

Rosenbloom, J. Lab. Invest. 51,605,1984.

Ross, M.H., and Reith, E.J. Histology, Harper and Row, Publishers, J.P. Lippincott Co., New York, Chapters 11 and 15, 1985.

Salonen, V., Roytta, M. and Peltonen, J. Acta Neuropathol. 74,13,1987.

Sandberg, L.B., Soskel, N.T., and Wolt, T.B. J. Invest. Dermatol. 79,128A,1982.

Scott, J.E. and Orford, C.R. Biochem. J. 197,213,1981.

Silbert, J.E. J. Invest. Dermatol. 79,31A,1982.

Silver, F.H. Biological Materials: Structure, Mechanical Properties, and Modeling of Soft Tissues, NYU Press, New York, Chapter 3, 1987.

Silver, F.H., Christiansen, D. and Buntin, C.M. CRC Critical Reviews, Biocompatibility In Press, 1989.

Simpson, J.W. Handbook of Experimental Aspects of Oral Biochemistry, edited by E.P. Lazzari, CRC Press, Inc. Boca Raton, Florida, p. 213,1983.

Slavkin, H.C. In: The Chemistry and Biology of Mineralized Tissues, edited by W.T. Butler, Ebsco Media, Inc., Birmingham, Alabama, p. 237,1985.

Stanley, J.R., Woodley, D.T., Katz, S.I., and Martin, G., J. Invest. Dermatol. 79,69s,1982.

Tajima, S., Nishikama, T., Hatono, H., and Nagai, Y., Arch. Dermatol. Res. 273,115,1982.

Termine, J.D. In: The Chemistry and Biology of Mineralized Tissues, Ebsco Media, Inc., Birmingham, Alabama, p. 94,1985.

Timpl, R. and Martin, G.R. Immunochemistry of the Extracellular Matrix, Vol. 2, edited by H. Furthmayr, CRC Press, Boca Raton, Florida p.119,1982.

Timpl, R., Weiderman, H., VanDelden, V., Furthmayr, H. and Kuhn, K. Eur. J. Biochem. 120,203,1981.

Unsworth, D.J., Scott, D.L., Almond, T.J. Beard, H.K., Holborow, E.J., and Walton, K.W. Br. J. Exp. Pathol. 63,154,1982.

Ushiki, T. and Ide, C. Arch. Histol. Jap. 49,553,1986.

Vogel, K.G. and Heinegard, D. J. Biol. Chem. 260,9298,1985.

Yurchenco, P.D., Tsibibary, E.C., Charonis, A.S. and Furthmayr,H. J. Histochem. Cytochem. 34,93,1986.

Chapter 3

Response to Injury: Pathobiology

3.1. Introduction

The biomaterials scientist and engineer is concerned with replacing biological tissues and organs. Since many of the tissues and organs are diseased or injured, it is essential that he/she understand cellular changes that represent abnormal as well as normal responses. In many cases the effects of a biomaterial or medical device can only be determined after cells are brought in contact with it. Therefore the biomaterials scientist must be able to evaluate the effects of biomaterials on cells, tissues, and organs. In addition, as part of the requirements for approval of medical devices by the Federal Food and Drug Administration (FDA) materials must be evaluated for safety and efficacy. Safety evaluation requires understanding the interaction of cells and tissues with implant materials.

It is necessary to briefly review the components of cells that are examined to assess the compatibility of materials and devices. Such cellular structures are observed and analyzed using transmission electron microscopy (TEM). Changes that are associated with cellular adaptation and injury include induction, secretion, adaptation, and cell injury and death. Several factors are involved in these changes.

3.2. Cell Components: Structural Considerations

Cells are exposed to a multitude of chemicals, physical forces, as well as biological agents during a lifetime. The successful use of biomaterials and artificial organs will require that such intervention will not result in additional cell injury and death. In the design stage, the biomaterials scientist and engineer builds a device prototype that is initially tested in animals and in cell culture. It is essential that the prototype be evaluated for any potentially harmful effects on cells. In this section we discuss those cellular and subcellular components that are studied by the biomaterials scientist during evaluation of safety.

3.2.1. The Normal Cell

Biocompatibility of materials and devices with tissues is most specifically seen by examining changes in cell structure and function. The components of cells that are examined as evidence of normal cell function include cell membrane, cytoplasm, endoplasmic reticulum, mitochondria, ribosomes, Golgi apparatus, lysosomes, and nucleus (Table 3.1 and Figure 3.1).

3.2.1.1. The Cell Membrane

The cell or plasma membrane is a lipid bilayer that contains proteins on the inner and outer surface. These proteins are either partially or totally embedded in the membrane as diagrammed in Figure 3.2. The cell membrane (1) maintains ionic and chemical concentration gradients, (2) carries cell-specific surface markers and receptors for a variety of compounds including glycoproteins, hormones, growth factors, and histocompatibility antigens (HLA), (3) participates in intercellular communication, and (4) regulates cell growth and proliferation.

Figure 3.2 is a diagram of the fluid-mosaic model of membrane structure introduced by Singer and Nicolson, which incorporates a discontinuous lipid bilayer composed of phospholipids and cholesterol. Proteins and glycoproteins are embedded or attached to the membrane in this model. Peripheral proteins (such as red cell spectrin) are found attached to the cell membrane and play a structural role in maintaining cell shape, while integral proteins are embedded in the cell membrane and are involved in transport and cell recognition (class I and II histocompatibility antigens). Transplantation of tissue from one host to another leads to rejection if the cell recognition glycoproteins (histocompatibility antigens) are different, i.e., the cells of the transplanted tissue have different glycoproteins than do the cells of the recipient's tissues.

The cell membrane is also involved in interaction between cells. This function is necessary for adhesion, intercellular electrical and chemical communication, and tissue organization. Cell-to-cell attachment via specific types of junctions is also required for exchange of proteins and ions. For example, epithelial cells exhibit tight junctions, gaps junctions, and desmosomes (Figure 3.1).

Exposure of cells to chemicals used to process biomaterials or break down products of biomaterials can cause changes in the cell membrane. Typically, cells that take up the dye trypan blue have defective cell membranes.

Cells exhibit a polarity in their organization (Figure 3.1). Most cells present an apical pole (rich in microvilli) and a basal pole that is normally in contact with basement membrane. As described below, the cell membrane is involved in release and uptake of materials by the cell.

Table 3.1 Components of Normal Cell and Their Function

Cell component	Composition	Function
Cell membrane	Lipid bilayer, contain ing surface proteins (peripheral proteins), proteins totally embed-ded in the membrane (integral proteins), and glycoproteins par-tially embedded in the membrane	Maintains ionic and chemical concentration gradients, cell-spe-cific markers, inter-cellular communication, regulates cell growth and proliferation
Cytosol or cyto-plasm	Water,ions, soluble proteins	Contains enzymes and structures for: generation of energy (ATP) in the absence of oxygen (TCA cycle), activation of amino acids, carrying out specialized cell functions
Endoplasmic reticulum	Membrane enclosed channels	Involved in transport of proteins for extracel-lular secretion, and modification or detoxi-fication of chemicals
Mitochondria	Contains membrane lined channels to which en-zymes are attached that generate ATP from glu-cose	Involved in: TCA cycle, respiratory chain and oxidative phosphoryla-tion
Ribosomes	Small and large subunits composed of ribonucleic acid (rRNA);strand of messenger ribonucleic acid (mRNA) forms a complex with large and small subunits	Involved in synthesis of enzymes and structural components and proteins for extracellular re-lease

Table 3.1 (Continued)

Components of Normal Cell and Their Function

Cell Component	Composition	Function
Golgi apparatus	Membrane-lined tubular system that forms stacks	Packaging of proteins in vesicles for extracellular release
Lysosomes	Membrane lined vesicle containing hydrolytic enzymes	Involved in break down of intracellular and extracellular material
Nucleus	Membrane limited area of cell containing nucleolus and chromatin	Site of synthesis of ribonucleic acid and chromatin, involved in cell division

3.2.1.2. Cell Cytoplasm

The gellike material contained within the cell membrane is termed the cytoplasm or cytosol. It is composed of water, ions, and soluble proteins and contains enzymes that are involved in generation of energy (ATP) in the absence of oxygen (Figure 3.3, TCA cycle) and activation of amino acids for protein synthesis (Table 3.1 and Figure 3.3). Specialized structures termed organelles (see below) are found within the cytoplasm.

In addition to organelles, the cell cytoplasm contains cytoskeletal filaments (Figure 3.1) that control shape. Actin, myosin, and α-actinin are involved in cell contraction. Actin may be involved in cell attachment to extracellular matrix through the cell membrane. Other filaments such as intermediate filament, tubulin, calmodulin, and spectrin form networks that modify cell and organelle mobility and shape.

3.2.1.3. Endoplasmic Reticulum

Endoplasmic reticulum (ER) is a branching system of membrane limited channels that appear inside the cell (Figure 3.1). These channels are 40-70 nm in diameter and are enclosed by a membrane similar to the plasma membrane. In some cases the channels are covered with 20nm spheres; in this case the endoplasmic reticulum is termed rough. These spheres are termed ribosomes and are the site of protein synthesis. In the absence of ribosomes, the ER is termed smooth.

Proteins that are synthesized (i.e., collagen and elastin) on the ribosomes pass into the center of these channels and are transported into the Golgi apparatus where they are packaged into vesicles for release from the cell. Membranes that line the smooth endoplasmic reticulum are involved in modification and detoxification of chemicals released by biomaterials.

Biomaterials that release chemicals that stimulate modification or detoxification result in proliferation of smooth ER. In addition, biomaterials

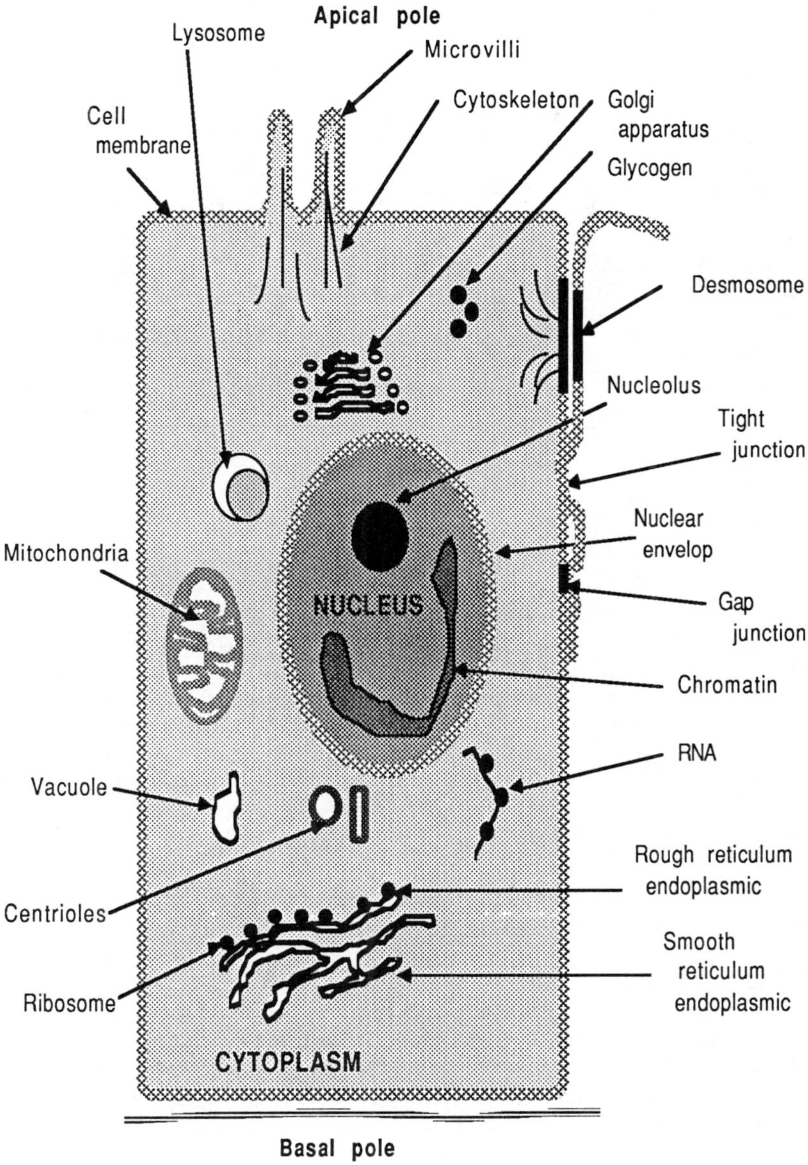

Figure 3.1. Schematic Presentation of a cell. Cells are surrounded by a cell membrane that contains cytoplasmic components such as nucleus, mitochondria (energy source); endoplasmic reticulum (transfer of molecules); ribosomes (synthesis of proteins) centriole (mitosis), Golgi appparatus (concentration and synthesis of molecules); and other organelles. Cells (particularly epithelial cells) have a polarity (basal and apical poles). Cell-cell contact involves several cell junctions where transport of molecules or electrical signals can occur.

that stimulate tissue deposition and protein synthesis result in proliferation of rough ER. Therefore, active proliferation of the ER alone is not a sufficient

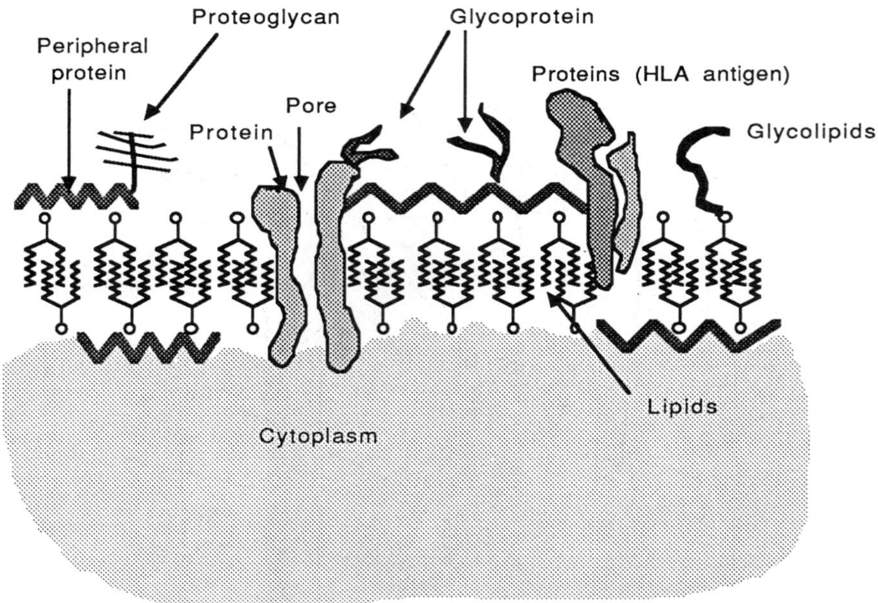

Figure 3.2. Cell membrane structure. Cell membranes are composed of two layers of lipids containing integral proteins with pores that allow communication between the exterior and the cytoplasm. Other proteins, proteoglycans, glycolipids, and glycoproteins can be attached to the lipids of the cell membrane. Some proteins such as the receptors for growth factors and the antigens of the major histocompatibility complex (HLA in human) have an intramembrane site and an external site.

indicator of biocompatibility.

3.2.1.4. Mitochondria

Mitochondria are cigar-shaped organelles (Figure 3.1) that are separated from material found in the cell cytoplasm by a double membrane. Both inner and outer mitochondrial membranes are similar to the plasma membrane. The inner membrane is connected to a series of folded channels (cristae) upon which the enzymatic reactions of the tricarboxylic acid cycle (TCA), respiratory chain, and oxidative phosphorylation occur (Figure 3.3). These reactions are required to generate energy (ATP) from glucose in the presence of oxygen.

Cells actively involved in enzyme synthesis required for modification or detoxification of chemicals or involved in synthesis of structural proteins (fibrotic capsule around an implant) such as collagen contain numerous mitochondria.

3.2.1.5. Ribosomes

Ribosomes are particles that contain ribonucleic acids in the form of large and small subunits (Figure 3.3). The small subunit is shaped like a donut that has been cut in half. The large subunit is spherical and contains a notched groove on

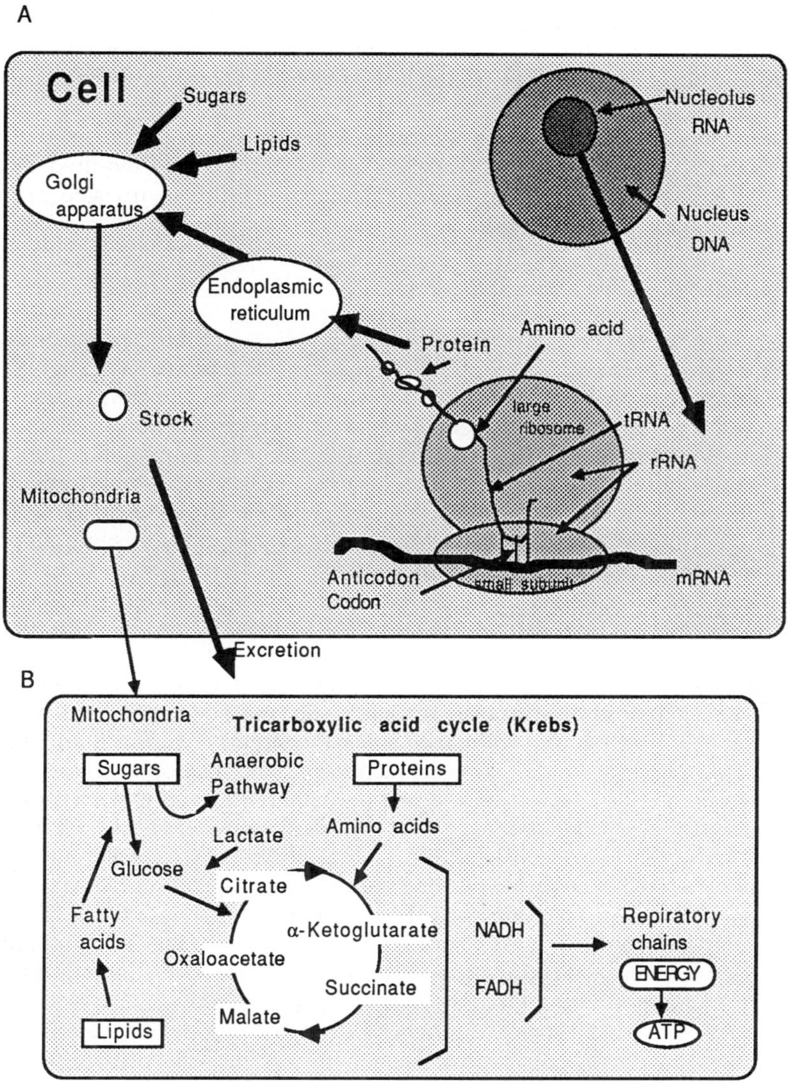

Figure 3.3. Cellular Mechanisms for energy and protein synthesis. (A). The nucleus contains the genetic code (deoxyribonucleic acid or DNA) that alows synthesis of proteins using ribosome (rRNA), a message (mRNA) and a transfer molecule (tRNA) that carries an amino acid. Proteins are then processed through the endoplasmic reticulum and transferred to the Golgi apparatus where lipids and/or sugars can be attached. Molecules are then stored in vesicles and can be excreted to the cell environment. (B). These processes require energy, which is produced in mitochondria from amino acids, lipids and sugars that are converted into adenosine triphosphate (ATP).

the top surface. A strand of messenger ribonucleic acid (mRNA) is found between the notched groove of the large subunit and the hole in the center of the small subunit; mRNA in association with the large and small ribosomal subunits acts as a template for protein synthesis.

Ribosomes found free within the cell cytoplasm are the site of synthesis of enzymes and structural proteins that stabilize the cell and are required for cell division. Ribosomes attached to endoplasmic reticulum synthesize proteins destined for extracellular release.

3.2.1.6. Golgi Apparatus

The Golgi apparatus is a membrane-bound system of tubes that is connected to and may derive from the endoplasmic reticulum. The tubular system is in the form of individual tubules that make up a winding system of cisternae that form stacks termed a dictyosome. Proteins synthesized on ribosomes attached to the inner membrane of the endoplasmic reticulum are transported into the Golgi apparatus where they are packaged into vesicles. Vesicles 40-80 nm in diameter at the outside of the cisternae are used to release these proteins extracellularly. Cells synthesizing connective-tissue components as part of the healing response have a well-developed Golgi apparatus.

3.2.1.7. Lysosomes

Lysosomes are membrane-bound vesicles found intracellularly that are involved in break down of proteins, nucleic acids, and sugar polymers or other debris that arise either intracellularly or extracellularly. Cells found in the blood, termed polymorphonuclear cells, monocytes, and macrophages, are rich in lysosomes and are involved in breakdown of biomaterials and materials that are derived from implants. Lysosomes are 0.1-0.8 μm in diameter ($1\mu m = 10^{-6}m$) and contain a variety of hydrolytic enzymes. Extracellular release of these enzymes is associated with cell death and bioincompatibility.

Lysosomal enzymes (particularly acid phosphatase) are synthesized on the rough endoplasmic reticulum and packaged into vesicles in the Golgi apparatus. At this stage they are termed primary lysosomes (Figure 3.4). Implant material is ingested by neutrophils and macrophages by a process termed phagocytosis. A membrane-coated vesicle containing material ingested by phagocytosis is termed a phagosome. A phagosome in turn fuses with a primary lysosome forming a secondary lysosome. Autophagy (degradation of cellular components) is a process by which damaged cell organelles are digested by fusion with a lysosome. Undigested debris may remain within vacuoles in the cell and are termed residual bodies. It is not uncommon to see residual bodies in cells involved in either degrading implants or wear particles that are derived from implants. Cells containing residual bodies surrounding an implant are a sign of a chronic inflammatory response to the implant.

Other vesicles termed microbodies are seen within the cell. These vesicles include peroxisomes and glyoxysomes and generate hydrogen peroxide and catalase (peroxisomes), respectively, and enzymes that generate energy from acetate and fatty acids (glyoxysomes).

Cytoplasmic inclusions observed following specific cellular activity include glycogen, lipids, crystalline inclusions, and secretory products.

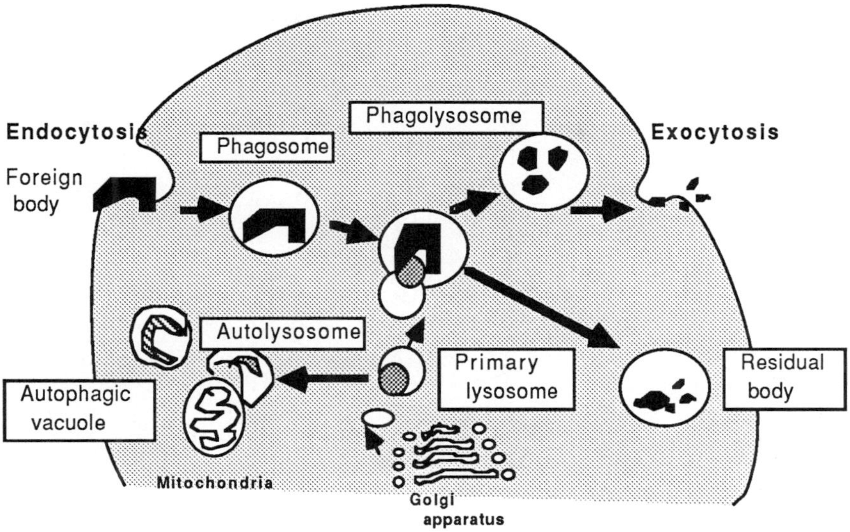

Figure 3.4. Phagocytosis of extracellular material. Foreign particles are introduced into the cytoplasm of leukocytes and macrophages by forming cell membrane around the particle. The foreign body is then progressively digested by lysosomal enzymes and can be either excreted or remain in the cytoplasm as a residual body.

3.2.1.8. Nucleus

The nucleus is composed of a porous nuclear membrane, a nucleolus, and soluble materials. The nucleolus (Figure 3.1) contains ribonucleic acids (RNA) and genetic material also termed chromatin. The nuclear membrane is continuous with the outer membrane of the endoplasmic reticulum. Chromatin containing deoxyribonucleic acid (DNA) and proteins termed histones are also found adherent to the inner surface of the nuclear membrane. Figure 3.5 shows diagrammatically the relationship between DNA and histones that is found within the chromatin.

The nucleolus is the site for synthesis and packaging of ribosomal RNA into preribosomal particles. The nucleic acids that are synthesized within the nucleus pass through pores in the nuclear membrane into the cytoplasm. After staining with special dyes for light microscopy, the nucleolus appears as a dense spot in the nucleus. In comparison, chromatin is acidic and stains darkly with basic dyes. Loosely coiled chromatin (euchromatin) stains lightly with basic dyes while tightly coiled DNA (heterochromatin) stains darkly with basic dyes. Stimulation of cell replication results in increased amounts of heterochromatin while resting cells exhibit increased amounts of euchromatin.

During cell division (M phase, Figure 3.5) heterochromatin is separated and arranged into distinct clumps of genetic material termed chromosomes. A finger print of the genetic material of a cell, a karyotype, is obtained by preventing termination of cell division. This allows the observation of the size and shape of the genetic material found in any cell type. In this manner, effects of a biomaterial or implant on chromosome structure can be evaluated by examining the cell karyotype in the presence (and absence) of the material. Biomaterials

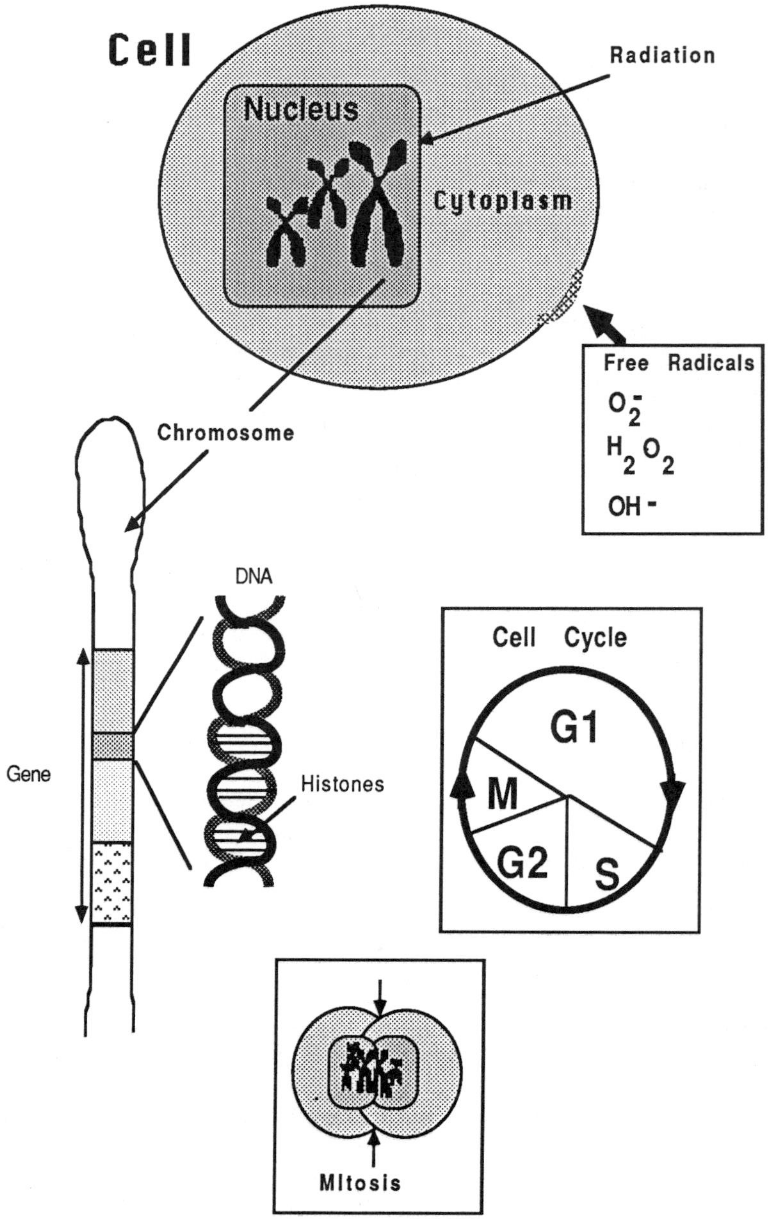

Figure 3.5. Cell cycle, DNA, and genetic material. Cells undergo a cycle during replication and pass through a step of mitosis (M phase) where the genetic material is duplicated and the cell is divided. Genes are present in the chromosome and consist of chains of DNA and histones. DNA has the genetic code that allows synthesis of specific molecules. Free radicals such as H, radiation, and viruses induce chromosomal damage and genetic mutation. The G1 phase is a resting step; S and G2 phase cells are synthesizing structural molecules in preparation for mitosis during M phase.

that actively enhance healing encourage cell division of fibroblasts, endothelial cells, and epithelial cells.

3.2.2. The Normal Cell: Summary

The normal cell contains many components that are essential for survival. Changes in these structures are associated with cell injury and contact with debris from implants. In Section 3.3 mechanisms of cellular adaptation and injury will be examined.

3.3. Cellular Adaptation

Events that lead to cell injury, including mechanical insult, contact with a foreign material, or exposure to microorganisms in some instances lead to cell degeneration. Cell degeneration is reversible; however, continual exposure to the insult leads to degeneration and cell death (necrosis).

Signs of reversible degeneration include cell swelling, compression of intercellular junctions, appearance of intracellular vacuoles, dilatation of endoplasmic reticulum and Golgi apparatus, proliferation of ribosomes, and degeneration of the cell membrane.

Signs of irreversible degeneration or cell death include loss of intercellular attachments, contraction of the nucleolus, rupture of the nuclear membrane, evidence of autophagy, and appearance of lysosomes. Degeneration of mitochondria and disorganization and disintegration of endoplasmic reticulum is accompanied by accumulation of lipids, glycogen, and protein in the cytoplasm (Figure 3.6).

As the human life expectancy increases, cells of the body are exposed to an increasing variety of stimuli that can disrupt cellular components or lead to changes in cell structure. As discussed in Section 3.2 the factors leading to cell injury include genetic, nutritional, physical, infectious, chemical, immune, and aging (Figure 3.7 and Table 3.2). Each of these factors will be discussed in detail in Section 3.40.

Before we deal with cellular responses to specific stimuli, it is necessary to consider general cellular adaptation mechanisms that pertain to one or more of these classes of stimuli. From the morphological point of view it is difficult to distinguish between normal, adapted, and injured cells in many instances. In many cases only cells that are destined to die can be distinguished from normal cells. Transmission electron microscopy is a useful tool to study cellular degeneration. The response of each cell line to a specific stimulus varies as well as the stimulus intensity required to bring adaptation.

3.3.1. Induction of Endoplasmic Reticulum

Chemicals such as carbon tetrachloride are detoxified by endoplasmic reticulum in the liver (Table 3.3). As carbon tetrachloride enters the liver enzymes are synthesized by the endoplasmic reticulum. These enzymes theoretically will convert the chemical moiety to a harmless substance that can be excreted. However, the reaction that takes place results in the formation of free radicals ($CC1_3$. and C1.) that are themselves toxic to cells. Carbon tetrachloride is an example of a chemical that by itself is nontoxic but is converted by enzymatic modification into a toxic material. Free radicals

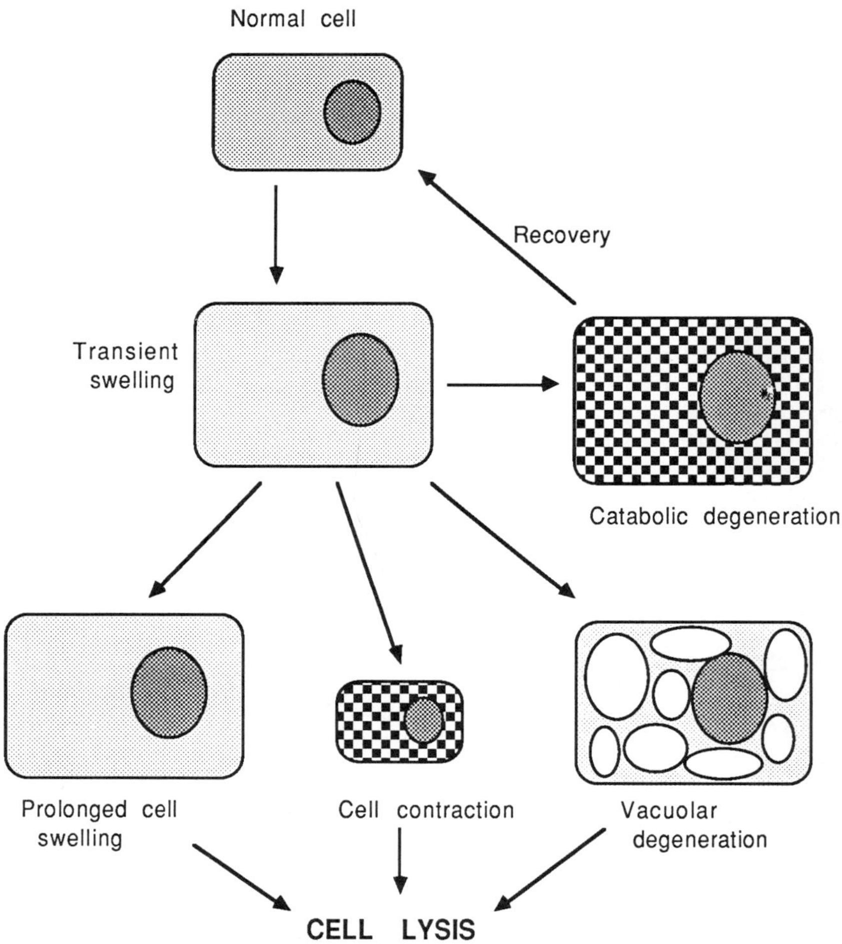

Figure 3.6. Cell degeneration. Diagrammatic representation of cell degeneration by swelling, contraction, and catabolic degeneration resulting in cell lysis.

produced by breakdown of any polymeric material result in decreased cellular biocompatibility.

3.3.2. Change in Cell Size

A stimulus may result in an increase or decrease in cell size. Increase in cell size without an increase in the number of cells that make up a tissue is a process termed hypertrophy. For example, in the heart increased work load that results from vascular disease results in synthesis of cellular components (i.e., enzymes, mitochondria, and endoplasmic reticulum) that result in an increase in the size of muscle cells.

A cell may shrink and cause an organ to get smaller (atrophy) as a result of decreased use or function. For example, loss of electrical stimulation results in a

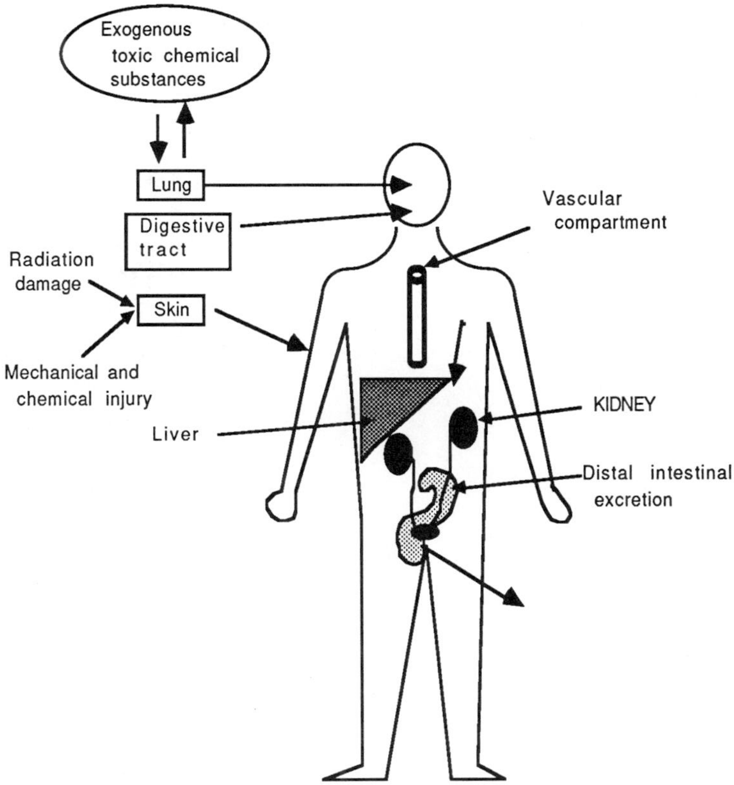

Figure 3.7. Sites of entrance of toxic substances. Toxic substances can be either inhaled into the lung, or swallowed into the digestive tract, or pass through the skin or enter through open wounds. The substance or foreign materials can be toxic at the site of entrance or to tissues as they pass through the vascular system and then into different peripheral organs including kidneys and liver.

decrease in the size of a muscle cell. This can result from immobilization of a limb or decreased blood supply. At the cellular level decreased cell mass is a result of breakdown of cell components and an increase in the number of autophagic vacuoles.

Materials used as replacements for blood vessels should not decrease the blood supply to an organ. If the blood supply to an organ is decreased then, at the cellular level, a decrease in cell size as well as autophagic vacuoles will be seen.

3.3.3. Increase in Cell Number (Hyperplasia)

When certain types of cells are stimulated, they increase in number resulting in a volume increase of the associated organ. This process is termed hyperplasia and can result from contact of a material with the cells on the surface of the skin.

Table 3.2 Classes of Stimuli Leading To Cell Injury

Stimulus	Effect
Genetic	Disfunctional protein
Nutritional	Delayed or inadequate wound healing
Physical	Loss or necrosis of tissue
Infectious	Tissue necrosis and delayed wound healing
Chemical	Detoxification by liver or kidney,or cellular injury
Immune	Vascular leakage, antibody production
Aging	Impaired healing

Table 3.3 Cellular Adaptation Processes

Process	Cellular changes
Induction of endoplasmic reticulum	Dilated endoplasmic reticulum
Autophagy	Removal of worn out organelles
Hypertrophy	Increased amount of cell components
Hyperplasia	Increase in cell number
Metaplasia	Change in cell type
Dysplasia	Change in nuclear size and shape

For example, abnormal healing of the skin can result in the formation of repair tissue during the proliferative phase of wound healing which is characterized by an increased cellularity. After surgical removal of part of the liver an increased cellularity of the remaining organ is observed.

3.3.4. Change in Cell Types (Metaplasia)

Under the influence of a stimulus (chemical, physical, or viral), a reversible change can occur in which one adult cell type is replaced by another cell type. The respiratory tract is normally lined by cells that are rectangular in cross section (columnar epithelial cells). Under the stimulus of chemicals found in cigarette smoke these cells change to elongated epithelial (squamous) cells. The process of replacing one cell type by another is termed metaplasia.

3.3.5. Change in Cell Characteristics (Dysplasia)

In response to chronic irritation, changes are observed in differentiated (adult) cells. These changes include variation in cell size, shape, and loss of organization and result in a change in cell staining characteristics. It most commonly involves epithelial cells that line the entrance to the uterus. The number of dysplastic cells increases especially in tissues where dividing cells are not normally seen.

3.4. Cellular Injury

Cellular injury can be induced by a variety of factors that lead to alteration in the genetic material, a change in cell characteristics, a degeneration of certain organelles and even cell death. These factors include genetic, nutritional, physical, infectious, chemical, and immune and are described in the following sections

3.4.1. Genetic Factors

During replication of the genetic material found in a cell a change in the sequence of nucleic acids (mutation) can occur that leads to an alteration of the messenger ribonucleic acid and abnormal protein synthesis. The classical example is that of a disease termed sickle cell anemia. In sickle cell anemia a mutation occurs that results in substitution of a valine for a glutamic acid residue. As a result of this mutation the hemoglobin molecule cannot maintain its shape at low oxygen tension and the red cell sickles (collapses). This change results in anemia, kidney, heart, and lung problems, and chronic skin ulcers.

Genetic mutations can also occur as a consequence of contact of cells and tissues with chemicals and radiation. Of particular importance to the biomaterials scientist are the chemicals and solvents used in the processing and preparation of a biomaterial or medical device.

3.4.2. Nutritional Factors

Nutritional deficiences or excesses can lead to many disease states. A variety of low- and high-molecular-weight molecules are required by humans. Nutritional injury is only of indirect importance to the biomaterials scientist. It is important that the biomaterials scientist be aware of the nutritional requirement of any species used in animal experiments. For instance, guinea pigs require high levels of ascorbic acid. If the biocompatibility of a dermal dressing is to be evaluated in a guinea pig model then the guinea pig diet must always be supplemented with ascorbic acid or else the wound healing response is impaired. In humans, low levels of essential nutrients are required for normal wound healing.

3.4.3. Physical Factors

Physical factors that result in cellular adaptation and injury include (a) thermal contact, (b) excessive pressure, (c) radiation exposure, and (d) electrical injury.

Contact with hot surfaces results in destruction of the epidermis, dermis, muscle, and even bone. Even contact with surfaces that are only 10° above body temperature can induce cell injury by accelerating cellular metabolism (inactivation of cellular enzymes) and by inducing vascular injury. The duration of exposure to elevated temperature is important in determining the extent of injury. The injured area becomes reddened as a result of dilatation of blood vessels and release of vascular components into tissue spaces.

Exposure to low temperature results in vasoconstriction of blood vessels and decreased blood flow. This may lead to cell death due to decreased oxygen supply. In comparison to thermal injury, freeze injuries of skin do not damage the connective tissue. Therefore, freeze injuries are not characterized by redness and increased local circulation.

Excessive pressure produced by inflation of a balloon catheter or movement of an implant can produce mechanical injury to cells. Patients required to undergo prolonged bed rest, may lose epidermal and dermal tissue of skin due to excessive pressure induced by their weight.

Radiation such as cosmic rays, ultraviolet and visible light, microwaves, or radio waves can cause cell injury depending on the frequency, duration, and type of radiation. Ionizing radiation leads to ejection of an electron from the target cell (or cells) and leads to the creation of positively and negatively charged ions. Electromagnetic waves with sufficient energy result in ionization; these waves include x rays, gamma rays, and cosmic photons. Ionizing radiation involves two subclasses: (1) oscillating electromagnetic waves including x rays and gamma rays, and (2) high-energy subatomic particles such as an α-particle. Ionizing electromagnetic waves penetrate deeply into tissues because of their low masses and transfer energy to electrons in cellular molecules. Charged particle radiation penetrates only a few centimeters but transfers more energy to tissue because of the higher mass. The unit of absorbed ionizing radiation is the rad. One rad is that amount of radiation that deposits 10^{-2} joules of energy per kilogram of tissue. For x rays and gamma rays, the unit of exposure dose is termed the rotengen. A dose of one rotengen to living matter results in energy absorption of about one rad. A dose of 500 rads is sufficient to destroy a cell by causing polymer chain cleavage in DNA, cel lmembrane proteins, and phospholipids.

Cells that divide often (high mitotic activity) such as the epithelium in skin and GI tract, bone marrow cells, and lymphocytes in the lymph nodes are very sensitive to radiation. Differentiated mature cells that have a low mitotic activity including cells found in muscle and brain, require high doses of radiation to be altered.

Many biomaterials are sterilized by exposure to a dosage of 2.5 mega (million) rads of gamma radiation (Co^{60}). Exposure to gamma radiation kills bacteria, viruses and fungi. In addition, irradiation can alter the chemical structure and biocompatibility of a material.

3.4.4. Infectious Agents

Infectious agents such as bacteria, viruses, and fungi are a major complication of surgery. Biomaterials processed under nonsterile conditions are contaminated and must be sterilized before implantation or evaluation in cell culture. Infectious agents are associated with tissue inflammation and cellular hyperplasia and hypertrophy.

Viruses must gain entrance into the cell to become infectious. However, bacteria and fungi can be infectious when present extracellularly. Toxins from bacteria also have the potential to destroy cells.

Viruses contain either ribonucleic acid (RNA) or deoxyribonucleic acid (DNA) in single- or double-stranded forms within a glycoprotein envelope that is coated on its inner surface with a protein layer termed capsid. Viruses are 100-300nm in diameter and have enzymes necessary to replicate. They infect and enter the host cell after which they direct the host cell to replicate the viral membrane capsid and nucleic acid or part of the viral genetic material.

Bacteria may be infectious (pathogenic) if they invade host tissues and release factors that are toxic to cells. A variety of bacteria release toxins that result in cell injury.

Infection is a side effect of procedures involving sterilized biomaterials and medical devices, in part, because biomaterials are implanted through or into previously contaminated wounds. It is hypothesized that conventional sterilization procedures are not able to penetrate all surfaces and crevices of the implant.

3.4.5. Chemicals

Chemicals found in our environment or used to process materials are known to produce cell injury. In most cases chemicals enter the body through the lungs, skin, and gastrointestinal tract (Figure 3.7). The chemicals can be eliminated by excretion or can be detoxified by cells in the liver and kidney. Cellular components, such as lysosomes containing enzymes, are synthesized by cells in contact with a chemical in an attempt to degrade or inactivate it. However, the metabolic products of ingested chemicals can be toxic.

In many cases the plasticizers and solvents normally used to process synthetic polymeric materials induce cell injury. For this reason polymers used for biomedical applications are processed by minimizing solvent contamination or by use of aqueous solutions.

3.4.6. Immune Reactions

When a foreign material is introduced into the body, immune cells immediately recognize the foreign surface and respond by either secreting or synthesizing antibodies. In addition, other immune cells help antibody-producing cells to produce higher levels of antibodies.

Immune disorders that arise from contact with natural collagenous biomaterials include immediate (also termed anaphylaxis) and delayed hypersensitivity. Some portion of the population has been exposed to collagen from beef and pork products and have circulating antibodies to collagen, which results in the formation of antigen-antibody complexes. Immediate hypersensitivity occurs on contact with a collagenous biomaterial derived from cow or pig tissues such as skin. Initial contact with the antigen (i.e., collagen from meat products) results in differentiation of blood-borne B-lymphocytes into producing plasma cells situated in the wall of the gut that produce a class of antibodies termed IgE. These antibodies become bound to specific receptors on the surface of mast cells found in connective tissues and basophils found in blood. Reexposure to the antigen (collagen in a biomaterial) causes mast cells and basophils to release histamine causing leakage of fluid from vessels and swelling of the surrounding tissues.

During the response to injury resulting from implantation of a biomaterial, inflammatory cells are seen surrounding the implant for short (acute) or long (chronic) periods depending on the extent of irritation. During the inflammatory reaction, inflammatory cells undergo structural changes. Most cells release the contents of their dense, round granules into the extracellular space. Endothelial cells swell and lose their intercellular attachments. Neutrophils ingest microorganisms as well as particulates derived from biomaterials. However, in some cases leukocytes are destroyed by the irritant. Leukocytes involved in phagocytosis exhibit surface projections, cel membrane invaginations, extend long tentacles, and cytoplasmic granules.

Other blood cells such as eosinophilic leukocytes, basophilic leukocytes, lymphocytes, plasma cells, and monocytes are involved ininflammatory reactions. On activation, blood monocytes are transformed into macrophages which destroy cell debris, fibrin, collagen, and biological biomaterials in general. The number of lysosomes, mitochondria, and pinocytotic vesicles increases when monocytes are transformed into macrophages. In addition, the Golgi apparatus is enlarged and the amount of intracellular enzymes (hydroxylases) increases. Several macrophages collected at a site of chronic irritation is termed as a granuloma. A collection of macrophages fuse together form a single cell (giant cell) with many nuclei. Giant cells are frequently observed in reactions to foreign bodies.

In response to solid implants "fibrotic tissue" envelops the implant in the form of a capsule. A foreign body capsule is formed by fibroblasts that actively synthesize collagen and other connective tissue components. Active fibroblasts are characterized by a well developed endoplasmic reticulum, prominent nucleoli, and the presence of cytoplasmic microfilaments rich in actin filaments.

Delayed hypersensitivity occurs at some time after exposure to the antigen. This type of response usually involves T-lymphocytes that have been previously exposed to the antigen.

3.5. Summary

The biomaterials scientist must be familiar with cell components and pathobiological changes that occur on contact with a biomaterial or medical device. Biomaterials are not inert. They induce formation of new connective tissue, cellular hyperplasia, and even cell death (necrosis). It is important for the biomaterials scientist to recognize early in a biocompatibility analysis the difference between cellular hyperplasia due to an irritant from that associated with rapid healing. However, this will require some familiarity with the field of histopathology, which is beyond the scope of this text.

Bibliography

Cheville, N.F. Cell Pathology, Iowa State University Press, Ames, Iowa, 1976.

King, D.W., Fenoglio, C.M. and Lefkowitch, J.H. General Pathology, Principles and Dynamics, Lea & Febiger, Philadelphia,1983.

Robbins, S.L. and Kumar, V. Basic Pathology, W.B. Saunders, Philadelphia, 1987.

Sheeler, P. and Bianchi, D.E, Cell Biology: Structure, Biochemistry, and Function, John Wiley and Sons, New York,1980.

Chapter 4

Wound Healing: *In Vivo* Response to Biomaterial Implantation

4.1. Introduction

An implant is used to replace the function of a natural tissue that is lost due to injury. If the injury such as ulceration of the skin is superficial, then installation of the implant requires little or no surgery. However, if the device is to be used internally, such as to replace a blood vessel, then surgical intervention is necessary to install the device. In either case the device is used in existing or surgically created wounds. The tissue reaction that occurs around the implant also involves the process of wound healing. Therefore it is important to understand the events associated with wound healing.

Wound healing and tissue repair are a set of processes that are automatically triggered by tissue injury (Figure 4.1). These processes include an inflammatory phase, a proliferative phase, and a remodeling phase. Tissue injury normally results in cell death, denaturation of extracellular connective tissue components, and loss of blood vessel integrity. Blood-vessel disruption leads to blood cells and proteins escaping into the injured area. Vasoconstriction occurs in intact vessels preventing continued blood loss into the injured area (Figure 4.2).

The inflammatory phase involves blood-vessel dilation and leakage of blood components into extravascular spaces as well as local accumulation of inflammatory cells including neutrophils, lymphocytes, and macrophages (Figure 4.3). The proliferative phase consists of formation of new blood vessels and the laying down of new connective tissue termed granulation tissue to fill the wound (Figure 4.4). Remodeling consists of enzymatic removal of granulation tissue and deposition of aligned connective tissue in the form of a scar (Figure 4.5).

Each of these three phases of wound healing are dependent on the previous phase and can be prolonged or shortened depending on how the wound is treated and the state of health of the individual. The presence of an infectious agent, for instance, when wounds are open to air, the inflammatory phase can be prolonged extensively. The presence of infectious bacteria in the wound results in prolongation of the inflammatory phase, varying in length from a period of several days to as long as several months. In comparison, wound closure using sutures shortens the inflammatory phase to several days. This assumes that the

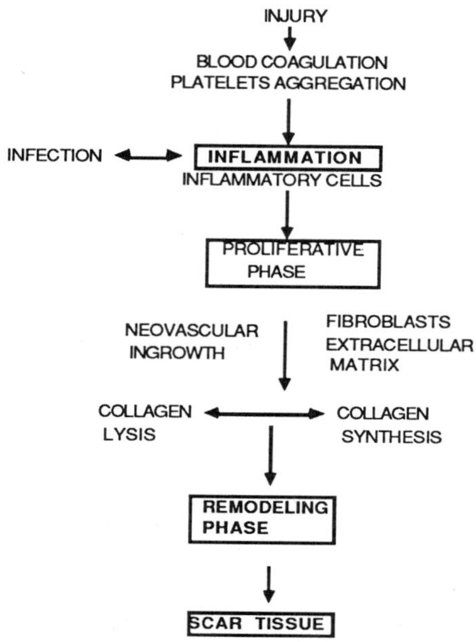

Figure 4.1. Diagram of events associated with wound healing. Injury to tissue is followed by blood coagulation and platelet aggregation and leads to attraction of inflammatory cells into the wound area. In the absence of infection, the inflammatory phase ends and proliferation of connective tissue cells leads to the deposition of new vessels and extracellular matrix. During remodeling collagen lysis is offset by new collagen synthesis resulting in the deposition of scar tissue.

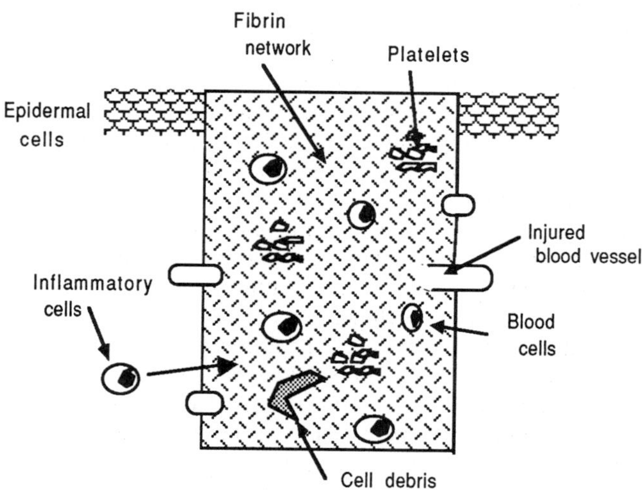

Figure 4.2. Fibrin clot formation. Diagram showing a skin defect filled with blood components that have formed a fibrin network containing platelets, injured blood vessels, and blood cells. Migration of inflammatory cells into the wound triggers removal of the clot.

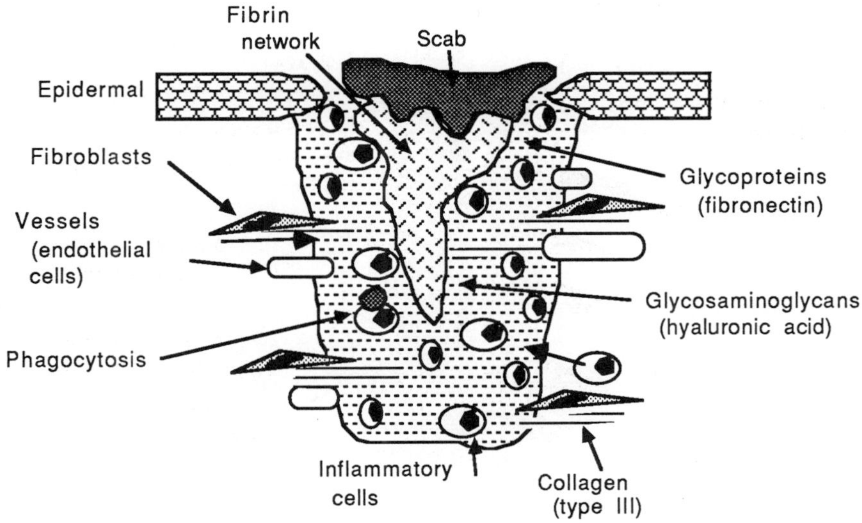

Figure 4.3. Inflammatory phase of wound healing. As clot digestion occurs by inflammatory cells (Figure 4.2), a new extracellular matrix is synthesized containing fibronectin, hyaluronic acid, and collagen that is produced by fibroblasts and endothelial and epithelial cells. Cell migration and proliferation occurs in response to release of chemoattractants.

edges of the wound are close and can be brought together by applying tension. In other cases the healing process is incomplete or ends with the formation of scar and fibrotic tissue. Patients that have suffered extensive burns sometimes do not form stable tissue but experience the formation of abnormal scar tissue, including hypertrophic scar tissue and keloids.

4.2. Biological Systems Involved in Wound Healing

Normal wound healing involves several biological homeostatic systems that maintain equilibrium, i.e., normal functioning of the physiological systems described in Table 4.1. These include the intrinsic and extrinsic blood clotting cascades, complement, kinin, immune and fibrinolytic systems. In addition, the inflammatory process is the key to triggering some of these systems.

4.2.1. Inflammation: Primary Response to Injury

Inflammation is associated with the redness, heat, swelling, and pain that occur as a result of cellular and tissue injury. Tissue injury can be caused by pressure due to an implant, chemicals, or the products of infectious organisms released from a biomaterial. Any of these factors sets into motion a series of events that results in some cases in healing and repair of damaged tissues while in other cases results in uncontrolled inflammation, which can lead to tissue destruction. Ingrowth of newly synthesized connective tissue is an example of

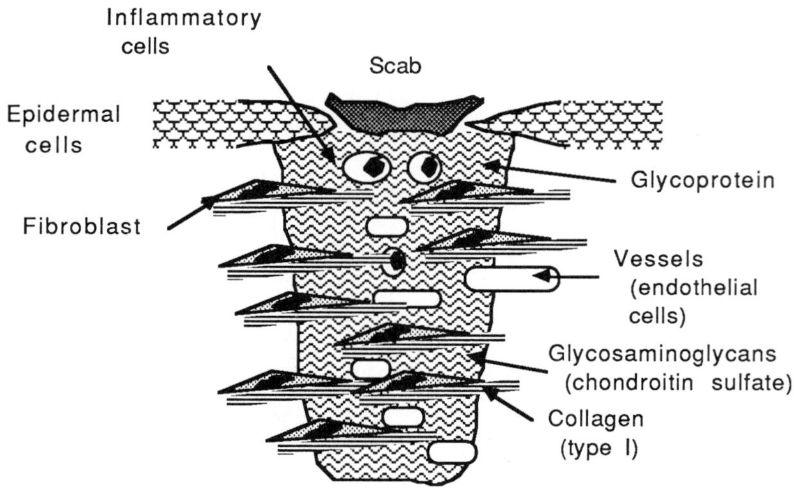

Figure 4.4. Proliferative phase of wound healing. Migration of cells into the wound is followed by cell division and maximization of the number of fibroblasts and endothelial cells in the wound. Extracellular matrix contains thin, unorganized collagen fibers and large amounts of interfibrillar matrix.

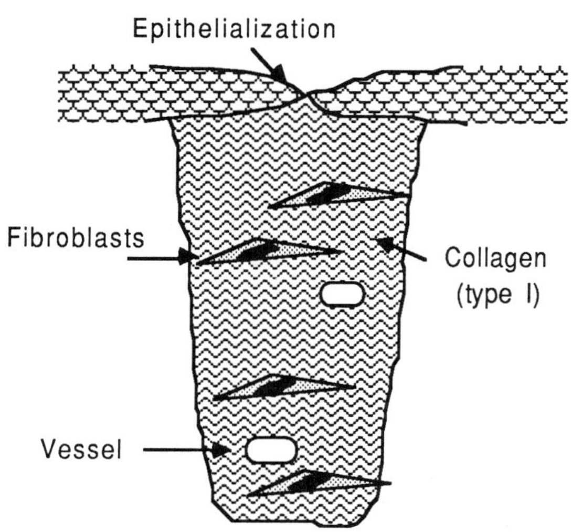

Figure 4.5. Remodeling phase of wound healing. As granulation tissue is remodeled a decrease in the number of fibroblasts as well as vessels occurs in the wound. Collagen is remodeled into large-diameter fibers and the concentration of proteoglycans changes. Aligned type I collagen fibers form a scar that is approximately parallel to the surface of the skin.

Table 4.1. Biological Control Systems Involved in Homeostasis

Control System	Components	Significance
Complement	Serum proteins	Lyse bacterial and viral membranes
Fibrinolysis	Soluble blood proteins-initiated by activated Hageman factor	Digests fibrin clots in tissues
Immune, humoral	Soluble blood proteins produced by plasma cells	Marks antigens for processing by cells and tissues
Immune, cell mediated	T lymphocytes and natural killer cells	Cytolytic destruction of tumor and other cells
Intrinsic and extrinsic clotting	Hageman and tissue factors, other blood proteins, and platelets	Prevents blood loss when vessels are damaged
Kinin	Serum proteins	Increase vascular permeability

tissue repair while graft rejection or implant extrusion are examples of destruction caused by inflammation.

4.2.1.1. Blood Coagulation Systems

Normal blood fluidity is required to maintain cardiac output to pump oxygen and nutrients to all tissues and organs. Leaks in the cardiovascular system, from an engineering point of view, lead to fluid loss and inadequate flow of blood to the peripheral tissues. For this reason loss of blood must be minimized via mechanisms that do not result in blockage of the vessel lumen. Blood coagulation is a mechanism to prevent excessive blood loss from the cardiovascular system and occurs via a complex mechanism (Figure 4.6). The first step involves contact of blood with collagen fibers in the media of a cut vessel wall; this part is termed the intrinsic clotting cascade. Contact of a cell extract termed thromboplastin, also can trigger clotting through the extrinsic pathway. Below we will discuss the connection between these pathways.

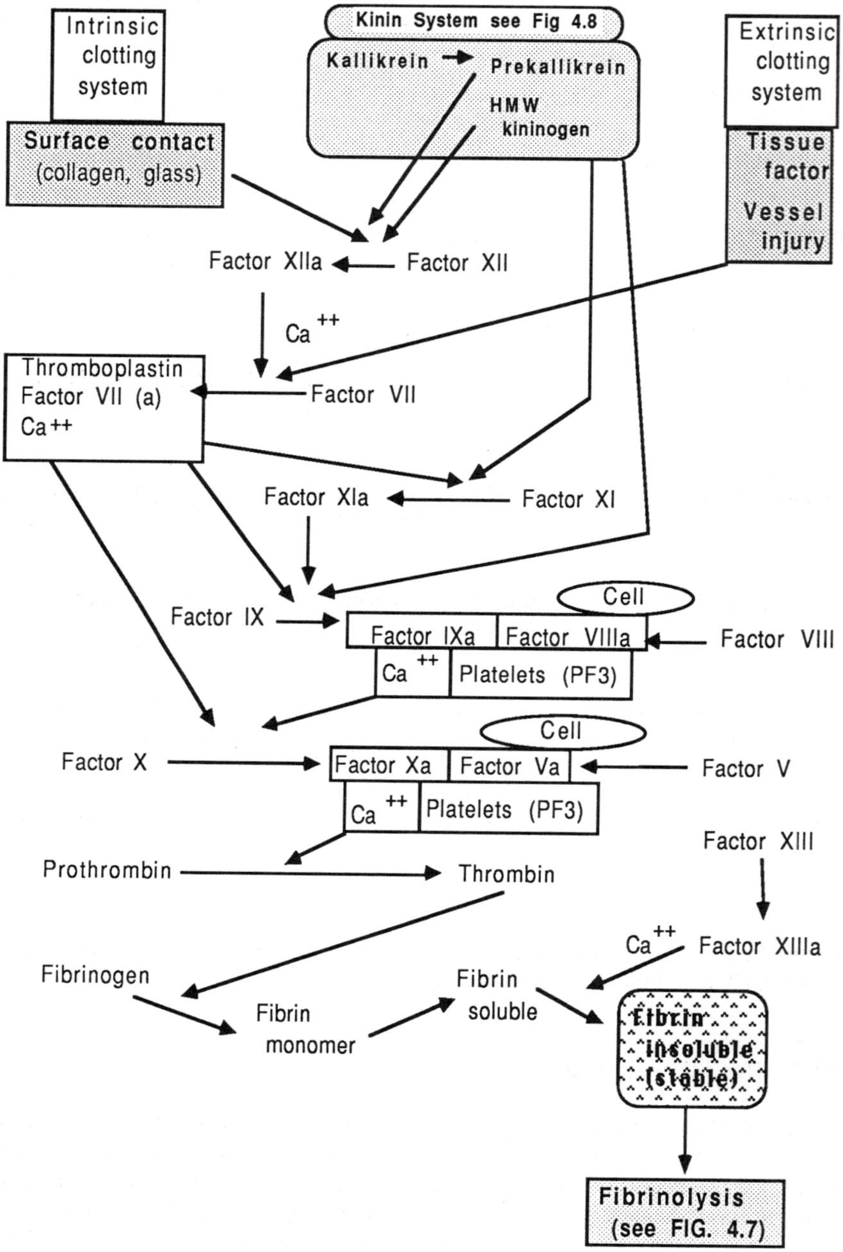

Figure 4.6. Coagulation system. Intrinsic and extrinsic blood coagulation pathways are triggered by contact with collagen, activation of the kinin system, or via tissue injury. Activation of either of these two pathways leads to formation of a fibrin clot that is insolubilized in the presence of activated factor XIII. Fibrin clot formation initiates fibrin network dissolution via the fibrinolysis pathway.

4.2.1.1.1. Intrinsic Coagulation Activation

Biomaterials are believed to trigger blood clotting by activating the contact factor (factor XII, Table 4.2 and Figure 4.6). In addition to factor XII, prekallikrein, factor XI, high molecular weight (HMW) kininogen, and C1 inhibitor are other major factors involved in control of the activation of the contact phase of coagulation (Schmaier and Coleman, 1985). Only a deficiency in factor XI results in an abnormal bleeding state suggesting that the other factors are involved in regulation of clotting. Factor XIIa, prekallikrein, high molecular weight kininogen, and Factor VIIa, interact in the activation of factor XI. Activated factor XI (XIa) sets in motion a cascade of reactions that are enzymatically triggered. Factor XIIa is inhibited by the first component of complement (C1 inhibitor) (Schreiber *et al.*, 1973). Antithrombin III also binds and inactivates fragments of factor XII (Stead *et al.*,1976). At this point further progress toward the formation of a clot requires the contribution of the extrinsic blood coagulation system (Figure 4.6).

4.2.1.1.2. Extrinsic Blood Coagulation System

For a long time it was conceived that the intrinsic blood clotting cascade and the extrinsic blood coagulation system could operate independently of each other (Prydz, 1985). As illustrated in Figure 4.6 the two systems activate each other and therefore are interdependent.

In the presence of activated factor VII, calcium, and a tissue factor termed thromboplastin, factor X is activated. Factor VII is activated in the presence of activated factor XII and its degradation fragments, XII_f.

Thromboplastin is an integral membrane protein that has been isolated from brain (Bjorklid *et al.*,1973) and requires phospholipid for coagulation activity (Deutsch *et al.*, 1964; Hvatum and Prydz,1969; Bjorklid and Storm, 1977).

Factor VII is a protein that requires vitamin K for its synthesis. Produced in the liver, this factor has a half-life of a few hours in plasma. Factor VII forms complexes with phospholipids, which increases its activity (Prydz, 1985). In the presence of thromboplastins-activated factors VII and XI and Ca^{2+}, factor IX is activated. Factor IX is a glycoprotein found in plasma that has binding sites for Ca^{2+} that link it to phospholipid on platelet membranes.

Factor VIII is activated in the presence of thrombin to VIIIa and combines with factor IXa, Ca^{2+}, and platelet phospholipid to activate factor X . Subsequent events involve conversion of prothrombin to thrombin by factors Xa and Va, platelets, and Ca^{2+}in a complex. Activated factor Va can be inactivated by protein C, in the presence of thrombin. Protein S stimulates inactivation of-factor Va by Ca^{2+}.

In the presence of thrombin, fibrinogen is converted to fibrin monomer and subsequently polymerizes into soluble fibrin. Thrombin also activates factor XIII and in the presence of Ca^{2+} forms insoluble fibrin (Fibrin-I). Fibrin networks in a clot play an important role as a scaffold for cell migration during the inflammatory response.

Activation of the intrinsic and extrinsic clotting pathway by biomaterials occurs through direct activation of factor XII and/or by tissue injury and generation of tissue thromboplastin. For biomaterials that are designed as wound

Table 4.2. Factors Involved In Intrinsic and Extrinsic Coagulation Pathway

Factor	Source	Structure (molecular weight)	Properties	Reference
XII (Hageman)	Blood plasma	Polypeptide (80,000)	Activated by negatively charged surfaces via kallikrein	Griffin and Cochrane, 1976; Cochrane et al., 1973
C1 inhibitor	Blood plasma	Gycoprotein (220,000)	Inhibits factors XIIa and XIa	Donaldson and Evans 1963; Schreiber et al., 1973
Antithrombin III	Blood plasma	----	Inhibits fragments of XII	Stead et al., 1976
Prekallikrein	Blood plasma	Polypeptide (88,000)	Converted by factor XIIa to kallikrein	Weupper and Cochrane, 1972
High Molecular Weight Kininogen (HMWK)	Blood plasma, platelets	Polypeptide (110,000)	Binds to kallikrein and activates factor XI	Jacobsen and Kriz, 1967
Factor XI	Blood plasma	2 chains polypeptide (175,000)	Activated by Factor XIIa complexed with HMWK in blood	Kurachi and Davie, 1977
Factor VII	Blood plasma	Protein (48,000)	Activated by XIIa and its fragments	Kissel and Davis, 1977
Thromboplastin	nembrane extract	Membrane protein (50,000)	Activates factor X in presence of VIIa and calcium	Bjorklid et al., 1973

Table 4.2. (cont.)

Factor	Source	Structure (molecular weight)	Properties	Reference
Factor IX	Blood plasma	Glycoprotein (56,000)	Possesses binding sites for Ca^{+2} for linkage to phospholipids	Katayama et al., 1979
Factor VIII	Blood plasma	?	Activated by thrombin-binds to complex of Xa + Ca^{+2} + platelets	Houghie and Baugh, 1984
Factor X	Blood plasma	2 chain polypeptide (55,000)	Activated by thromboplastin complex with Ca^{+2} and VIIa	Jackson and Hanahan 1968
Factor V	Blood plasma	Glycoprotein (330,000)	Activated by thrombin	Esmon, 1979; Nesheim et al., 1979
Protein C	Bood plasma	---	Ceaved by thrombin into CA which inactivates Va	Esnouf, 1984
Protein S	Blood plasma	Protein (65,000)	Stimulates inactivation of Va by CA	Kisiel, 1975
Prothrombin	Blood plasma	Glycoprotein (70,000)	Converted to thrombin in presence of factor Xa, Ca^{+2} and platelets	Esnouf, 1984
Fibrinogen	Blood plasma	Protein (340,000) dimer of three polypeptide chains	Converted to fibrin monomer in presence of thrombin	Caspary and Kekwick, 1957
Factor XIII	Blood plasma	Protein (320,000) tetramer of alpha and beta chains	Activated by thrombin in presence of Ca^{+2}, crosslinks fibrin	Schwartz et al., 1973

dressings, activation of clotting prevents unnecessary bleeding. In contrast, activation of clotting is an undesirable property of materials that are designed for blood conduits.

4.2.1.1.3. Role of Platelets in Hemostasis

Platelets are cells that lack a nucleus and are derived from megacaryocytes found in the bone marrow. These cells are actively involved in preventing bleeding via their ability to stick to ruptured vessel walls and their role in blood clotting.

Platelets adhere to gaps between endothelial cells in areas of exposed vascular basement membrane. Exposure of subendothelial connective tissue to platelets results in rapid adhesion and build up of several layers of cells. It is believed that collagen fibers in the vessel wall and ADP released from injured cells stimulate this process (see Born *et al.*, 1984). Platelet-to-platelet aggregation follows initial platelet adhesion and occurs until the defect is filled by a plug.

Platelet aggregation is correlated with binding of fibrinogen (Peerschke *et al.*, 1980), thrombin, serotonin, thromboxanes, and phospholipids. In addition, platelet factor 3 (a phospholipoprotein) is released from platelets in the form of micelles and is associated with acceleration of thrombin formation (Figure 4.6).

When a surgical incision is made in a tissue or mechanical trauma occurs, a sequence of events is triggered involving the biological systems discussed in Section 4.2.

Proteins in blood involved in the clotting cascade (Figure 4.6) that come in contact with defects in vascular connective tissue components trigger a sequence of events that leads to the formation of a clot. These events include activation of the contact factors, conversion of prothrombin to thrombin, and conversion of fibrinogen to a crosslinked fibrin network in the presence of fibrin stabilizing factor. In this cascade factor XII is activated by contact with collagen fibers found in connective tissue. Activated factor XII then acts to activate the next factor (factor IX), which in turn activates the next step until a fibrin clot is formed. Platelets are required as a cofactor for fibrin network formation.

Factor XII also triggers fibrinolysis that causes removal of the blood clot from the wound (Figure 4.7). This process involves the kinin system (Figure 4.7) and the complement system (Figure 4.8). Both of these systems involve activation of proteins found in blood plasma.

Injury to connective tissue found in th eextracellular environment causes release of collagen-derived peptides that attract fibroblasts, endothelial cells, and inflammatory cells through a process termed chemotaxis. These peptides as well as the complement components (C3a and C5a) and bradykinin (from the kinin pathway) are involved in initiating the inflammatory phase that results from tissue trauma.

4.2.2. Fibrinolysis

Once clots containing fibrin networks and platelet aggregates are formed to prevent excessive blood loss, the fibrinolytic system begins to dissolve fibrin clots

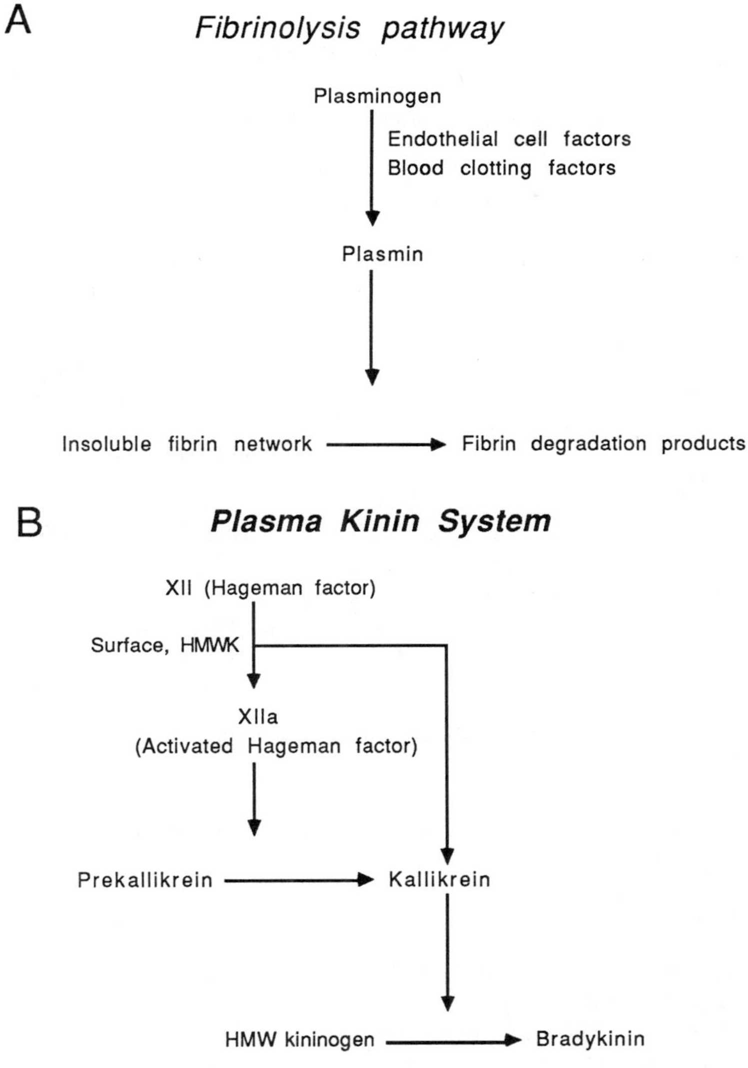

Figure 4.7. Fibrinolysis/kinin systems. Diagram illustrating (A) fibrinolysis pathway for generation of fibrin degradation products from fibrin, (B) plasma kinin system that generates bradykinin, a vasoactive agent.

to prevent or limit intravascular thrombosis. This process also aids in the removal of fibrin from extravascular tissues, which is the first step in the repair process.

Fibrinolysis involves conversion of plasminogen to plasmin, an enzyme that actively degrades fibrin into fibrin degradation products (Bennett *et al.*, 1984). Plasmin's fibrinolytic activity is rapidly neutralized by α_2-antiplasmin (Table 4.3 and Figure 4.7). Degraded fibrin networks are also digested by inflammatory cells.

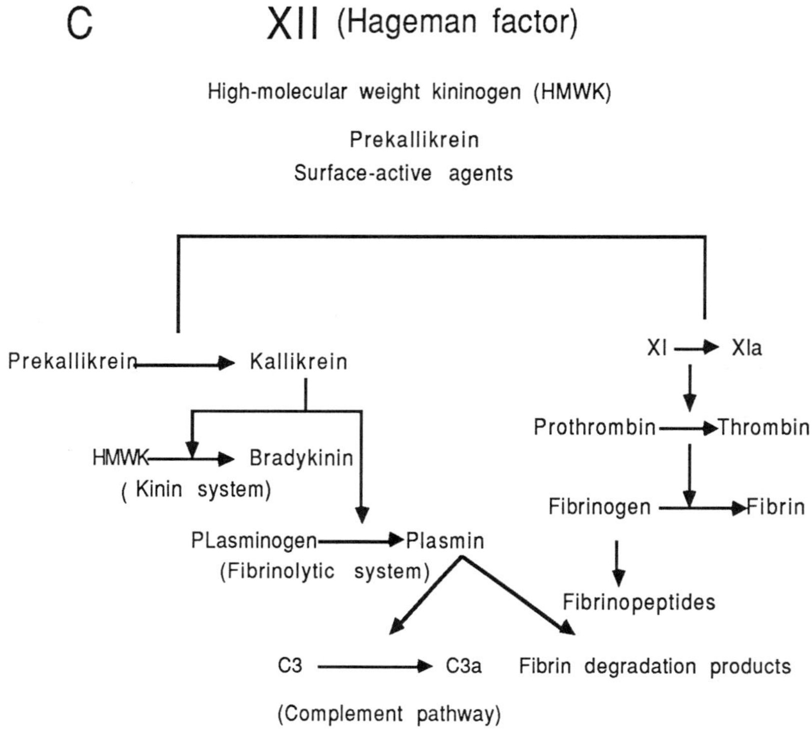

Figure 4.7 (cont.). (C) the interaction between blood clotting, kinin, complement, and fibrinolysis systems.

4.2.3. Complement System

The complement system involves about eleven soluble serum proteins that can lyse bacteria in the presence of an antibody response. Fragments of some of these proteins also contribute to inflammation. This system consists of two pathways, classic and properdin, that lead to membrane lysis of bacteria or target cells (Figure 4.8).

The classical pathway (see Mulle-Eberhard, 1983 for a summary) is triggered by an antibody antigen complex and is recognized by C1 which consists of subunits termed $C1_q$, $C1_r$, and $C1_s$. $C1_q$ is capable of recognizing antibody and also activating $C1_r$. $C1_r$ in turn activates $C1_s$ to provide a supply of enzymatically active Cl_s fragments. Next, $C1_s$ activates C_4 and C_2 to form a C_4-C_2 biomolecular complex that in an activated form termed C3 convertase, splits C3 into C3a (small fragment) and C3b (larger fragment). C3a acts as a mediator in inflammation by causing histamine release, smooth muscle contraction, and release of hydrolytic enzymes from neutrophils and mast cells. C3b associates with C3 convertase to form a trimolecular complex (C4-C-C3b) termed C5convertase. C5 in the presence of C5 convertase is cleaved into C5a (small fragment) and C5b (large fragment); C5b fuses with C6, C7, and C8 forming a complex that attaches to the membrane of the target cell. This complex

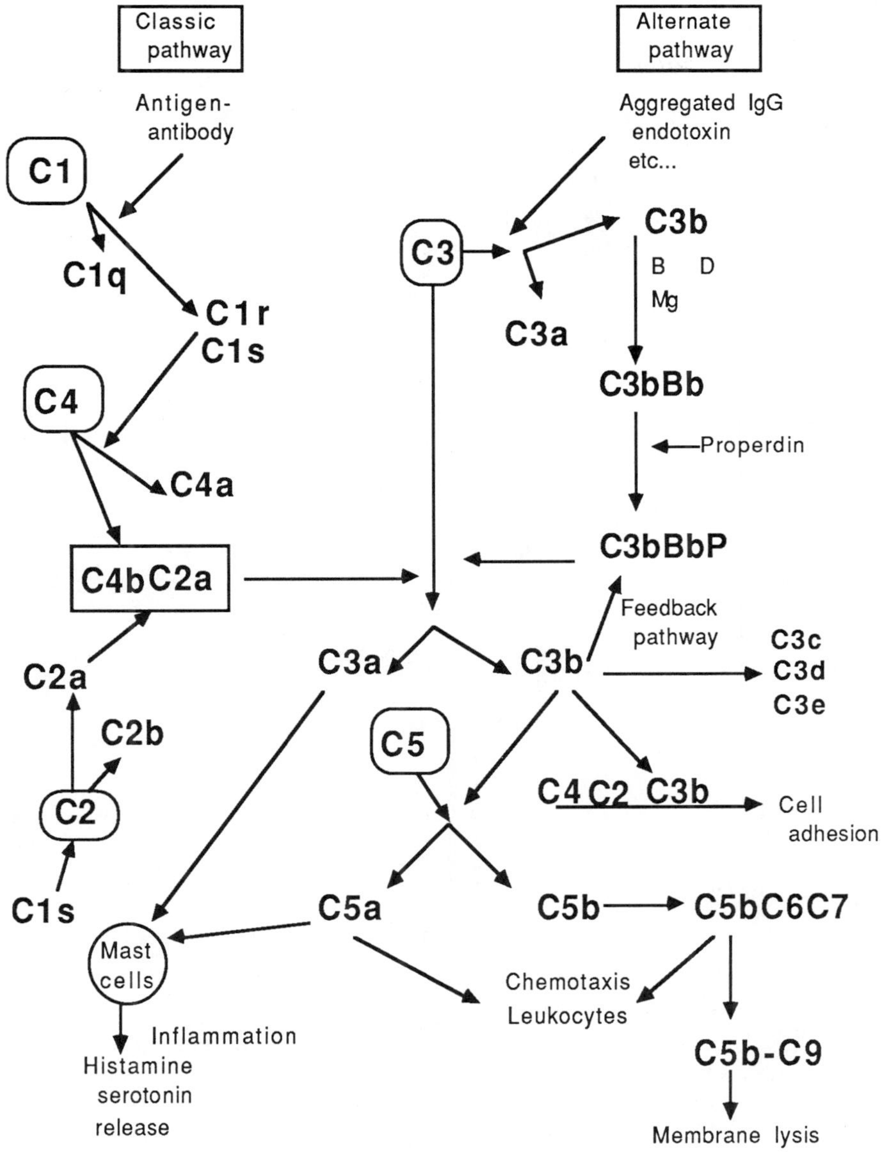

Figure 4.8. Complement system.
Formation of a C5C6C7C8C9 complex that results in membrane lysis of a target cell occurs through either the classic or properdin (alternate) pathways. The classic pathway is triggered via an antigen-antibody complex while the alternate pathway is triggered by aggregated antibody or bacterial endotoxins.

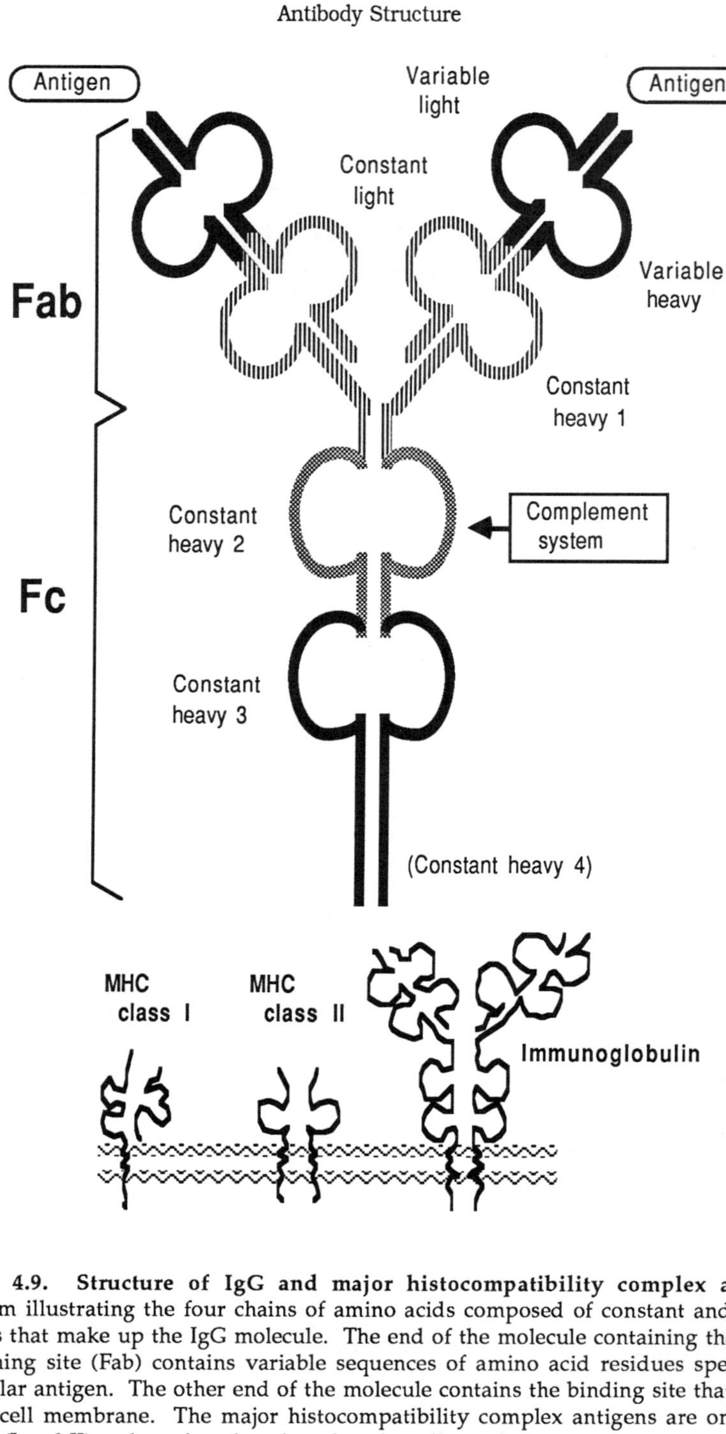

Figure 4.9. Structure of IgG and major histocompatibility complex antigens. Diagram illustrating the four chains of amino acids composed of constant and variable regions that make up the IgG molecule. The end of the molecule containing the antigen combining site (Fab) contains variable sequences of amino acid residues specific to a particular antigen. The other end of the molecule contains the binding site that attaches to the cell membrane. The major histocompatibility complex antigens are one of two classes (I and II) and are found anchored to the cell membrane.

Table 4.3

Factors Involved In Fibrinolytic Pathway

Factor	Source	Structure (molecular weight)	Property	Reference
Plasminogen	Blood plasma	protein (80,000-90,000)	in presence of activators, cleaved to plasmin	Robbins, 1977
Plasmin	blood plasma	two polypeptide chains crosslinked with disulfide bonds	digests fibrin	Wiman, 1978
Plasminogen activators	endothelial cells	protein? (65,000)	activates plasminogen into active plasmin	Aokiand and von Kaulla, 1971

activates the polymerization of C9 into a tubule that spans the full thickness of the cell membrane and permits ion and water flow into and out of the cell which leads to cell death.

The properdin or alternate pathway (Figure 4.8) involves hydrolysis of C3 into a C3b-like molecule that binds to another factor (B) and is activated by D in the presence of magnesium, yielding a smaller complex C3bBb. Properdin binds to the complex, which generates active 3b. The classic and alternate pathways join at this point.

The complement pathway is a mechanism for destruction of foreign bacteria and other targeted cells during graft rejection, in the presence of soluble antibody, and results in the generation of fragments C3a and C5a that are active in inflammation.

4.2.4. Kinin System

The plasma kinin system is an enzymatic system that is triggered by activated factor XII and the intrinsic blood clotting cascade and converts prekallikrein to kallikrein which converts high-molecular-weight kininogen to bradykinin. Bradykinin induces leakage of fluid from the vascular system during the process of inflammation.

The kinin, blood clotting, fibrinolytic, and complement systems all interact. This interaction contributes to the complexity of the inflammatory process.

4.2.5. Immune System

Over the course of a lifetime humans come in contact with a wide variety of foreign substances both cellular and acellular. Cellular organisms that invade the body are destroyed via complement-dependent lysis via a humoral-mediated response involving antibodies and B lymphocytes or via a cell-mediated response involving white blood cells termed T-lymphocytes. Acellular substances (antigens) that gain entry into the body are combined with proteins termed antibodies or immunoglobulins and removed in tissue by free tissue macrophages. Circulating antigens are removed from the blood by the reticuloendothelial system consisting of monocytes in the blood, Kupffer cells in the liver, and endothelia of the sinusoids that are found in the spleen and bone marrow.

4.2.5.1. Antibody Structure and Function

Each antibody molecule is composed of two high-molecular-weight (heavy) chains and two lower molecular weight (light) chains with disulfide crosslinks holding the heavy chains and ligh chains into a Y-shaped molecule (Figure 4.9). The light and heavy chains are composed of about 214 and 440 amino acids, respectively. Each chain consists of regions of variable (V) and constant (C) sequences of amino acids. The amino terminal portions of the light and heavy chains show different amino acid sequences and give rise to the site where antibody and antigen combine (antibody combining site or Fab fragment). Proteolytic cleavage of an antibody molecule with the proteolytic enzymes papain and pepsin result in formation of fragments that combine with complement (Fc fragment) and antibodies (Fab fragments), respectively.

Serum contains five types of immunoglobulins, IgG, IgA, IgM, IgD, and IgE, that vary in molecular weight and number of antigen combining sites. High-molecular-weight forms result from aggregation of the basic Y-shaped molecule. Antibodies of the IgM class represent the primary antibody that circulates in the blood. IgG antibodies are directed against a specific antigen.

Antibodies are produced by plasma cells in response to contact with an antigen. Plasma cells are derived from activated B lymphocytes found in peripheral lymphoid organs including spleen, tonsils, lymph nodes, and gastrointestinal-associated lymphoid tissue.

At birth pre-B cells have IgM antibodies on their surface. On migration to peripheral lymphoid tissue they acquire IgD on their surface. Binding of antigen to B cells triggers differentiation into plasma cells that synthesize IgM, IgD, IgG, IgE, and IgA, and secrete into the extracellular space IgM,IgG, IgE, and IgA, (Leder, 1983), and release chemicals (lymphokines) that stimulate other lymphocytes (Figure 4.10) and other cells. In addition, replication of B cells leads to synthesis and excretion of additional antibodies. Mature B cells exposed to an antigen (memory B cells) help other cells respond to the same antigen.

Other thymus-dependent lymphocytes termed T-cells are effective in suppressing (suppressor T-cells) or maximizing (helper T-cells) antibody production by plasma cells after exposure to antigen.

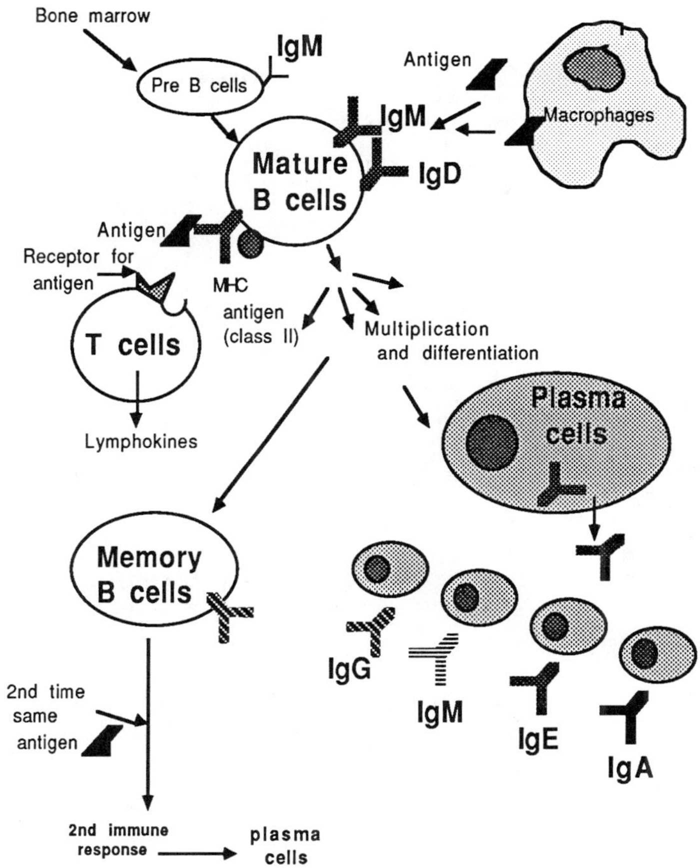

Figure 4.10. Humoral mediated immune response. In contact with an antigen presented by macrophages, bone marrow B cells mature and transform into plasma cells that secrete antibodies specific to the antigen. Memory B cells are formed that differentiate and produce antibodies immediately on contact with a second exposure to the antigen. T cells are stimulated to release lymphokines by antigen-antibody complexes attached to B cells.

4.2.5.2. Cell Mediated Immunity

T lymphocytes are involved in the killing of foreign cells by a process that determines the chemistry of cell-membrane-associated glycoproteins termed histocompatibility antigens. Several types of cytotoxic lymphocytes are present in tissues, including cytotoxic T-cells, natural killer (NK), or killer cells.

Cytotoxic T lymphocytes are a product of T-cell responses to surface membrane antigens and are involved in graft rejection. In contrast, K-cells achieve cytolytic specificity after target cells are coated with antibody. K-cell attachment to the Fc antibody receptor is followed by lysis of target cells. T cells also help (helper T cells) or inhibit (supressor T cells) the response of B cells

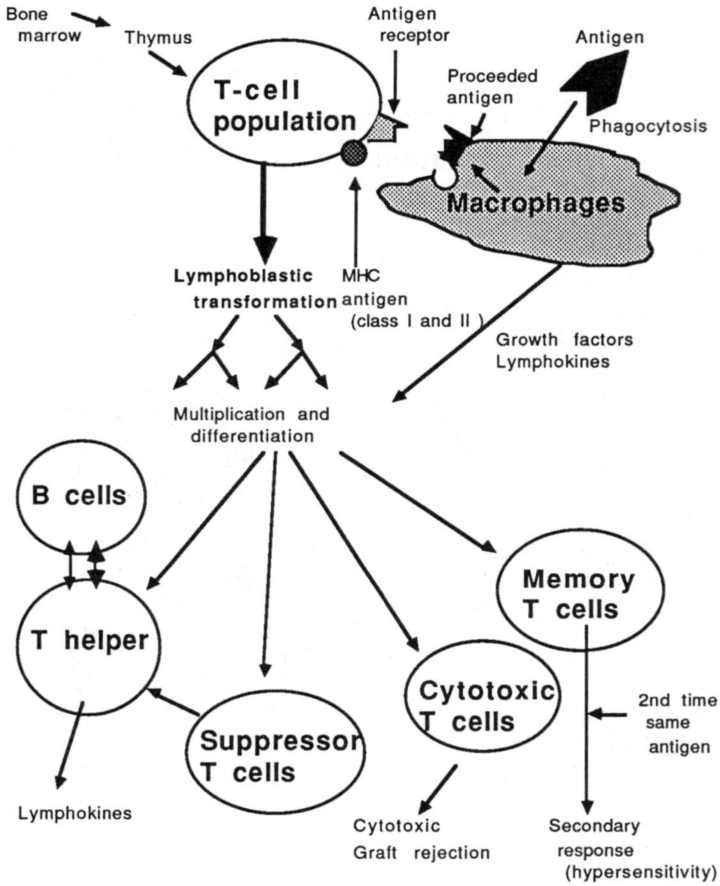

Figure 4.11. Cell-mediated immune response. T cells derived from bone marrow are activated by antigen presented by macrophages and class I and II MHC antigens. Activated T cells proliferate and are transformed into memory, helper, suppressor, and cytotoxic T cells.

to an antigen. Memory T cells exposed to an antigen for the second time result in a hypersensitive response.

4.2.5.3. Human Lymphocyte Antigens (HLA)

T-cell recognition of cell surface antigens leads to cell-mediated transplant rejection. Class I antigens (Figure 4.9) present on most cells are coded by the major histocompatibility complex (MHC) located on chromosome 6 in the human. This site codes for membrane antigens that recognize cell as either "foreign" or "self". In addition, class II antigens are coded for by the MHC and are found on lymphocytes, platelets, monocytes, endothelial cells, and Langerhan's cells in the skin. The MHC codes for a series of membrane antigens termed A, B, C, and D related (D_R). These antigens are grouped into class I (A, B, and C) and class II (D_R).

Class I antigens are glycoproteins consisting of a heavy chain (M_r 44,000) and a light chain(M_r 11,600). Class II antigens are composed of two heavy chains (M_r 29,000 and 34,000) (Figure 4.9).

Graft rejection involves a complex series of events (Figure 4.11) that are triggered by class I and II antigens on donor cells. These antigens are recognized as foreign and trigger activation of cytotoxic T cells in the presence of macrophages and antibodies to these antigens, production of lymphokines, production of antibodies by B cells, and eventual lysis of donor cells.

4.3. Inflammatory Phase

At the microscopic level tissue alterations seen during the inflammatory phase are due to vascular and cellular events (Figures 4.12 and 4.13). Vasodilation of blood vessels is due to release of prostaglandins, histamine, and serotonin. Histamines and serotonin are released from mast cells and platelets present in blood and tissue in the injured area. These agents cause loss of intercellular contacts between endothelial cells that line blood vessels and result in escape of proteins and serum into the vessel wall.

A wide variety of mediators of inflammation have been isolated (Table 4.4). Contact of blood cells (mast cells, basophils, and platelets) with collagen, thrombin, lysosomes from neutrophils, complement components (C3a and C5a), and physical agents induce secretion of histamine, serotonin, and platelet activating factors (platelet factor 3) by these cells. These factors stimulate vasodilation.

During degradation of fibrin networks (fibrinolysis), fibrin degradation products are stimulators of vasodilation. Prostacyclin secreted by endothelial cells inhibits platelet aggregation and vasodilation and serves to limit the effects of vasodilators released by blood cells.

Contact of factor-XII with collagen and bacterial endotoxins activates kallikrein and bradykinin (the kinin system). These activated plasma proteins increase chemotaxis of neutrophils and monocytes (kallikrein) and vasodilation (bradykinin). Vasodilation is also caused by the release of prostaglandins from a variety cells; however, prostaglandin activity is rapidly lost. Lymphokines and leukotrienes, secreted by lymphocytes and neutrophils, respectively, are other cellularly derived products that cause vasodilation.

Some components such as C5a from the complement system, besides increasing vascular permeability, also are chemotactic to neutrophils and monocytes. This factor attracts neutrophils and monocytes to the site of injury where these cells begin to clean up dead tissue and bacteria. Neutrophils and monocytes migrate from vessels near the wound and collect in the area of the wound within hours after injury has occurred. Chemotactic agents released from cells include leukotrienes (leukocytes), kallikrein (kinin system), prostaglandins (cell membrane metabolism), collagen derived peptides, fragments of fibronectin, and bacterial products.

Removal of dead cells, clotted blood proteins, and denatured tissue components is accomplished via phagocytosis. During phagocytosis, particles

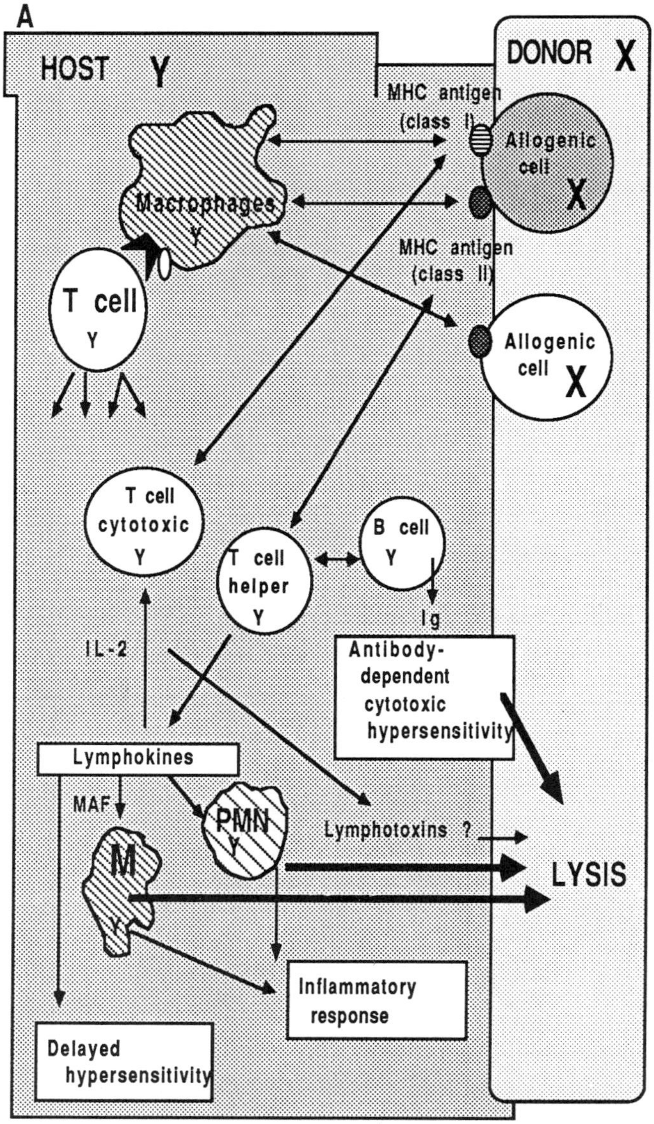

Figure 4.12. Mechanism of graft rejection. (A). Diagram showing interaction of class I and II MHC antigens from a donor tissue and host T and B cells, macrophages, and inflammatory cells. Allogenic cells activate macrophages, and cytotoxic and helper T cells to stimulate release of lymphokines from inflammatory cells as well as antibodies from B cells. Lysis of allogenic cells occurs via attack by neutrophils and monocytes or via antibody dependent cytotoxicity.

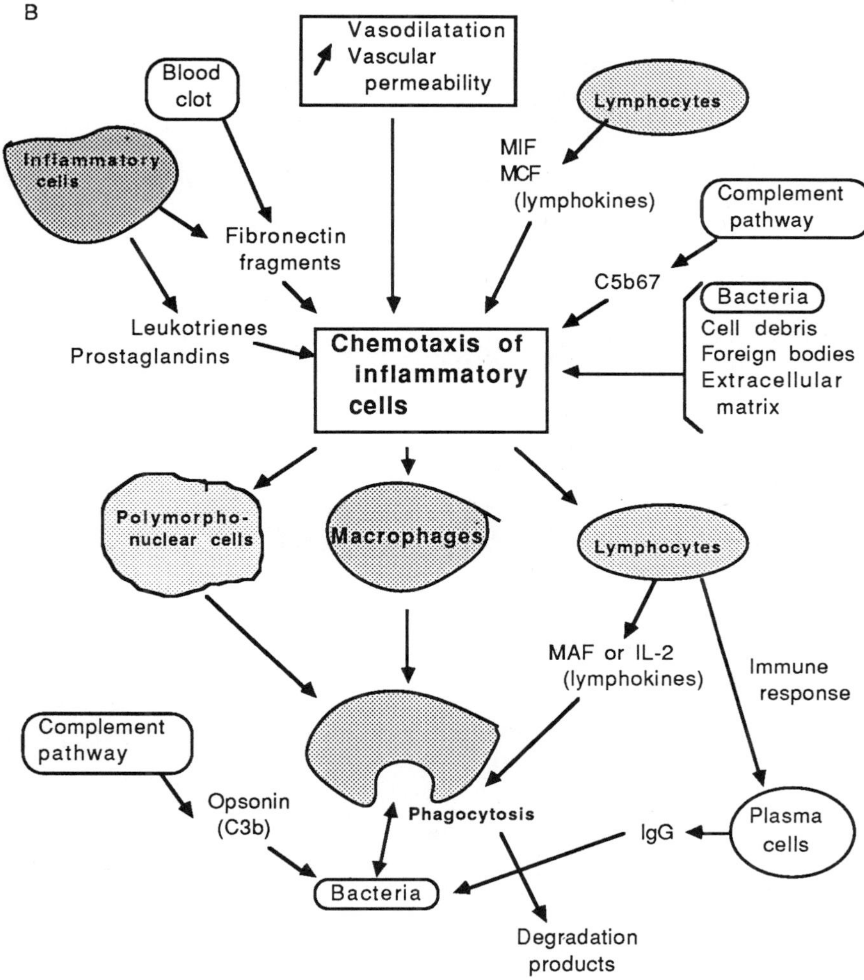

Figure 4.12 (cont.). (B). Diagram illustrating the sources of chemotactic stimuli that cause chemotaxis of inflammatory cells and destruction of foreign cells.

containing dead cellular material or denatured tissue macromolecules are engulfed by the cell membrane and then incorporated into the cytoplasm where they are incorporated into a membrane-lined vesicle (Figure 3.4). Lysosomal fusion with the phagocytic vesicle results in the formation of a single vesicle. Enzymes and peroxides present within lysosomes degrade the material within the vesicle. If the material resists digestion then "residual bodies" remain within the cell.

The interactions between monocytes and macrophages and other cells is diagrammed in Figure 4.14. Release of interleukin-1 affects neutrophils, fibroblasts, endothelial cells, and B lymphocytes while T cells modify factors that influence macrophage function. The complex interactions among these cells

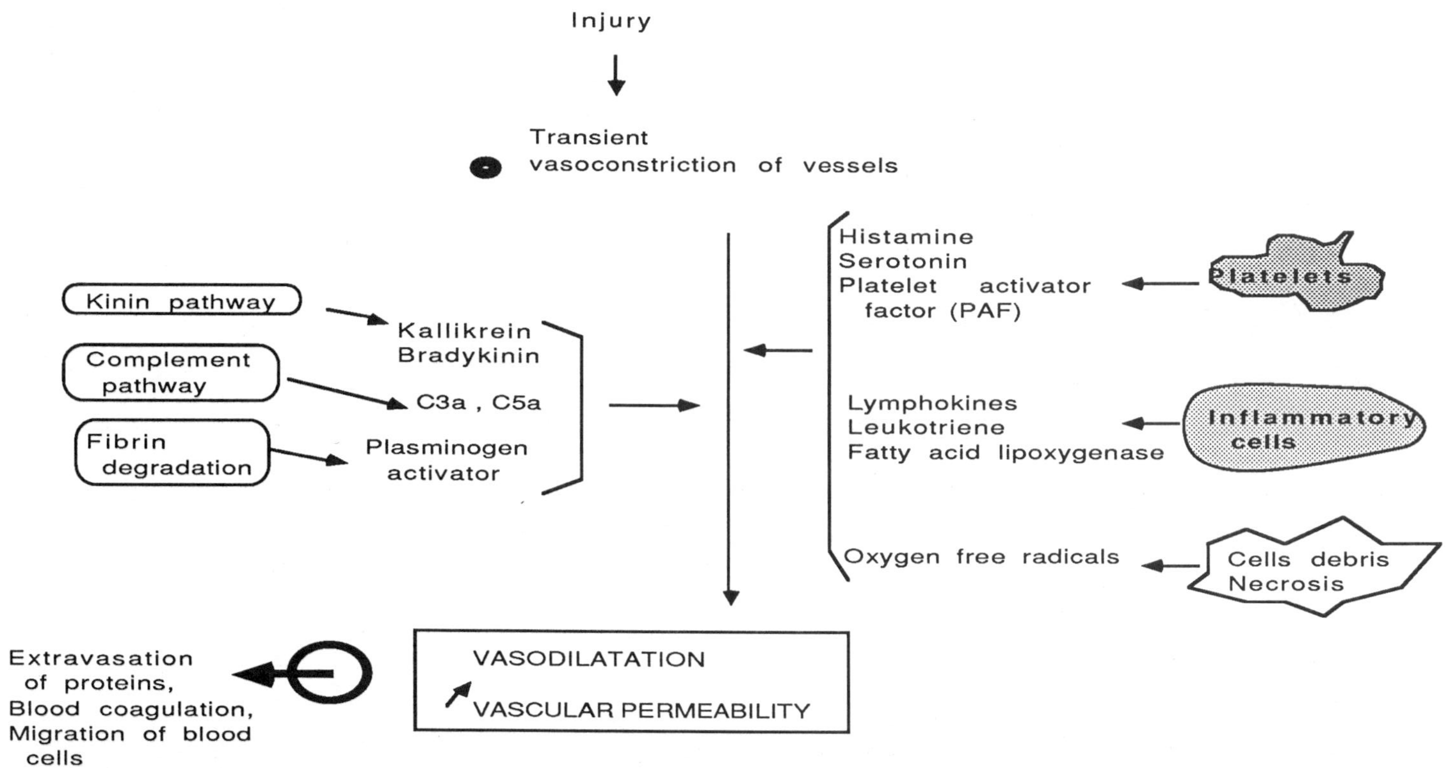

Figure 4.13. Vascular events associated with injury. Associated with injury is a transient vasoconstriction that is a result of release of products of platelets, inflammatory cells, tissue necrosis, and products of the kinin, complement, and fibrinolysis systems. Following this transient vasodilation is a period of vasodilatation of vessels in tissues surrounding the defect.

is beyond the scope of this text.

4.4. Proliferative Phase

The inflammatory phase is aimed at preventing bleeding and attracting white blood cells to the wound area to clean up damaged tissue, bacteria, and dead cells. However, the process of repair requires the involvement of cells that lay down new connective tissue components in the area of injury. The proliferative phase involves the migration of cells including fibroblasts, endothelial cells, and epithelial cells into the wound area and is followed by synthesis and deposition of collagen fibers and interfibrillar matrix.

It is triggered by events that occur during the inflammatory phase. Chemical mediators such as platelet activating factor cause platelet aggregation and release of platelet-derived growth factor (PDGF). PDGF stimulates fibroblast and endothelial cell migration to the site of injury. Other factors, including fibrin, plasma fibronectin, plasminogen activator, lymphokines, macrophage secretory products (interleukin-1 and fibronectin), and inflammatory cell secretions (hyaluronic acid), attract and cause proliferation of fibroblasts and endothelial cells (Figure 4.15).

Fibroblasts initially synthesize and deposit connective tissue extracellularly in a random form. Granulation tissue (Figure 4.4) consists of fibroblasts surrounded by extracellular matrix and endothelial cells involved in capillary formation. Synthesis and deposition of extracellular matrix is enhanced by fibroblast growth factor (FGF), and epidermal growth factor, (EGF), ascorbic acid (vitamin C), oxygen, glycosaminoglycans, proteoglycans, and glycoproteins. In contrast, interleukin-1 and prostaglandins inhibit synthesis of extracellular matrix.

Biochemically the composition of the extracellular matrix changes during the proliferative phase. Initially, the collagen typeIII/type I ratio of the extracellular matrix is high and decreases with increased healing. Hyaluronic acid is initially found in high concentrations in dermal wound tissue and is subsequently replaced with chondroitin sulfate and finally dermatan sulfate.

Proliferating and migrating fibroblasts contain large actin bundles that are present in the cell cytoplasm near the cell membrane. These bundles lie parallel to the long axis of the cell and appear to be involved in adhesion, via heparan sulfate proteoglycan, to fibronectin found in the extracellular matrix as well as indirectly to collagen. These fibroblasts have been referred to as "myofibroblasts" and are thought to be responsible for cell mobility within wound tissue. In the past, it has been hypothesized that myofibroblasts induce wound contraction. Wound tissue contraction is especially observed in animals and is characterized by the movement of the edges toward the center of the wound.

New blood-vessel growth (angiogenesis) is triggered by factors released by macrophages, low O_2 content, the presence of lactic acid in wounds, FGF, vasoactive amines, and glycosaminoglycans (heparin and heparan sulfate). Migration and replication by capillary-bed formation of endothelial cells is directed toward the center of the wound.

Reepithelialization is another event that frequently occurs during the proliferative phase. This event corresponds to the movement of epidermal cells

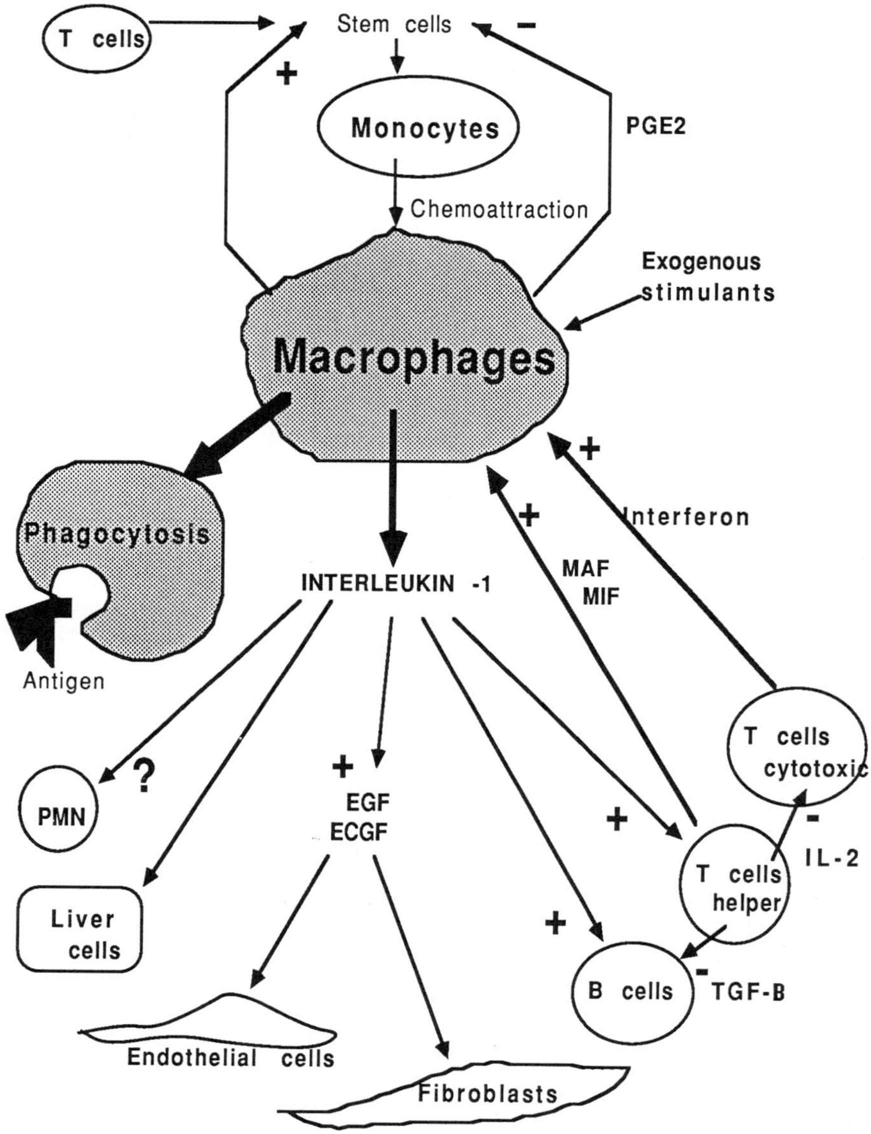

Figure 4.14. Macrophage/monocyte cellular interactions. Activation of monocytes into macrophages enables these cells to be involved in phagocytosis and results in stimulation of other cells via release of interleukin-1. Feedback inhibition occurs via the release of factors from T cells.

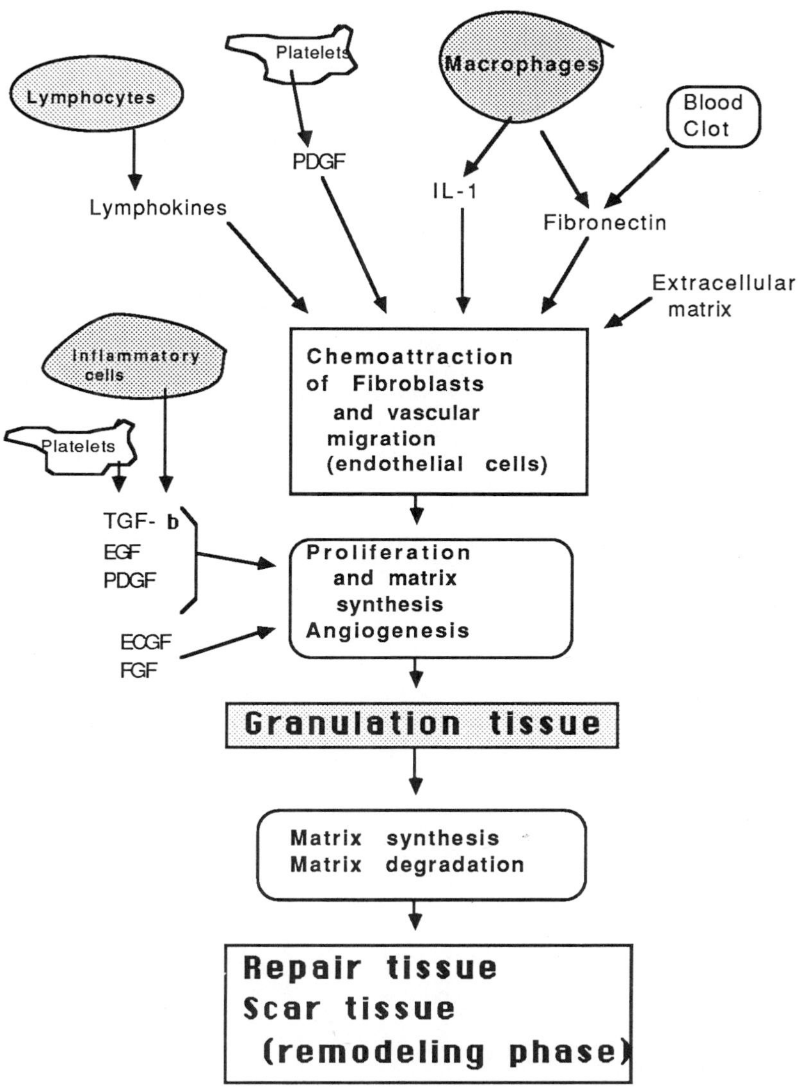

Figure 4.15. Interaction of inflammatory cells with fibroblasts and endothelial cells. Diagram illustrates release of lymphokines and other products from inflammatory cells and macrophages and the effects on synthesis and deposition of repair tissue.

(in skin repair) or epithelial cells (in other specific tissues) from the free edges of the tissue across the defect. Epidermal cells are activated at the wound edges and extend components of the cell membrane (pseudopodia) across the wound. These cells lose their polarity and become mobile, advancing on the wound surface by successive movement of the cell membrane. Chemotaxis and replication of epidermal cells are the result of chemical stimulation in which fibronectin plays the most important role. However, it is suggested that

Table 4.4. Summary of Mediators Involved in Inflammation

Mediator	Source	Action			
		Vascular Leakage	Chemotaxis		
			Neutrophils	Monocytes	Other
Histamine and serotonin	Mast cells, basophils, and platelets	+	-	-	
Kinins					
Bradykinin	Plasma substrate	+	-	-	Pain
Kallikrein		-	+	+	
Complement					
C3a	Plasma protein via liver,macrophages	+	-	-	Opsonic fragment (C3b)
C5a		+	+	+	
Prostaglandins	Most cells from membrane phospholipids	Potentiate other mediators	±	-	Vasodilation, pain, fever
Leukotrienes					
B_4	Leukocytes	-	+	+	Bronchoconstriction, Vasoconstiction
C_4,D_4,E_4	Leukocytes: mast cells	+	-	-	
Lysosomal components					
Cationic proteins	Leukocytes	+	-	-	Immobilization of neutrophils
Neutral proteases	Leukocytes	+			Tissue damage
Oxygen metabolites	Leukocytes	+			Endothelial damage, tissue damage
AGEPC	Mast cells, other cells	+	+	?	Bronchoconstriction

Table 4.4. (Continued)

		Action			
		Vascular Leakage	Chemotaxis		
Mediator	Source		Neutrophils	Monocytes	Other
Interleukins IL - 1 IL - 2 (MAF)	Activated macrophage Endothelial cells Keratinocytes Langerhans cells	N/A	N/A	N/A	Stimulate macrophage and lymphocytes (immune response)
Interferons IFN - α IFN - β	Leukocytes Fibroblasts	N/A	N/A	N/A	Stimulate natural killer cells (immune response)

N/A, not applicable.

AGEPC, acetylated glycerol ether phosphocholine platelet activating factor (PAF).

MAF, macrophage activating factor.

epidermal growth factor (EGF) is directly involved in epidermal replication. The basement membrane does not reform until after migration ceases. Epidermal cell migration always occurs from the living tissue (granulation tissue) to the repair tissue. Epidermal cells secrete enzymes that cause the formation of a passage between living and dead tissues.

4.5. Remodeling Phase

As described earlier, this phase overlaps with the previous phase. Remodeling corresponds to the formation of scar tissue. Scar formation requires both collagen synthesis and collagen catabolism (degradation). Wound collagen is degraded by a variety of enzymes including collagenase. At the same time new collagen is synthesized, deposited, and reorganized. Progressively, glycoproteins, glycosaminoglycans, and the content of type III collagen are diminished. During the remodeling phase large collagen fibers composed of thick fibrils are observed (mostly type I collagen). The thickness of the collagen fibers is observed to increase during remodeling although it never reaches the thickness observed in normal tissue except in wounds closed by sutures. The tensile strength of repair tissue from large dermal wounds is related to the amount of collagen and never reaches that of normal skin. It has recently been demonstrated that the organization of collagen into fibers is related to increased wound tensile strength in remodeled dermal tissue.

In the case of abnormal dermal healing, excessive remodeling results in hypertrophic scar tissue, which is characterized by whorls containing randomly deposited collagen that is eventually aligned parallel to flexion lines of the scar. Transmission electron microscopy shows slightly irregular-to-ovoid—shaped fibrils with smaller-than-normal diameters and excess glycosaminoglycans associated with collagen fibrils. The apparently reduced extensibility of hypertrophic scar tissue is a result of a partial pre-alignment of the collagen fibers.

4.6. Wound Healing in Different Organs and Tissues

4.6.1. Peripheral Nerve Repair

Nerves that are severed by mechanical means or via a pathological condition are surgically repaired. Wound healing in nerves involves Schwann-cell proliferation shortly after trauma occurs and endoneurial fibroblasts multiply and lay down collagen around and between the regenerating axons.

At the distal end, axons collapse, are compressed or obliterated by collagen deposition, and are resorbed during reinnervation. The distal portion of the nerve undergoes a Wallerian degeneration that results in disappearance of neurofibrils and myelin sheaths.

During the reinnervation of the distal part by the proximal part of the peripheral nerve, there is cellular bridge formation (if the two parts are close to each other) that leads to central migration and attachment of the ends. These new tunnels will be the scaffold for new axons to travel through. This process is an active phenomenon requiring proliferation of distal Schwann cells. In the proximal part, new axons move distally and then undergo remyelinization gradually. Initially, the number of axons exceeds the normal number by a factor greater than four and then decreases.

4.6.2. Vascular Repair

Large-and small-diameter blood vessels are repaired after the lumen is closed by atherosclerosis or after the lumen is damaged by disease. In most cases the subintima and media are exposed directly to blood and consequently they are covered by fibrin and blood cells. There cellularization (reendothelialization) varies as a function of the thickness of the initial platelet-fibrin thrombus (large thrombi induce slow reendothelialization and, subsequently, a thick, final neointimal layer). Calification of the internal elastic lamella occurs over time.

In the media, a neointima is thickened by an accumulation of collagenous tissue rich in elastin but is not organized like normal tissue. Calcified and noncalcified foci of inflammation are observed in the media, which are progressively resolved. When endothelial recovery is complete, the fibroblastic response stops. However, new endothelial cells exhibit a chronic alteration in permeability.

4.6.3. Tendon Repair

Tendons are severed as a result of occupational, home, and athletic injury. Repair occurs by fibrovascular tissue migration from the paratendineum (rich in proliferative fibroblasts) and results in formation of a callus that contains the epitendineum. Fibroblasts and newly formed fibers are oriented along the tendon axis but the fibers are thinner than normal and progressively thicken with time.

Cells of the epitendineum move longitudinally down into and across the area of tendon juncture and grow down into the tendon itself. The newly formed collagen is crosslinked to the remaining original collagen on each side of the injured area. However, the functional ability of the tendon depends on how much of the tendon is retracted into its sheath.

4.6.4. Bone Repair (Compact Bone)

Bone repair after traumatic injury involves a rapid regeneration of new Haversian systems. The osteoblasts form spearheads at the end of the Haversian canals, followed by osteoblasts and new vessels. New osteons traverse the fracture site. After the inflammatory response there is a stage in which a soft callus is formed, characterized by proliferation of osteoblasts. The periosteum plays an important role in supplying nutrients to undifferentiated cells that migrate into the wound. Progressively, the bony fragments are united by fibrous and/or cartilaginous tissue. Then, a hard callus is converted into bone with increased vascularization and mineralization.

4.6.5. Cartilage Repair

Cartilage repair associated with osteoarthritis is infrequent. Cartilage is unable to reconstitute its basic structure following injury, which is partially related to its lack of a blood supply.

Fibrocartilages are unable to regenerate (except at their periphery). Hyaline cartilage repair produces a fibrocartilage. Injury to the epiphyseal plate interferes with normal growth (particularly if the calcified zone of the cartilage is injured).

4.7. Summary

Response to injury by wound healing is a basic behavior that is also observed after implantation of biomaterials. Biomaterials behave like foreign bodies in which the degree of tissue reaction and wound healing depends on the biocompatibility of the implanted materials.

Implantation of biomaterials induces an inflammatory response as described earlier. This inflammatory response can be observed around and sometimes within the implant. Undigested implant-wear particles can induce chronic inflammation associated with a foreign-body reaction. A foreign-body reaction consists of an accumulation of macrophages that fuse into giant cells.

In the long term, biomaterials can be surrounded by a fibrotic capsule that corresponds to scar tissue. Alternatively, biomaterials are invaded by cells, and newly synthesized connective tissue is formed that resembles either repair or scar tissue.

References

Aoki, N. and von Kaulla, K.N. J. Lab. Clin. Med. 78,354,1971

Bennett, B., Ogston, D. and Douglas, A.S. In: Human Blood Coagulation, Haemostasis and Thrombosis, edited by R. Biggs and C.R. Rizza, Blackwell Scientific Publications, Oxford, p.415,1984.

Bjorklid, E., Storm, E. and Prydz, H. Biophys. Biochem. Res. Commun. 55,969,1973.

Bjorklid, E. and Storm, E. Biochemical J. 165,89,1977.

Born, G.V.R., Begent, N.A. and Cusack, N.J. In: Human Blood Coagulation, Haemostasis, and Thrombosis, edited by R. Biggs and C.R. Rizza, Blackwell Scientific Publications, Oxford, p. 319,1984.

Caspary, E.A., and Kekwick, R.A. Biochem. J 67,41,1957.

Cochrane, C.G., Revak, S.D., and Wuepper, K.D. J. Exp. Med. 138,1564,1973.

Deutsch, E., Irsigler, K. and Lomoschitz, H. Thromb. Diath. Haemorrh. 12,12,1964.

Donaldson, V.H. and Evans, R.R. Amer. J. Med. 35,37,1963.

Esmon, C.T. J. Biol. Chem. 254,964,1979.

Esnouf, M.P. In: Human Blood Coagulation, Haemostasis and Thrombosis, edited by R. Biggs and C.R. Rizza, Blackwell Scientific Publications, Oxford, p. 46,1984.

Griffin, J.H. and Cochrane, C.C. In: Method Enzymology, edited by L. Lorand, Academic Press, New York, p. 56,1976.

Houghie, C. and Baugh, R.F. In: Human Blood Coagulation, Haemostasis and Thrombosis, edited by R. Biggs and C.R. Rizza, Blackwell Scientific Publications, Oxford, p. 1,1984.

Hvatum, M. and Prydz, H. Thromb. and Diath. Haemorrh.19,217,1969.

Jackson, C.M. and Hanahan, D.J. Biochemistry 7,4506,1968.

Jacobsen, S.and Kriz, M. Br. J. Pharmacol. 29,25,1967.

Katayama, K., Ericsson, L.H., Enfield, D.L., Neurath, H., Davie, E.W. and Titani, K. Proc. Natl. Acad. Sci. USA 76, 4990,1979.

Kisiel, W. and Davie, E.W. Biochemistry 14,4928,1975.

Kurachi, K. and Davie, E.W. Biochemistry 16,5831,1977.

Leder, P. In: The Biology of Immunologic Disease, edited by F.J. Dixon and D.W. Fisher, Sinauer Associates, Inc., Sunderland, Massachusetts, p. 3,1983.

Muller—Eberhard, H.J. In: Biology of Immunologic Disease, edited by F.J. Dixon and D.W. Fisher, Sinauer Associates, Inc., Sunderland, Massachusetts, p.128,1983.

Nesheim, M.E., Myrmel, K.H., Hibbard, L.S. and Mann, K.G. J. Biol. Chem. 254,508,1979.

Peerschki, E.I., Zucker, M.B., Grant, R.A., Egan, J.J. and Johnson, M.M. Blood 55,841,1980.

Prydz, H. In: Blood Coagulation and Haemostasis, edited by J.M. Thomson, Churchill Livingstone, New York, p. 1,1985.

Robbins, K.C. In: Haemostasis: Biochemistry, Physiology and Pathology, edited by D. Ogston and B. Bennett, John Wiley and Sons, New York, p. 208,1977.

Schmaier, A.H. and Coleman, R.W. In: Blood Coagulation and Haemostasis, edited by J.M. Thomson, Churchill Livingstone, New York, p. 22,1985.

Schreiber, A.D., Kaplan, A.P. and Austen, K.F. J. Clin. Invest. 52,1402,1973.

Schwartz, M.L., Pizzo, S.V., Hill, R.L. and McKee, P.A. J. Biol. Chem. 248,1395,1973.

Stead, N.W., Kaplan, A.P., and Rosenberg, R.D. J. Biol. Chem. 251,6481,1976.

Weupper, K.G. and Cochrane, C.G. J. Exp. Med. 135,1,1972.

Wiman, B. In: Fibrinolysis: Current Fundamental and Clinical Concepts, edited by P.J. Gaffney and S.M. Balkav-Ulutin, Academic Press, London, p. 47,1978.

Chapter 5

Chemistry and Physics of Polymeric Materials

5.1. Introduction

Biomaterials scientists are involved in developing prototype medical devices that contain components made from synthetic and/or natural polymers. Prototype medical device development requires selection, fabrication, and testing of devices containing polymers and is based on an understanding of the chemistry and physics of polymers. The purpose of this chapter is to introduce the basic concepts needed for polymer selection in medical device design.

The physical properties of polymers depend on the chemistry of the units that make up this class of molecules. A variety of backbone chemistries are observed in both natural and synthetic polymers. Examples of repeat units found in polymers are given in Table 5.1.

Natural structural polymers include nucleic acids, proteins, polysaccharides, and complexes of proteins and polysaccharides. Synthetic polymers used in implants include polyamides, polyesters, poly (siloxane), polymers of methacrylates, poly (urethanes) and natural rubber. These polymers fall under the classifications of plastics and rubbers.

All polymers are composed of long chains of atoms linked end-to-end. It is not uncommon to have in excess of 10,000 atoms linked together in a single polymer molecule. Synthetic polymers derived from oil can be processed into a wide variety of biomaterials in a cost-effective manner. Natural polymers because of the expense involved in manufacturing, tend to be less cost effective; however, because of their biological properties, they are the materials of choice in some applications.

In this chapter an attempt will be made to correlate the chemical structure of polymers with their physical properties.

Chemical Structure of Polymer Repeat Units

The physical properties of polymer chains reflect the properties of the constituent repeat units. Monomer molecules can add together to form polymers

Table 5.1. Repeat Units of Typical Synthetic Polymers

Polymer	Repeat Unit
Polyethylene	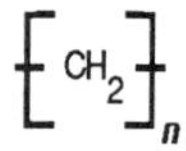
Polytetrafluoroethylene	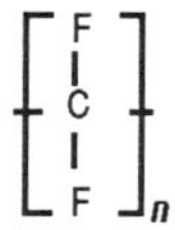
Polyethylene terephthalate (Dacron)	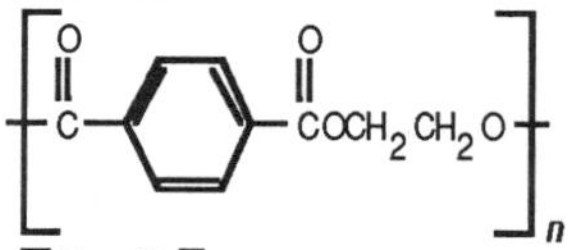
Polyvinyl chloride	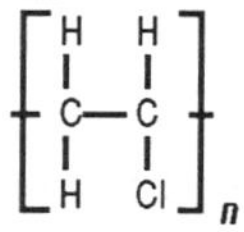
Polyvinyl flouride	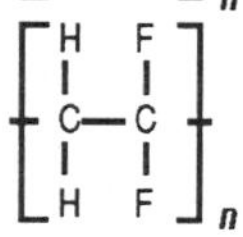
Polymethyl methacrylate (Plexiglas)	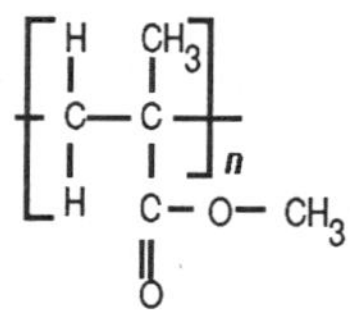
Poly(2-hydroxyethyl methacrylate) (HEMA-hydrogel)	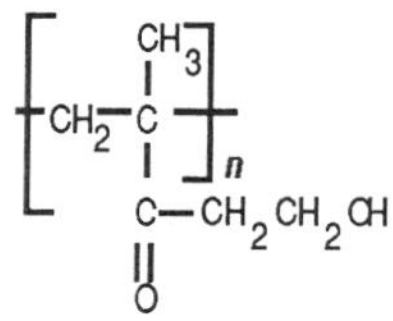
Polydimethyl siloxane	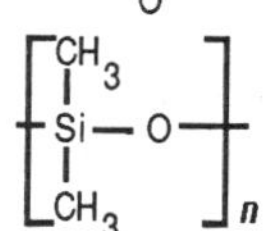
Polysulfone	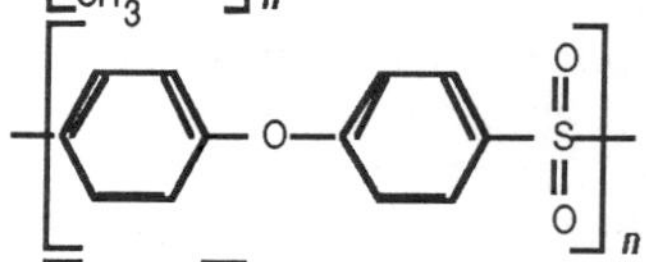
Polyamide	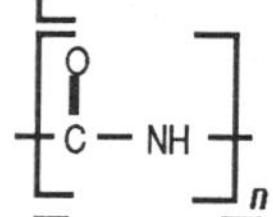
Polyurethane	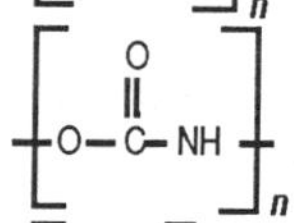
Polyester	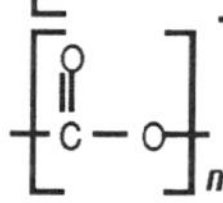

Table 5.1 (cont.) Polyurethanes

Polyurethanes

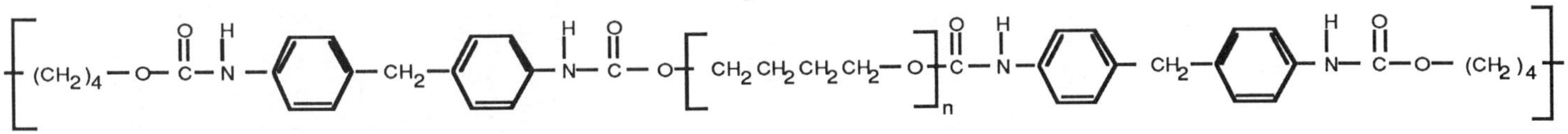

Pellethane (Upjohn product)

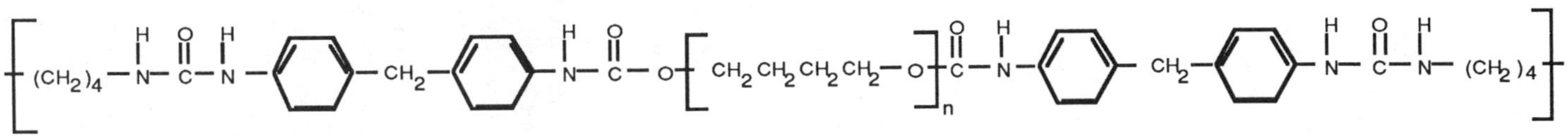

Biomer (Ethicon product)

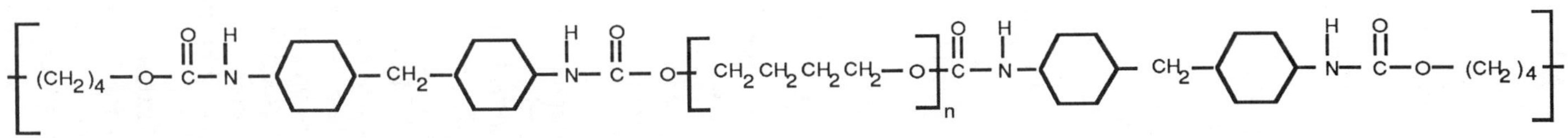

Tecoflex (Thermedics product)

Table 5.1 (Cont.) Nucleic acid, polypeptides, polysaccharides

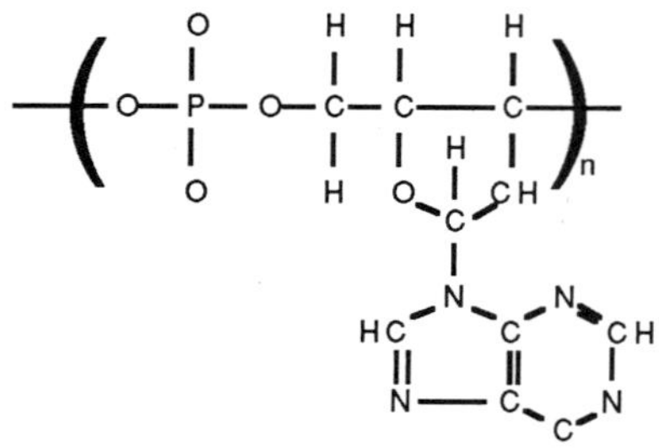

nucleic acid

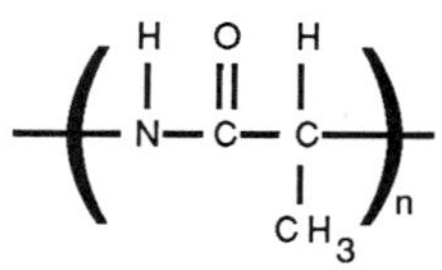

polypeptide

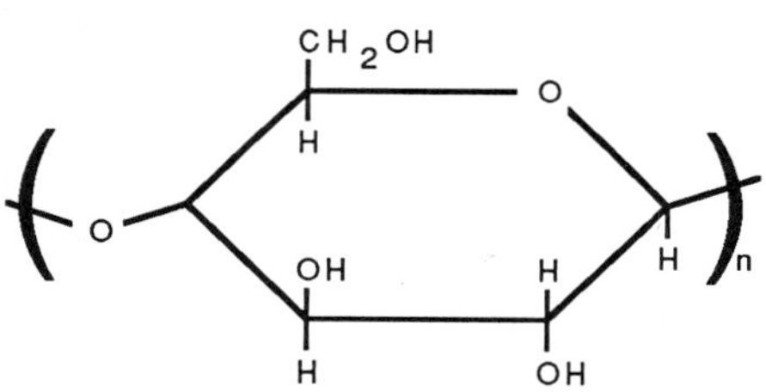

polysaccharide

via addition or condensation polymerization. Monomers containing double bonds such as ethylene can add together by addition polymerization in the presence of a free radical (such as that generated by peroxides and a catalyst) to form long chains (Figure 1.9A). Addition polymerization occurs without the release of by-products. In comparison, condensation polymerization occurs when two bifunctional monomers react to release a by-product such as water (Figure 1.9B). Formation of a protein chain occurs by the condensation of two amino acids to form a peptide repeat unit.

Condensation polymers that are formed yielding water as a by-product tend to depolymerize in an aqueous environment. Therefore, biomaterials that make use of condensation polymers may degrade when implanted, releasing free monomers. The monomers may or may not be toxic to tissues depending on their chemistry.

Addition polymerization requires metallic catalysts as well as free radicals for initiation of a growing chain. As discussed in Chapter 3, free radicals can induce cell injury, therefore biocompatibility of addition polymers may be adversely affected by small amounts of contaminants although the polymer itself maybe compatible.

The chemistry and size of the atoms in the backbone of the polymer chain dictate how the chain folds. Unless stated otherwise, discussion of polymer structure in three-dimensional space refers to linear, unbranched chains.

5.2. Physical Structure of Polymer Chains

Polymers can exist in several macroscopic forms at room temperature. These forms include hard rigid states, the rubbery state, and a gellike state. Examples of these forms include plexiglass (a rigid, clear material), polyethylene (a rigid, opaque plastic), natural rubber (a rubbery material), and polyhydroxyethyl methacrylate, HEMA, (a water swollen hydrogel).

In the absence of solvent molecules, polymer chains can exist in either structureless (amorphous) or ordered (crystalline) states (Figure 1.11). In an amorphous state (i.e., natural rubber) the polymer chains are randomly folded and do not have long-range order. This means that if we stand at a point along the polymer chain, we would not be able to predict the exact location of the chain at any other spot. In comparison, some polymer chains such as polyethylene form crystals with long range order, such that the exact location in space of atoms within a polymer chain is known.

Physical properties of polymers depend on the amount of long-range order (i.e., degree of crystallinity) exhibited by a polymer as well as the chain flexibility (amorphous regions of polymer chains). For example, flexible chains that do not form crystals require less force to deform than do rigid polymer chains that form crystals. The relationship between chain structure and physical properties of polymers will be discussed further below. The point to be made in this section is that the stiffness (force per unit area per unit of deformation) of a polymer chain is inversely related to the degree of randomness. This means that, at room temperature, amorphous polymers tend to stretch easily (low stiffness) compared to crystalline polymers. However, as we will see below, the inverse relationship between randomness and stiffness is complicated by the effects of temperature and interactions between side chains that are attached to the polymer backbone.

At absolute zero, all polymer chains are rigid (high stiffness). As the temperature is increased, polymer chains with small atoms attached to the backbone (side chain) begin to exhibit rotation around the backbone. If the force of attraction between polymer molecules is high, then a crystal is formed. Polymer molecules in a crystal are stiffened by the interactions between atoms on the side chains (secondary interactions). If the largest dimension of a polymer crystal is larger than 1/20th of the wavelength of visible light, then the material appears translucent. When a crystalline polymer is heated above its melting temperature, the individual polymer chains become highly mobile and the polymer changes from translucent to transparent.

Polymer chains that do not interact to form crystals behave as isolated, entangled, flexible chains. The rubbery state is characterized by flexible polymer molecules that stretch by undergoing backbone reorientation, which requires little force. At a higher temperature a rubbery polymer will begin to flow like a liquid unless the chains are held together by chemical crosslinks.

Many industrial plastics are composed of polymer molecules in crystalline and amorphous forms. Amorphous plastics such as poly (styrene) and poly (carbonate) are used to make petri dishes for tissue culture, while crystalline polymers such as poly (tetrafluoroethylene) are used to make vascular prostheses.

5.2.1. Physical States of Polymer Chains

As mentioned earlier, polymer chains can exist in amorphous and crystalline states. Amorphous polymers can exist as glasses below a temperature termed the glass transition temperature. Thermal transitions of amorphous glassy polymers and crystalline polymers can be followed by determining the stiffness as a function of temperature (Figure 5.1). The stiffness (tensile modulus) can be determined from the force required to extend a sample of a polymer to a fixed strain at a constant rate of strain. The stiffness (modulus) is calculated as the force per unit area of cross-section divided by the change in length over the original length of the specimen.

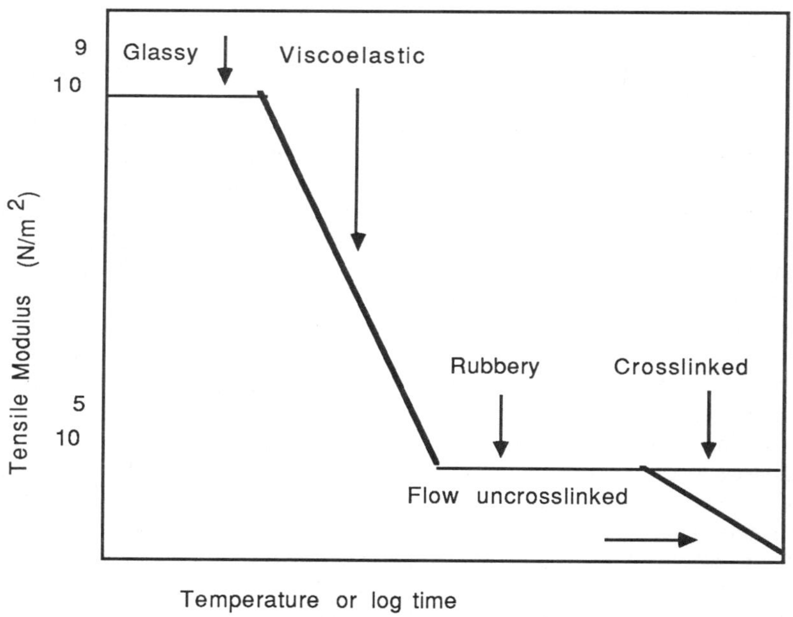

Figure 5.1. Temperature and time dependence of tensile modulus. Schematic diagram illustrating the temperature and time dependence of the tensile modulus (stiffness). At low temperatures and short times amorphous polymers have moduli of approximately 10^9 N/M^2 . As the temperature and time increases the polymer chains become flexible and the modulus decreases in the viscoelastic, rubbery, and flow regions. Crosslinked polymer chains have a modulus of approximately 10^5 N/M^2 and decompose on further heating.

As the temperature is increased, the modulus of a glassy polymer such as plexiglas (PMMA) decreases from about 10^3 MPa (1MPa = 10^6 N/m^2) to a plateau value of about 0.1 MPa. The plateau value is equivalent to the modulus of a rubbery material (rubberband). The transition region between the glassy and rubbery regions is a region that is termed the viscoelastic region. Uncrosslinked amorphous polymers, when heated further, flow like liquids and exhibit a further decrease in modulus while crosslinked rubbery materials retain a stiffness of 0.10 MPa until they decompose upon raising the temperature.

Crystalline polymers have a modulus in excess of 10^3 MPa at temperatures below their crystalline melting temperature. The modulus does decrease somewhat as the temperature is raised; however, a sharp decrease in modulus to that of a rubber, 0.1MPa, is only seen at the melting temperature. The behavior of crystalline and amorphous polymer molecules is similar at temperatures above the crystalline melting temperature.

5.2.2. Molecular States of Polymer Chains

In the previous section, the mechanical property of stiffness was used to describe the glassy, crystalline, viscoelastic, rubbery, and rubbery flow regions of polymer behavior. This behavior of a polymer reflects the behavior of individual polymer chains. For instance in the glassy region, amorphous polymer chains are frozen into a random (amorphous) three-dimensional organization. The property of forming a random chain orientation derives from the rotational freedom of the polymer backbone. If the polymer chain is composed of amino acid residues (see Silver, 1987) two degrees of free rotation exist between each peptide unit (Figure 5.2) and therefore a chain can theoretically adopt a large number of sets of the dihedral angles, ϕ and ψ. Figure 5.2 is a plot of the sterically allowed dihedral angles for amino acids that contain a side chain. Figure 5.2 demonstrates that many sets of ϕ and ψ are not found within the allowed regions (dotted and solid lines) for a general polypeptide chain. Note that the areas enclosed by the solid lines and the areas within the dotted lines are the combinations of ϕ and ψ that are sterically allowed. When the number of allowed conformations (combinations of ϕ and ψ) is large, the chains are in continuous motion and the polymer chain is flexible, assuming that the force of attraction between polymer molecules is low. The number of allowable dihedral angles is dependent on the temperature, solvent composition, and intermolecular forces between chains. Usually at high temperatures (above T_M and T_g) all polymer chains are ideally flexible and undergo continuous motion. Figure 5.2 illustrates the combinations of ϕ and ψ found commonly in proteins.

As the temperature is lowered, free rotation around the angles ϕ and ψ is hindered because of energy barriers. Energy barriers exist because of attraction between atoms that arises from hydrogen, ionic, and secondary forces (see Silver, 1987, for a more complete discussion of attractive forces). Table 5.2 lists the types of bonds that form and the free energy changes associated with bond formation. A highly mobile group of polymer chains in the "rubbery" state freezes into a glass at T_g, the glass transition temperature. The glass transition temperature is higher for polymers with large bulky side chains that prevent free rotation around the polymer backbone than it is for polymers with smaller ones. Table 5.3 summarizes the glass transition temperature of different polymers.

An empirical relationship, Eq. (1), has been developed to show the dependence of T_g, on chain molecular weight. Molecular weight, M, equals degree of polymerization times repeat unit molecular weight,

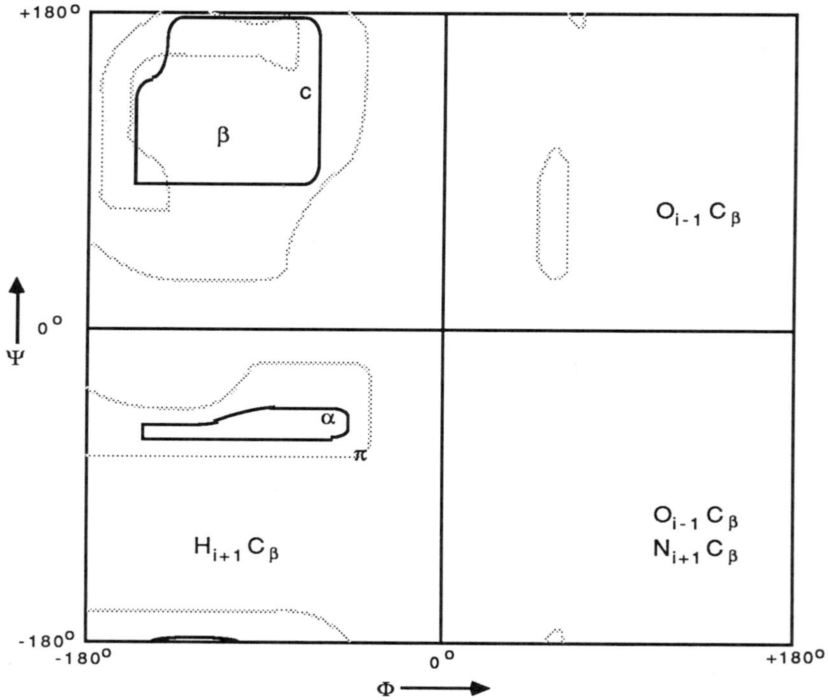

Figure 5.2. Flexibility of polymer chains. Conformational plot of totally (____) and partially (............) allowed dihedral angles between neighboring peptide units based on normal and minimum interatomic distances. Fully allowed conformations observed in proteins include the α-helix, B-sheet, π-helix, and collagen triple-helix. Adapted from Schulz and Schirmer (1979).

$$T_g = T_{g,\infty} - K_g / M \tag{1}$$

where $T_{g,\infty}$ is the glass transition temperature of an infinitely long chain and K_g is a constant. For poly(methylmethacrylate) $K_g = 2.1 \times 10^5$ °Cg/mol and $T_{g,\infty}$ =114°C/mol for molecular weights greater than 10,000 (Beevers and White, 1960). For polystyrene $K_g = 1.7 \times 10^6$ °Cg/mol and $T_{g,\infty}$ = 100°C above molecular weight 3,000 (Bueche, 1962).

Polymeric materials can be completely amorphous, partially crystalline, or highly crystalline. Polymer crystals are usually 5-50 nm in length and contain polymer chains that are folded within the crystal as diagrammed in Figures 5.3 and 5.4. The schematic diagram shown in Figure 5.3 is the widely accepted fringed micelle model for polymer chains in a crystal. A model of the polyethylene crystal (Figure 5.4) indicates that the covalent axis of the polymer chain is perpendicular to the longest dimension of the crystal.

As the temperature is raised the polymer chain will unfold at the melting

Table 5.2. Types of Bonds that Limit Free Rotation of Polymer Chains [a]

Bonds	Example	Free energy change (kcal/mole)
Dispersive	Lennard-Jones 6-12 potential	-0.03
Electrostatic	Salt bridge $-COO^{-}\text{-}H_3N^{+}-$	-5.0
Hydrogen Bond	Ice	-4.0

[a] From Schulz and Schirmer, 1979.

temperature. The crystalline melting temperature for different polymers varies from -54°C for poly (dimethylsiloxane), commonly known as silicone rubber, to 267°C for poly (ethylene terephthalate) (Table 5.3).

Since glass transition and melting involves physical changes within the polymer chains these phenomena are both related to the chemistry of the polymer chain.

A generalized relationship between T_g and T_M has been established. The ratio T_M/T_g is always greater than 1.4 but less than 2.0. The glass transition temperature of a polymeric material can be lowered by adding low-molecular-weight materials termed plasticizers that add to the rotational freedom of the chain. Figure 5.5 illustrates how the flex temperature at which polymeric compounds become brittle is lowered with increased plasticizer content. The plasticizer molecules prevent or lower secondary bonding between polymer chains and allow polymer molecules to adopt a variety of allowed conformations.

5.2.3. Mechanical Properties of Polymer Chains

The mechanical properties of polymers can be determined by stretching a sample to a predetermined length and by measuring the stress as a function of time. A diagram of the events that occur in this "stress relaxation" experiment is shown in Figure 5.6. The stress is seen to be time dependent, indicative that polymeric materials exhibit viscoelastic behavior, i.e., their behavior is characteristic of both viscous fluids and elastic solids. Stresses applied to a viscous fluid are dissipated during fluid flow while stresses applied to an elastic solid are reversibly stored as strain energy in the deformed solid. Therefore, the recoverable part of the stress stored in a polymer is the elastic component while the stress lost as polymer chains slide by each other is the viscous component. In designating the stiffness (ratio of stress to strain) of a polymer, one must indicate the time of loading.

As diagrammed in Figure 5.1, the stiffness in tension for a polymeric

Table 5.3

Effects of Different Polymeric Side Chains On The Glass Transition and Melting Temperatures

Polymer	TM (^{0}C)	*Tg* [Transition Temperature at 1 HZ(^{0}C)]
Polypropylene		0 [a]
Polystyrene	137	116 [a]
Poly(vinyl chloride)	212	87 [b]
Poly(vinyl acetate)	---	29 [b]
Poly(methyl methacrylate)	---	105 [b]
Poly(ethylene terephthalate)	267	69
Poly(dimethyl siloxane)	-54	-123 [a]
Polyethylene	137	-120 [a]
Cis-Poly(isoprene)	28	-70 [a]

[a] Vincent (1965)

[b] Nielsen (1983)

material depends on the time of observation. For short observation times the polymeric material has a stiffness of about 10^9 N/m^2 which decreases to 10^5 N/m^2 with increased time. The stiffness of all amorphous crosslinked polymers falls within these limits. Increased temperature or plasticizer content has the same effect as increased time on the stiffness of a polymeric material.

As is indicated earlier, the chemistry and size of the side chains attached to the repeat unit dictate the flexibility and bonding interactions available to polymer chains. Chains containing repeat units with large bulky groups such as styrene, do not form interchain bonds and remain as amorphous glasses below the glass transition temperature while above this temperature they are rubbery materials. Chains containing repeat units with small groups such as polyethylene, interact with neighboring chains to form crystalline materials.

The behavior of crystalline polymers above the melting temperature (T_M) is

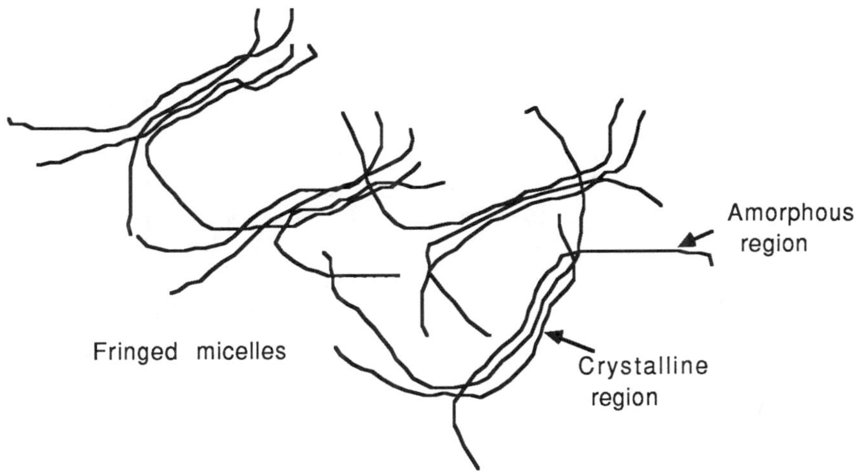

Figure 5.3. Fringed micelle model of semi-crystalline polymers. Diagram showing fringed micelle model of polymer crystalline regions. Polymer molecules contain amorphous and crystalline regions within the same polymer chain. Crystalline regions are formed from regions on several closely associated chains. These regions are responsible for melting behavior that is observed while the amorphous regions undergo a glass transition.

similar to that of rubbery materials. Williams et al, (1955) empirically developed a relationship for the viscosity (n) and modulus (G), at the glass transition temperature (T_g) and at some temperature T which is close to T_g. The relationship is as follows:

$$\log\left(\frac{N}{N_g}\right) = \log\left(\frac{G}{G_g}\right) = \frac{-17.44\,DT}{(51.6 + DT)} \tag{2}$$

where DT is $T\text{-}T_g$. This relationship holds between T_g and T_g +100°C and suggests that the stiffness of amorphous polymers depends only on the temperature of interest (T) and the glass transition temperature (T_g)

5.5.3. Stress-Strain Relationships for Polymers

The mechanical behavior of materials is often characterized by placing a tensile force on a long thin specimen at a constant rate of strain. This results in force-extension measurements that are required to formulate a stress (force per unit area) —strain (change in length over original length) curve (Figure 5.7). For polymeric materials the resulting stress-strain curve can have one of three general shapes. Below the glass transition temperature amorphous polymers have linear stress-strain curves (Figure 5.8, curve C) while cross-linked rubbery materials are characterized by curve A (Figure 5. 8) and crystalline materials by curve B. Biological tissues such as skin and tendon have stress-strain curves that

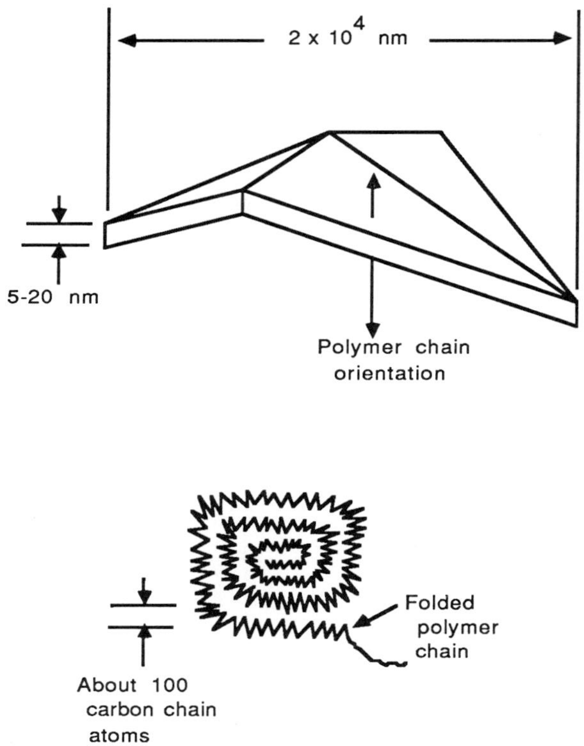

Figure 5.4. Structure of a polyethylene crystal. Diagram showing large crystal of polyethylene and the orientation of individual chains. Growth of the crystal occurs by folding and twisting of polymer chains.

curve upward as the strain is increased (see Figure 5.8). The shape of the stress-strain curve is important since the amount of force required to reach a certain level of strain is one of the design criteria for a biomaterial. For instance, a bandage used to cover a skin wound over a joint must be able to double its length as the joint is bent and should not exhibit any residual strain after the force is removed.

Stress-strain relationships (also known as constitutive relationships) for polymers in the rubbery state have been developed based on thermodynamic principles.

Generalized stress-strain relationships based on molecular phenomena are more complicated because of the time dependence of the mechanical behavior of polymeric materials.

5.3.1. Rubber Elasticity Theory

There have been several approaches to studying the mechanical properties of rubbery materials. In this section we will discuss a thermodynamic approach that relates the equilibrium force required to stretch a rubbery material as a

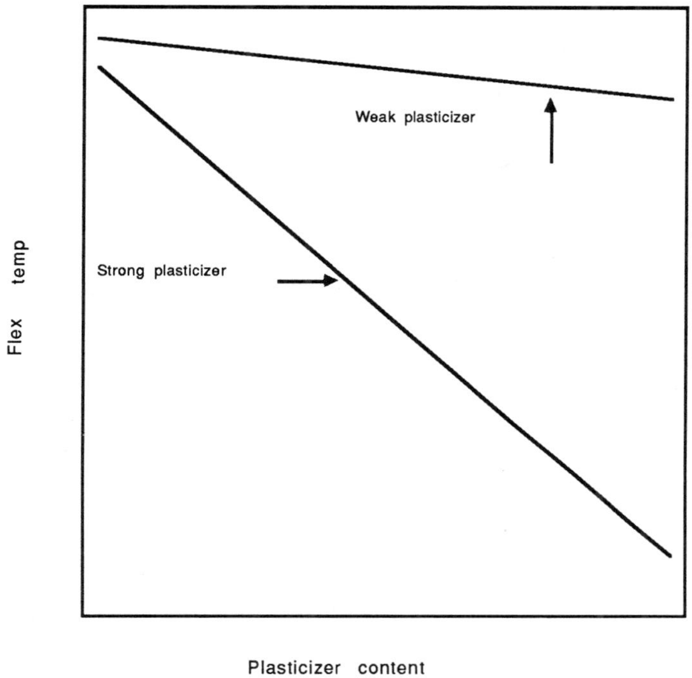

Figure 5.5. Role of plasticizers in lowering flex temperature. Diagram illustrating the decrease in flex temperature with increase in plasticizer content observed with weak and strong plasticizers. Plasticizers are used to prevent stiffening of plastics near the glass transition temperature.

function of extension ratio (equivalent to strain plus one) and changes in free energy.

The first law of thermodynamics states that the change in internal energy (ΔE) of a system of polymer molecules is equal to heat added to the system of polymer molecules (Q) minus the work (W) done by the system of polymer molecules.

$$\Delta E = Q - W \tag{3}$$

The second law of thermodynamics states that the free energy change (ΔG) of a system of polymer molecules is equal to the change in internal energy (ΔE) plus the change in the pressure times the volume of the polymer molecules ((PV)) minus the temperature (T) times the polymer-molecule entropy change (ΔS).

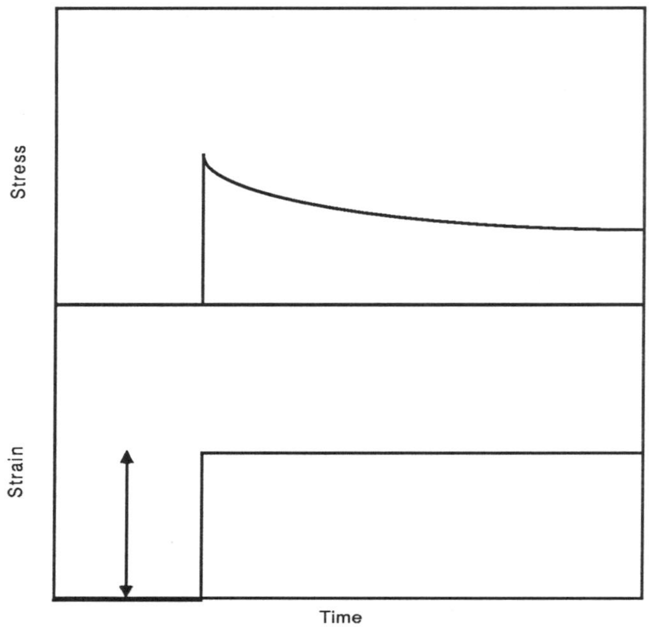

Figure 5.6. Stress relaxation test. Diagram illustrating stress and strain as a function of time for a viscoelastic material. When strain is fixed and the stress allowed to vary the experiment is termed stress relaxation.

$$\Delta G = Q - W + \Delta(PV) - T\Delta S \tag{4}$$

If the process of stretching a rubbery material is carried out reversibly, i.e., ΔG is zero for the cycle, and at a constant temperature (T), then no energy is lost as heat to the environment and the differential change in entropy (dS) times the temperature (T) is equal to the incremental heat added reversibly (dQ).

$$dQ_{rev} = TdS \tag{4a}$$

As we stretch a piece of rubber through a distance (dL) with a force (f) we do work (dW). Since the angle between f and dL is 180°, then the dot product of f and dL is fdL:

$$dW = f \circ dl = +fdl$$

$$dE = TdS + fdL \tag{5}$$

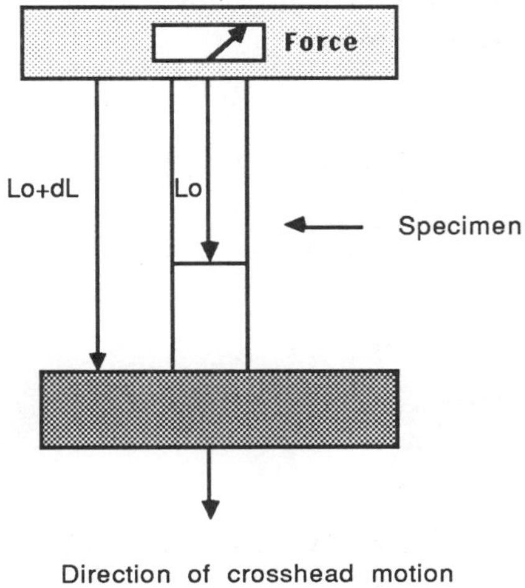

Figure 5.7. Stress-strain test. Diagram illustrating the relationship between force, stress, strain, and extension ratio for a stress-strain experiment. Definitions: Stress, Force/cross-sectional area, strain, dL/L_o, extension ratio, 1 + strain.

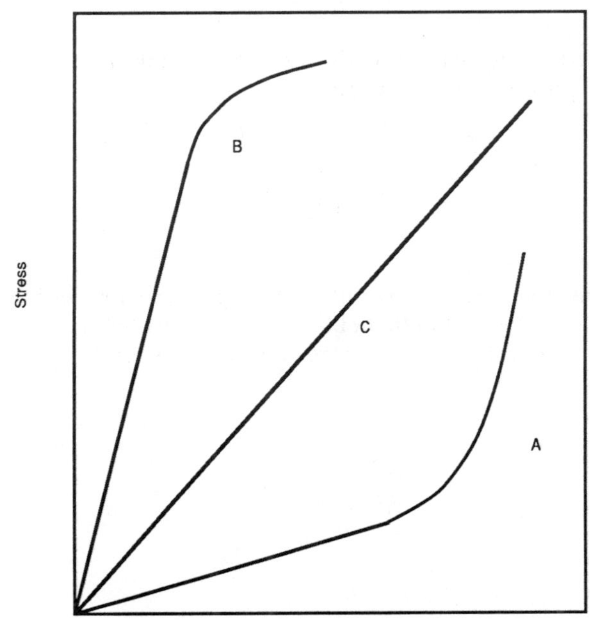

Figure 5.8. Stress-strain curves for polymers. Diagram illustrating stress-strain curves for biological tissues (A), polymer fibers (B), and glassy polymers (C).

After partial differentiation of Eq. (5) with respect to length (L) holding temperature and volume constant we get:

$$f = \left(\frac{\partial E}{\partial L}\right)_{T,V} + T\left(\frac{\partial S}{\partial L}\right)_{T,V} \tag{6}$$

Using the differential form of eq. (4) results in:

$$dG = dE + PdV + VdP - TdS - SdT \tag{7}$$

Substituting eq. (5) for dE we get:

$$dG = VdP - SdT + fdL \tag{8}$$

Taking the partial derivative with respect to length in eq. (8), holding temperature and pressure constant we get:

$$\left(\frac{\partial F}{\partial L}\right)_{T,P} = f \tag{9}$$

The second derivative obtained by differentiating $(\partial G/\partial L)_{T,P}$ with respect to T at constant P and L is identical to that obtained by differentiating $(\partial G/\partial T)_{P,L}$ with respect to L at constant T and P (Euler relationship). After this is done we obtain eq. (10).

$$\left(\frac{\partial f}{\partial L}\right)_{L,P} = \left(\frac{\partial S}{\partial L}\right)_{T,P} \tag{10}$$

Substituting into eq. (6) we get eq. (11):

$$f = \left(\frac{\partial E}{\partial L}\right)_{T,V} + T\left(\frac{\partial f}{\partial T}\right)_{L,P} \tag{11}$$

Based on eq. (11) it is observed that the retractive force is composed of an internal

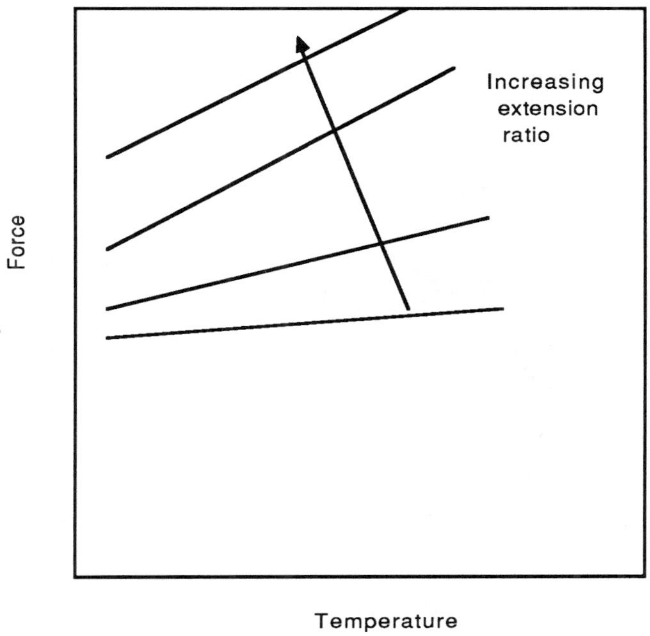

Figure 5.9. Force-temperature plot. Diagram illustrating the force required to stretch a rubbery material at increasing temperatures and as a function of increased extension ratios.

energy term that involves changes in bond energy and an entropy term due to changes in comformational states of a polymer chain. An "ideal rubber" is defined as a polymeric material that deforms with no change in bond energy, i.e., $\partial E/\partial L=O$. This can be evaluated for a polymer by plotting the equilibrium retractive force versus temperature at different extensions (Figure 5.9). As diagrammed in Figure 5.10, if the intercept of a force versus temperature plot (*C*) is much less than the product of temperature and the slope (*B*), then the material behaves as an ideal rubber.

Now that we have identified the properties of an ideal rubber, i.e., the retractive force is proportional to the change in entropy over the change in length (eq. 12), it is possible to develop a relationship between structural parameters (location of chain ends) and the refractive force (*f*). It is known that the entropy of a polymer chain

$$f = T\left(\frac{\partial S}{\partial L}\right)_{T,V} \tag{12}$$

is related to the number of available conformations, i.e., combinations of (ϕ, ψ) (Figure 5. 2) that a polymer chain undergoes as is indicated by equation (13)

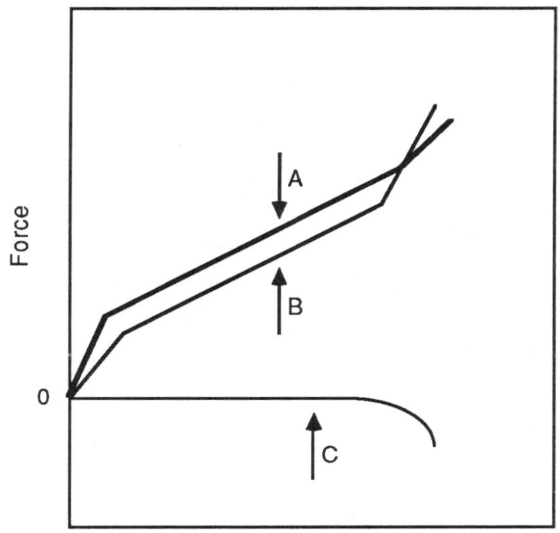

Figure 5.10 Energy and Entropy Components of Total Force. Diagram illustrating the total force (A), the slope (B), and intercept (C) of the force-temperature plot shown in Figure 5.9. Diagram illustrates that entropy component dominates the retractive force for natural rubber and that the energy component is only significant at high extensions.

$$S = k \ln \Omega \tag{13}$$

where k is Boltzmann's constant and Ω is the number of combinations of ϕ and ψ for a chain.

The interested reader is referred to Rodriguez (1982) for the details of this derivation. After differentiation and assuming the volume of the specimen remains constant during stretching, the force per unit of original area (σ) is related to the extension ratio (λ) using eq. (14):

$$\sigma = RTN\left(\lambda - \frac{1}{\lambda^2}\right)$$
$$\lambda = \frac{L}{L_0} \tag{14}$$

where the stress σ is the force per unit underformed area, R isthe gas contant, T is the absolute termperature, N the number of moles of polymer chains per unit volume, and λ is the extension ratio (L/L_o). Using eq. (14), the stress (σ) at any extension ratio (λ) can be calculated knowing the temperature and density of the rubber chains. This of course assumes that the deformation of a rubbery polymer occurs without any changes in internal energy, i.e., no loss of energy due to viscous effects.

The aorta is composed of elastin, a protein with rubber-like properties and other connective tissue components. At low extension ratios, the mechanical behavior of the aorta is dominated by the behavior of elastin. The number of polymer chains per cm^3 of elastin can be calculated knowing that, at an extension ratio of 1.2, the stress that the aorta can withstand at equilibrium is $2 \times 10^4 N/m^2$. If the density of elastin is about 1.39g/cc then the molecular weight of the chains that supports the stress can be calculated.

If the rubber is swollen in a solvent and crosslinked, then the chain density, N, must be corrected for chain ends in a polymer with a chain molecular weight of M_n (number average molecular weight). The new chain density is equal to N minus $2\rho / M_n$ where ρ is the polymer density.

The Mooney-Rivlin equation [see eq. 15] is a modified form of eq. 14) and applies to swollen polymers. This equation contains a constant C_2, and V_2, the volume fraction of polymer in the swollen state. The stress and extension ratio are based on the unswollen cross-section and the ratio of length to unstretched swollen length, respectively (Rodriquez, 1982).

$$\sigma = \left[\left(N - \frac{2\rho}{M_n}\right)RT + \frac{2C_2}{\lambda}\right]\left[\lambda - \frac{1}{\lambda^2}\right]\left[\frac{1}{V_2}\right]^{1/3} \tag{15}$$

5.3.2. Viscoelasticity

Polymers and biological tissues exhibit a time dependent change in specimen dimensions when a fixed tensile load is applied. This behavior is characterized by relaxation of the sample that is manifested by a continuous increase in the specimen length. Increased specimen length is a result of molecular rearrangement within the specimen primarily by molecular chain alignment along the tensile axis.

Rubber elasticity theory applies to polymers that exhibit rubbery behavior at equilibrium (time approaches infinity). However, we are most concerned with material behavior that occurs between the time a load is applied and infinite time. Therefore rubber elasticity theory applies to only a small subset of mechanical observations.

Polymers are viscoelastic as pointed out above and the stress-strain ratio termed stiffness is time dependent. If the log of the relaxation stiffness (obtained from a stress relaxation experiment, see Figure 5.6) is measured as a function of time at different temperatures it is possible to get a series of curves for stiffness (Figure 5.11). If the log modulus at a fixed time, i.e., one hour, is plotted as a function of temperature, we get a curve that shows a transition between high and low stiffness (Figure 5.12). The region with a tensile stiffness (E) equal to 10^9 or 10^{10} Pa is the glassy region. The rubbery region has a modulus of betwen 10^6 to 10^7 Pa. The temperature which is half way between these two regions is termed the glass transition temperature T_g.

If the curves of Figure 5.12 are shifted along the time axis using eq. (2), the WLF equation, using a shift factor that is equal to -17.44 DT/51.6+D T we get a stiffness-time master curve for temperatures between T_g and T_g+100°C (Figure 5.12). Time and temperature have equivalent effects on changing the stiffness of amorphous polymers.

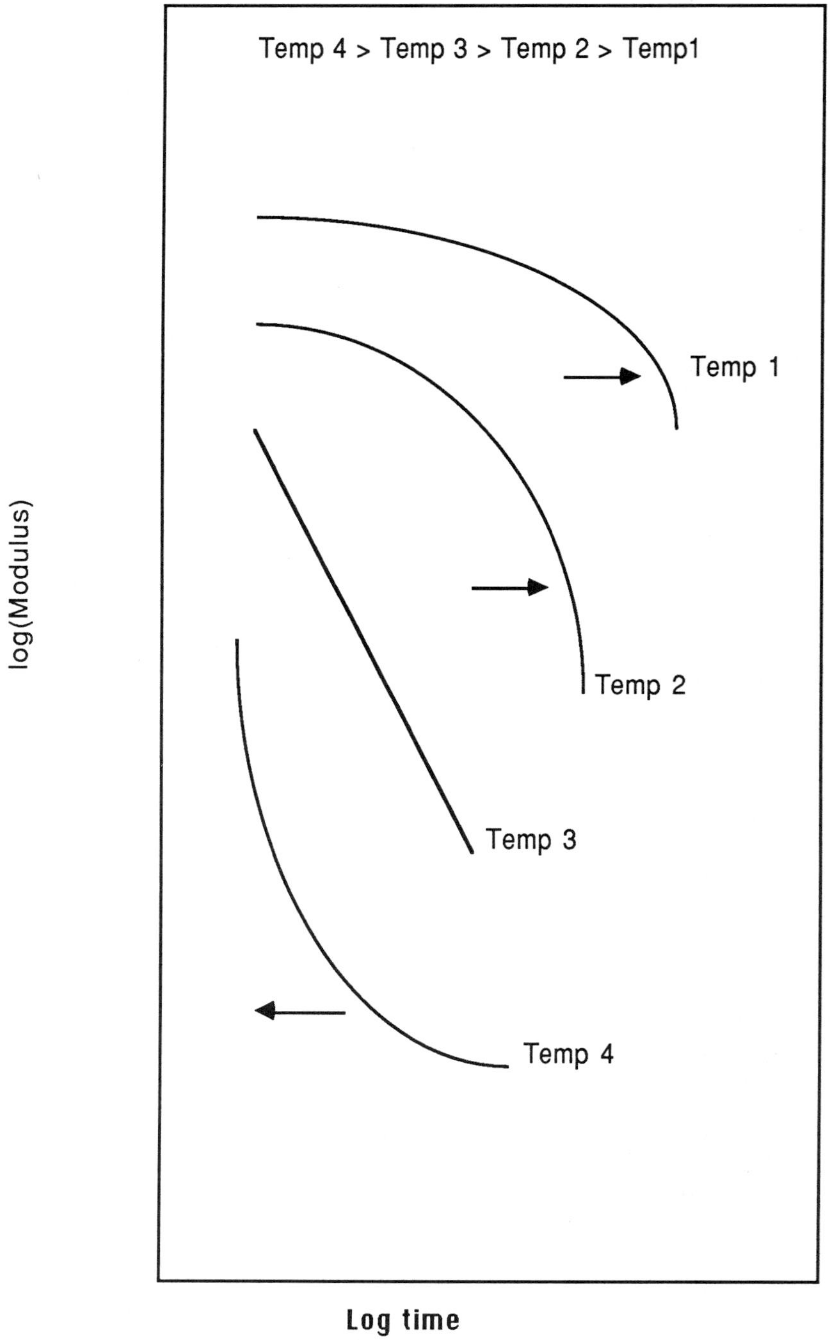

Figure 5.11. Temperature dependence of modulus. Diagram illustrating the temperature dependence of modulus measurements as a function of time for an amorphous polymer.

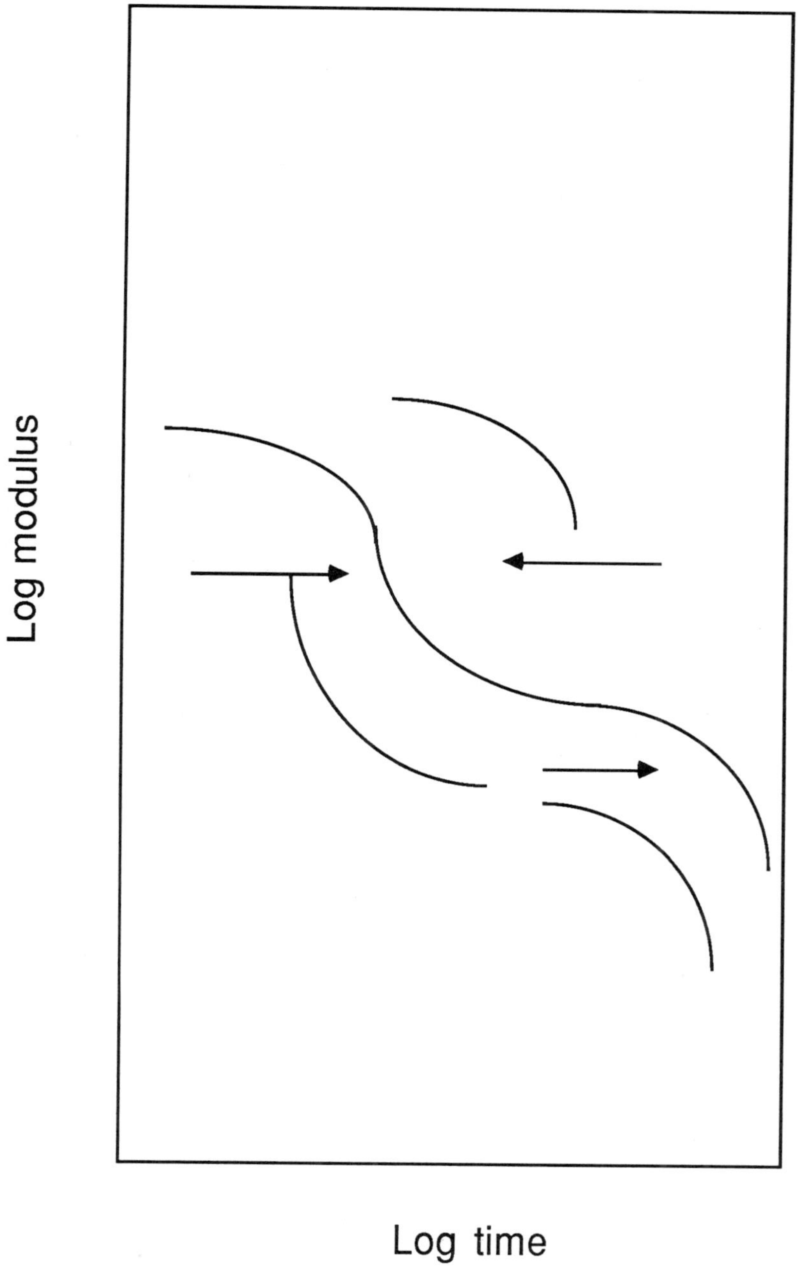

Figure 5.12. Shift diagram for modulus-time curves. Diagram illustrating the process of shifting log modulus-log time data for amorphous synthetic polymers. Curves are shifted to the right and left depending on the temperature of the reference curve.

The effects of increasing the polymer molecular weight and crosslink density on stiffness of amorphous polymers are shown in Figure 5.13. According to eq. (15) the stiffness is proportional to the gas constant (R), absolute temperature (T) and the number of moles of polymer chains per unit volume. If the molecular weight of the segment decreases (increased crosslinking) then the stiffness should increase.

For a polymer with a small degree of crystallinity, the glassy and rubbery plateau regions are essentially unchanged. However, at the melting point a sharp decrease in the stiffnessis observed. For highly crystalline materials the glass transition is less pronounced due to a large stiffness loss at T_M. The sharpness of T_M will depend on the distribution of crystalline sizes and melting temperatures. The melting temperature is usually proportional to the crystallite size.

Mechanical tests used to evaluate the properties of polymers include determination of the stress at which a material fails (ultimate strength or tensile strength). Amorphous polymers reach a point of maximum stress and they fail. Crystalline polymers may reach a point of maximum linear stress (yield point) and continue to deform "plastically" and absorb energy. Tensile-strength measurements are valuable when they can be correlated with the energy required to cause failure (toughness) or the number of cycles of force application required to cause failure (fatigue endurance limit).

Implants that fill structural requirements must be repeatedly loaded and unloaded during their useful lifetime. For these applications structural integrity is an important design criterion. For implants that have limited lifetimes by design, such as biodegradable drug-delivery systems, structural integrity may interfere with proper function. For this reason the degree of crosslinking of amorphous polymers and the degree of crystallinity of crystalline polymers is varied to achieve the desired mechanical properties.

The energy absorbed during deformation of a polymeric material may be reversibly stored in the form of strain energy in the rubbery state or it may be dissipated as heat in the viscoelastic transition region.

5.4. Types of Mechanical Tests

A variety of different mechanical tests are used to evaluate phenomenologically the mechanical properties of polymers. These include constant-strain-rate, breaking energy, creep, and fatigue tests. A tensile test involves the elongation of a dumbbell-shaped sample in hydraulically pressurized jaws at a predetermined strain rate (Figure 5.7). The stress is measured as a function of time. If the material to be tested deforms more than 10% before failure, then the cross-sectional area or at least the specimen width should be measured as a function of time.

A stress versus strain curve (Figure 5.8) is determined knowing the force, cross-sectional area, and strain. The strain is typically obtained by knowing the strain rate, the initial test length between the jaws (guage length) and the time of deformation. The area under the stress-strain curve is the energy required to cause failure (toughness) and the difference between the areas under the loading and unloading portions of this curve is the energy lost as heat.

Another method for testing toughness involves measuring the difference in

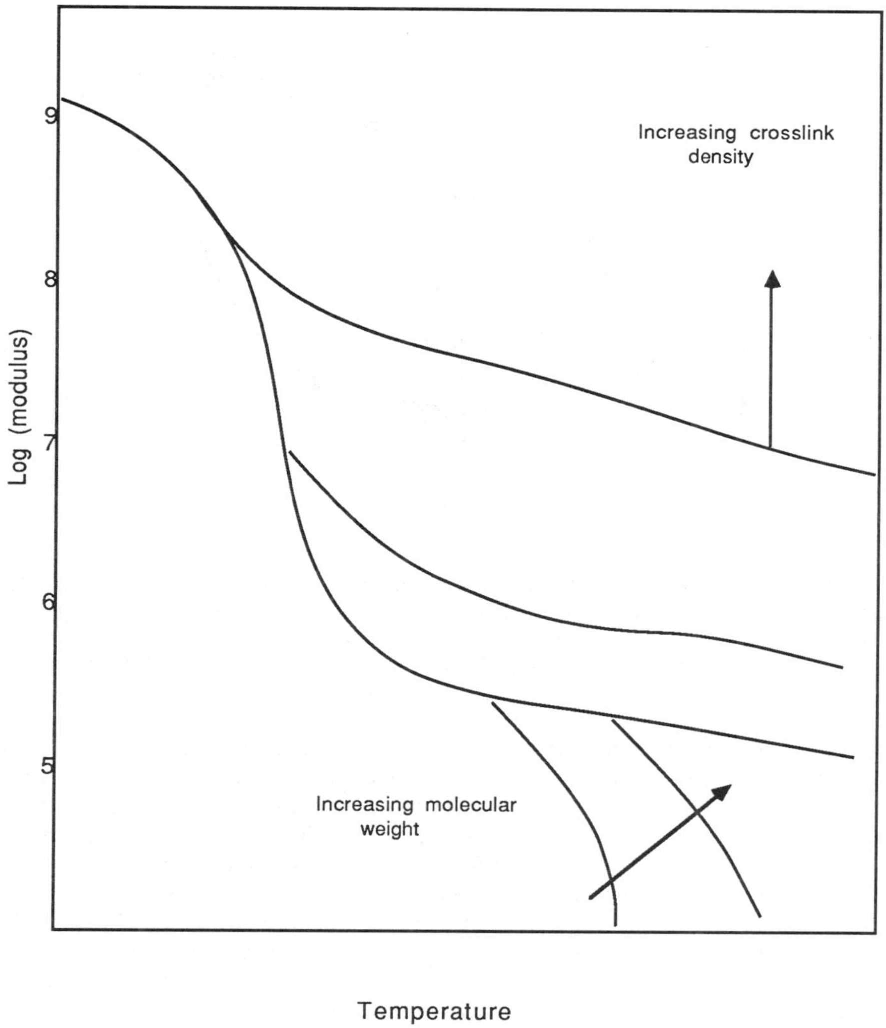

Figure 5.13. Dependence of modulus on molecular weight and crosslink density. Diagram illustrating the effects of increasing molecular weight and crosslink density on the modulus temperature curve for an amorphous polymer. Increased crosslink density or molecular weight increases the modulus.

height that a swinging pendulum reaches before and after causing a notched specimen to fail (Figure 5.14). The impact strength is the energy required per unit thickness to break the sample. It is equal to the difference in potential energy w $h)/d$, where d is the sample thickness. The impact energy or toughness changes discontinuously at the glass ransition (T_g) and melting (T_M) temperatures.

A creep test involves exposing a specimen to a fixed stress and measuring the change in length with time. Near the glass transition (T_g) and melting (T_M) temperatures, the creep rate (slope of strain versus time curve) increases rapidly. Creep tests are useful in predicting service life at a particular temperature.

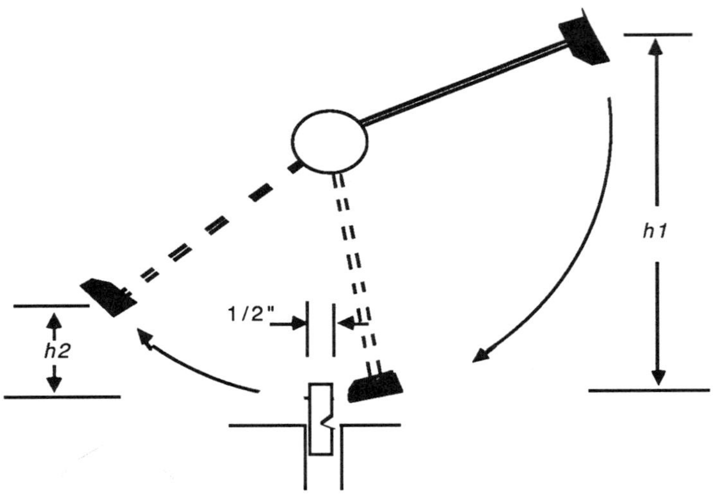

Figure 5.14. Fracture energy determination. Diagram illustrating a method for determining the energy required for fracturing a polymer specimen using a weight that falls from a distance h_1. The difference in heights (h_1-h_2) is directly proportional to the energy to cause failure. This energy decreases sharply at the temperature where the specimen becomes brittle.

Fatigue tests are conducted by cycling a polymeric material repeatedly to a predetermined stress and counting the number of cycles until failure occurs. Some polymers such as polytetrafluoroethylene and polymethylmethacrylate exhibit an endurance limit which is the stress below which a polymer will endure an infinite number of cycles (Figure 5.15). Polymers used to replace skeletal components need to be able to be cycled between 10^6 and 10^8 times before failure.

Rubbery materials have tensile strengths in the range of 10^5-10^7 Pa and will stretch from 1-10 times their original length. Glassy amorphous polymers have tensile strengths of about 10^9 Pa and fail at strains of several percent. Crystalline polymers can have strengths between 10^6-10^9 Pa and fail at strains as high as 200%.

5.5. Summary

Polymeric materials are composed of either long amorphous or crystalline chains produced by condensation or addition polymerization. Amorphous polymers behave like "window" glass at temperatures below the glass transition temperature. In the glassy state, amorphous polymer chains have moduli of

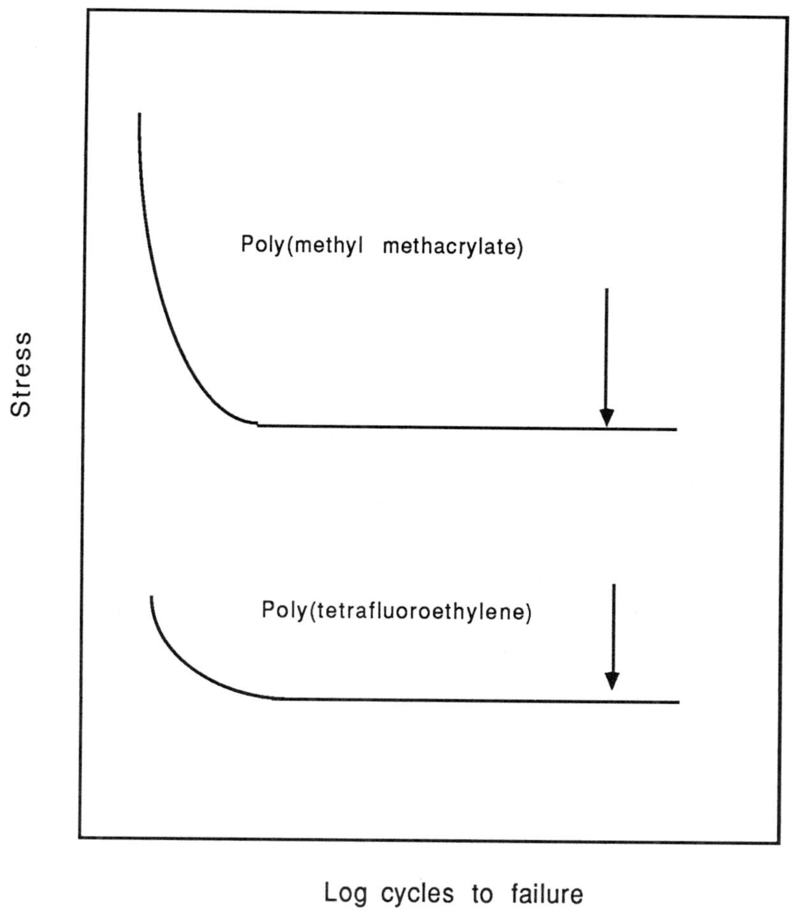

Figure 5.15. Endurance limit testing. Diagram illustrating the number of cycles to failure required as a function of stress for poly(methyl methacrylate) and poly (tetrafluoroethylene). The endurance limit is the stress below which an infinite number of cycles is required before failure will occur.

about 10^9 Pa. Increasing the temperature decreases the modulus of crosslinked chains to 10^5 Pa.

Crystalline polymers contain highly ordered regions in which polymer molecules are well organized in a three-dimensional lattice. The order of these regions is lost on heating above the melting temperature, T_M. Crystalline polymers have moduli in excess of 10^9 Pa at temperatures below T_M; however, above T_M their behavior is similar to that of amorphous polymers.

Polymers used to replace skeletal components must be able to withstand many cycles of loading and unloading while in comparison polymers used for drug-delivery systems may be required to bioerode within a period of days. For this reason the physical properties of a polymer used in construction of medical devices must be carefully considered in the preliminary design stage.

References

Beevers, R.B. and White, E.F.T. Trans. Faraday Soc. 56,744,1960.

Bueche, F. In: Physical Properties of Polymers. Interscience Publishers, New York, Chapter 5, 1962.

Nielsen, L.E. Mechanical Properties of Polymers and Composites, Dekker Publishing Co., New York,1962.

Rodriguez, F. Principles of Polymer Systems, McGraw-Hill Publishing Corporation, New York p. 2016,1982.

Schulz, G.E. and Schirmer, R.H. Principles of Protein Structure, Springer—Verlag, New York, Chapter 2,1979.

Silver, F.H. In: Biological Materials: Structure, Mechanical Properties and Modeling of Soft Tissues, NYU Press, New York, Chapter 2, 1987.

Vincent, P.I. In: The Physics of Plastics, edited by P.D. Ritchie, Van Nostrand, London, 1965.

Williams, M.L., Landel, R.F. and Ferry, J.D. J. Amer. Chem. Soc. 74,3701,1955.

Chapter 6

Characterization of Polymeric Materials Used in Medical Devices

6.1. Introduction

The development of a medical device is a long process. The steps include physical and chemical characterization of the raw materials and prototypes, cytotoxicity screening of prototypes in cell culture, safety and efficacy testing in animals, and finally human testing. It is important that any implant or device developed be carefully characterized so that the final product can be reproduced repeatedly and to document what parameters may be important in controlling the commercial production of the device. This chapter is devoted to presenting some of the tests used to characterize polymeric materials used in medical devices.

The properties of polymers depend on the chemistry of the chain backbone and the average number of repeat units in the chains that make up the polymer. The chemistry of polymer chains can be analyzed using several techniques including infrared and ultraviolet spectroscopy and nuclear magnetic resonance spectroscopy.

6.2. Chemical Analysis of Polymers

In principle, polymers are formed from monomers of defined chemistry via either free radical or condensation polymerization. However, to ensure that the polymerization process does not result in side reactions that alter the nature of the chemistry, the repeat unit of each polymer must be chemically identified.

One approach is to degrade the polymer chain into monomeric units and to compare the properties of the monomeric units to that of known chemicals. For instance polypeptides are polymers that breakdown, to the constituent amino

Table 6.1. Amino Acid Composition of Collagen α - Chains

	α1(I)	α2(I)	α1(II)	α1(III)	α1 (IV)	α2 (IV)	α1(V)	α2 (V)
Hyp	86	85	93	120	146	127	106	107
Asp	45	47	43	47	48	52	51	55
Thr	16	17	22	16	18	28	22	27
Ser	34	24	26	48	34	26	26	34
Glu	77	71	87	71	77	62	99	90
Pro	135	120	129	106	79	69	119	92
Gly	327	328	333	341	318	313	220	318
Ala	120	101	102	95	31	49	46	59
Val	18	34	17	16	30	24	19	30
Cys	-	-	-	2	-	-	-	-
Met	7	4	11	8	15	12	7	10
Ile	9	17	9	14	34	42	20	18
Leu	21	34	26	18	54	59	40	37
Tyr	4	3	1	4	5	6	1	-
Phe	12	16	14	9	24	37	12	12
Hyl	5	11	23	7	61	42	36	25
Lys	32	21	15	28	5	6	7	18
His	3	8	2	7	6	8	19	10
Arg	50	57	51	45	20	45	48	57

acids in the presence of strong acid or base. The amino acids are reacted with a dye, separated on a chromatographic column, and their light-absorbing properties are compared to the properties of standard amino acids. A typical amino acid analysis for type I collagen is shown in Table 6.1.

The chemical composition of the monomeric units can also be determined using a technique termed mass spectroscopy (Skoog, 1985; Willard *et al.*, 1981). Purified monomer units are converted into rapidly moving gaseous ions that are characterized on the basis of their mass-to-charge ratios. A mass spectrum is produced by plotting the relative abundance of the compound or its components against the mass to charge ratio. The compound or atom is identified by comparing the spectrum to that of known compounds or atoms.

Chemical identification of a polymer is also routinely made using infrared and ultraviolet spectroscopy. In the presence of infrared light, a polymer chain undergoes changes in vibrational energy of the individual atoms. Specifically, a bond such as that between carbon and hydrogen vibrates when light of specific wavelengths is absorbed. By measuring the amount of light absorbed as a function of the frequency, a characteristic fingerprint is obtained that can be used to identify the polymer. Infrared spectroscopy determines the types of bond-stretching and-bending vibrations that atoms in a polymer chain can undergo. The precise frequency of absorption of a bond between two atoms depends also on other adjacent atoms.

The fingerprint of a polymer is a plot of light absorbance (or percent light transmission through the sample) as a function of the inverse of the light wave length (wave number) (Figure 6.1). To identify an unknown polymer the infrared spectrum of the sample is compared to a library of spectra for different known polymers.

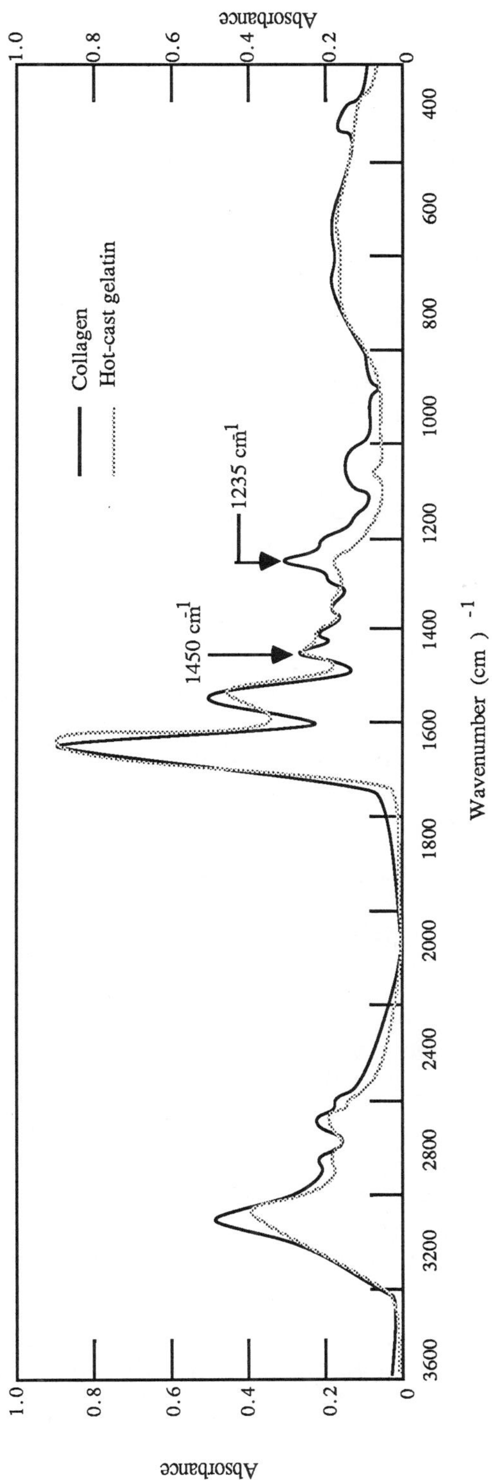

Figure 6.1 . Infrared spectra of collagen and gelatin. Diagram illustrating absorption peaks as a function of wave number (1/wavelength) of the triple helix. Adapted from Silver (1987).

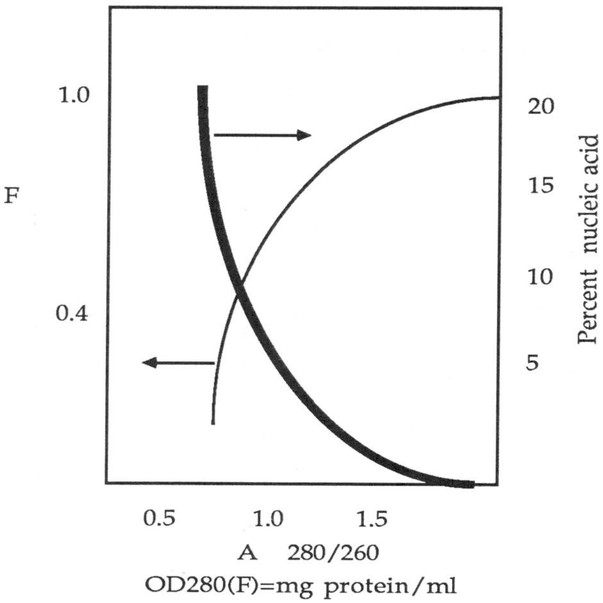

Figure 6.2. Determination of protein and nucleic acid concentrations. Protein and nucleic acid concentrations can be determined from the ratio of optical density at 280 nm to that at a wavelength of 260 nm.

Ultraviolet and visible light spectra are also used to characterize polymer samples. Light at these wavelengths causes electronic transitions in atoms and results in a series of peaks. For instance, it is well known that nucleic acids absorb ultraviolet light maximally at a wavelength of 260 nm while proteins absorb light at 280 nm. Therefore, from the absorption peaks at 260 and 280nm, the content of protein and nucleic acid, respectively, can be identified (Figure 6.2).

Nuclear magnetic resonance (NMR) is another method for characterizing a polymer. In an external magnetic field the spin orientation of an atom is changed. The exact frequency at which energy is absorbed is sensitive to the atomic environment of the proton. The characteristic fingerprint of the polymer is a scan of absorption versus frequency of the magnetic field.

6.3. Molecular Weight Distributions

Once the chemistry of a polymer has been characterized, it is necessary to define whether the chains have a single chain length or are characterized by a particular distribution of chain lengths (Billmeyer, 1984).

One gram of a polymeric material with an average molecular weight of 100,000 is composed of 6.023×10^{18} polymer molecules. In the case of natural polymers it is possible that each molecule has the same number of repeat units and therefore each molecule has an identical molecular weight. In the case of synthetic polymers there is a small probability that the molecular weights of all chains are identical. Therefore for synthetic polymers and some natural polymers, it is necessary to define the average molecular weight of a polymer molecule and the nature of the distribution of molecular weights. The distribution of polymer molecular weights dictates a number of physical properties such as the flow behavior required to form polymers into a variety of shapes used in implants.

If N molecules each have a molecular weight M_i, then the number average molecular weight, M, is equal to M_i. However, if there are N_i molecules where i can range from 1 to over 10^{23}, and each molecule has its own molecular weight, M_i, then the number average molecular weight becomes:

$$\overline{M}_n = \frac{\sum_{i=1}^{\infty} N_i M_i}{\sum_{i=1}^{\infty} N_i} \tag{1}$$

Eq. (1) is equivalent to the total weight of polymer molecules divided by the total number of molecules. Another way of calculating M_i is from the degree of polymerization, N_i. If one multiplies the degree of polymerization times the repeat unit molecular weight, M_0, then Mi can be calculated from Eq. (2).

$$M_i = M_0 N_i \tag{2}$$

Another manner by which the distribution of molecular weights can be assessed is by determining the weight average molecular weight, $\overline{M}_W$. This average molecular weight is calculated by weighing the molecular weight of each molecule by its weight, W_i. If the weighing factor, W_i, is introduced into eq. (1) we get a relationship for determining $\overline{M}_W$ [see eq. 3]. Recognizing that the weight of a molecule

$$\overline{M}_W = \frac{\sum_{i=1}^{\infty} W_i M_i}{\sum_{i=1}^{\infty} W_i} = \frac{\sum_{i=1}^{\infty} N_i M_i^2}{\sum_{i=1}^{\infty} N_i M_i} \tag{3}$$

is proportional to the molecular weight leads to a relationship between $\overline{M}_W$ and $N_i M_i^2$ [see eq. 3].

The ratio $\overline{M}_W/M_n$ is termed the polydispersity index and is used to assess the broadness of the polymer chain distribution. Properties involved in processing polymers into end—use forms such as flow behavior are related to the polydispersity index. For instance, some synthetic polymers with a

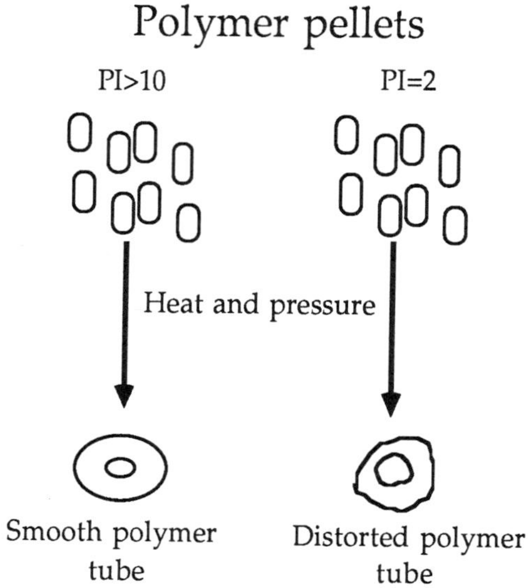

Figure 6.3 . Diagram illustrating relationship between polydispersity index and processing properties of polymers. Polydispersity index (PI) is defined as the ratio of weight and number average molecular weights ($\overline{M}w/\overline{M}n$). Polymers with a high polydispersity index form tubes with smooth contours when molded under heat and pressure. In comparison polymers with a low polydispersity indexes form distorted polymer tubes.

polydispersity index of two are more difficult to mold thermally into a variety of useful shapes than are synthetic polymers with a polydispersity index of twenty even though the average molecular weight of all the chains may be identical (Figure 6.3). Low- molecular-weight components tend to flow more uniformly and fill a mold better than high-molecular-weight components; however, high-molecular- weight components give a higher tensile strength and toughness. Therefore in many applications a distribution of low-and high-molecular weight chains are necessary to satisfy a variety of physical properties.

For more sophisticated analysis of polymer molecular weight distributions, a Z average molecular weight, $\overline{M}z$, is determined using eq. (4). The ratios

$$\overline{M}_z = \frac{\sum_{i=1}^{\infty} W_i M_i^2}{\sum_{i=1}^{\infty} W_i M_i} = \frac{\sum_{i=1}^{\infty} N_i M_i^3}{\sum_{i=1}^{\infty} N_i M_i^2} \tag{4}$$

$\overline{M}z/\overline{M}n$ and $\overline{M}w/\overline{M}n$ are used to give information concerning the shape of the molecular weight distribution.

6.3.1. Determination of Molecular Weight Distributions

There are several methods used to separate polymers into different molecular weight fractions. These include gel permeation and liquid chromatography, ultracentrifugation and differential salt or solvent precipitation. Once the polymer is fractionated into aliquots, the number average and weight average molecular weight of each fraction can be determined from ultracentrifugation, viscometry, osmotic pressure and light scattering (Table 6.2). If the number average molecular weight is determined for each fraction then a distribution can be directly plotted (Figure 6.4) and M_n obtained. This distribution can be easily converted into a weight average by plotting the weight of molecules in each aliquot versus the molecular weight of that aliquot. The weight average molecular weight is that molecular weight which separates equal weights of molecules as shown in Figure 6.5.

A Z average molecular weight is obtained by plotting the squared weight of each aliquot versus molecular weight of that aliquot. The Z average molecular weight is that molecular weight that separates equal areas under the Z distribution plot.

Table 6.2. Methods For Determination of Polymer Molecular Weight

Method	Molecular Weight Average
End group analysis	M_n
Equilibrium ultracentrifugation	M_w, M_n and M_z [a]
Light scattering	M_w
Osmotic Pressure	M_n
Viscometry	M_n

[a] obtained by calculation.

6.3.2. Polymer Fractionation

The terms gel filtration, gel permeation chromatography (GPC), and high pressure liquid chromatography (HPLC) refer to a process by which molecules are separated on a basis of size. These processes involve pouring or pumping under pressure (HPLC) a solution of polymer molecules through a column that is packed with porous beads (Figure 6.6). The beads can be made of porous glass, a crosslinked polymer such as a dextran, or any other surface active material.

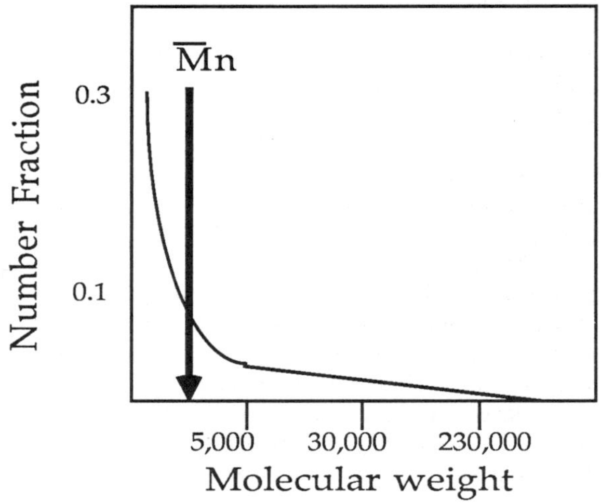

Figure 6.4. Diagram illustrating determination of the number average molecular weight. Number average molecular weight is determined as the molecular weight of the point that separates equal areas on a number fraction-molecular weight plot.

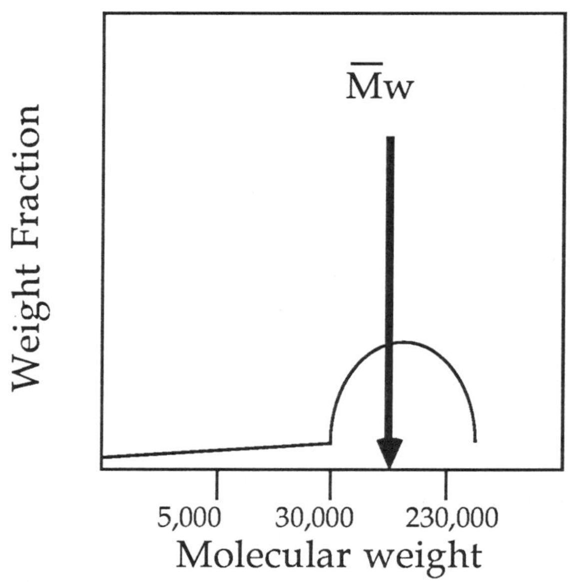

Figure 6.5. Diagram illustrating determination of the weight average molecular weight. Weight average molecular weight is determined as the molecular weight of the point that separates equal areas on a weight fraction-molecular weight plot.

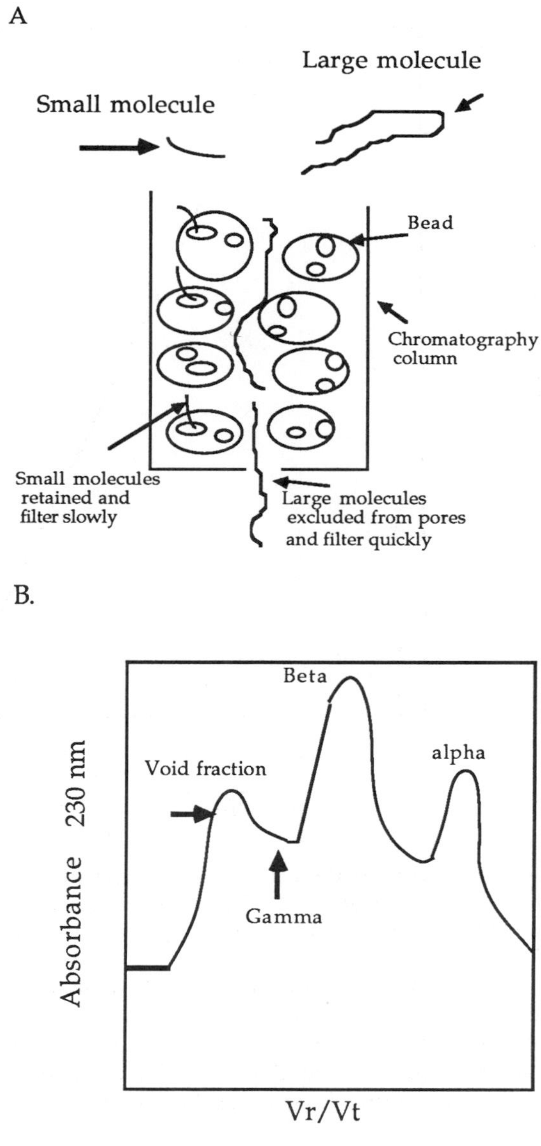

Figure 6.6. Separation of polymer chains by chromatography. (A). Large polymer molecules sterically cannot fit into the pores in the beads used to pack a chromatography column and therefore they flow between the beads and are eluted in the exclusion volume. The exclusion volume is the first volume to be obtained from the column. Lower molecular weight species require longer periods of time to travel through the bead pores and therefore are found in later volumes removed from the column. (B). A typical plot of absorbance (concentration) versus ratio of retention volume to total volume (V_r/Vt) for collagen derived gelatin chains with molecular weights of 99,000 (alpha chains), 189,000 (beta components), 296,000 (gamma components), and higher molecular weight species (void fraction). The separation was achieved using gelatin chains chromatographed over a column packed with Agarose A5M (from Silver and Trelstad, 1981).

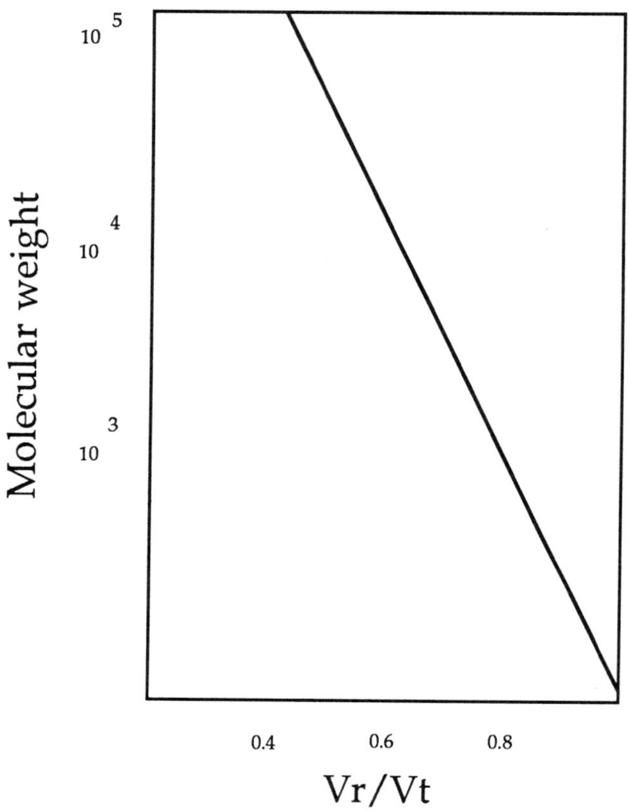

Figure 6.7. Diagram showing relationship between molecular weight and V_r/V_t. Log molecular weight is linearly related to Vr/V_t for gelatin chains derived from collagen by thermal denaturation (F.H. Silver and R.A. Berg, unpublished observations).

Separation is achieved when the smaller molecules penetrate the beads while the larger molecules flow around the beads. Continuous application of solvent to the column after the sample is applied causes the large molecules to flow through the column rapidly in what is termed the exclusion volume. The exclusion volume is the material that is "excluded" from passing through the beads.

The molecular weight species that is found in the exclusion volume is dependent on the size of the pores in the beads. Typically pore diameters of 10-250 nm are used to separate polymer molecules in the molecular weight range of 10^3 to10^7. The volume of solvent that must be eluted from the column before a molecule of a known molecular weight is found in the effluent, is determined for standards with known molecular weights. In some cases the log of the molecular weight is linearly related to the inverse of the fraction of solvent that is retained divided by the total volume through the colum V_r/V_t. An example of this relationship is given for gelatin (Figures 6.6 and 6.7).

Using a calibration curve of log molecular weight versus retention volume fraction, the molecular weight distribution of an unknown sample can be established. An example of a molecular weight distribution for a gelatin sample is shown in Figure 6.4.

In some cases the molecular weight distribution can be obtained directly from centrifugation data. Polymer solutions can be characterized using a technique termed equilibrium ultracentrifugation. The solution is placed in a sample cell at a known concentration and is rotated in a centrifuge at speeds of at least several thousand rpm. At these speeds, the polymer molecules are sedimented to the periphery of the rotor containing the sample. At the same time the molecules tend to diffuse towards the center of the cell due to a concentration gradient. At equilibrium, the centrifugal force due to rotation of the rotor is balanced by the frictional force. The optical density (as a function of position across the cell) of the solution is measured through a window and is converted to concentration using a calibration curve of concentration as a function of optical density. Yphantis (1964) developed a relationship between the concentration of the polymer at a distance *r*, *c* (*r*), and the molecular weight of the ith component (eq. (5)-(7)),

$$C(r)=\sum_{i=1}^{N} C_i(r)=\sum_{i=1}^{N} A_i \exp(\sigma_i r^2/2) \tag{5}$$

$$\sigma_i=\frac{M_i(1-\nu\rho)}{RT}\omega^2 \tag{6}$$

$$A_i=\frac{C_{i,0}\sigma_i(b^2-a^2)}{2\left[\exp\left(\sigma_i\frac{b^2}{2}\right)-\exp\left(\sigma_i\frac{a^2}{2}\right)\right]} \tag{7}$$

where *a* and *b* are the values of *r* at the meniscus and base of the cell, respectively. $C_{i,o}$ is the concentration of component i in the initial solution, *V* is the specific volume of the polymer, ρ is the polymer density, *W* is the angular velocity of the rotor, *R* is the gas constant, and *T* is the absolute temperature. The experimental distribution of C(r) obtained from the ultracentrifuge, is fit by varying $C_{i,o}$, *M*, and adding components i until a least- squares- fit of eq. (5) to the experimental data is obtained.

The meniscus is the point at which the optical density is seen to change discontinuously across the cell. For polymers that have very broad molecular weight distributions ultracentrifugation may underestimate the width of the distribution. This is due to poor separation of high molecular weight components at the meniscus.

6.4. Determination of Molecular Weights

For distributions of polymer chains there are several experimental approaches that can be used to estimate number, weight, and Z average molecular weights. Usually at least one technique described below is used to measure average molecular weight; however, it is often recommended that molecular weights be measured using several techniques.

6.4.1. Number Average Molecular Weight

End-group analysis is a method by which the number average molecular weight is determined. In this method the free ends of the polymer chain are reacted or counted by titration. For example, proteins or polymers with a peptide backbone and a free carboxylic acid group can be titrated with hydrochloric acid. From the number of moles of hydrogen ions added and the number of grams of polymer chains present, the molecular weight can be determined [see eq. 8]. Care must be taken to correct for titration of non-end groups.

$$\overline{M_n} = \frac{\text{weight of polymer}}{\text{moles of end group}} \tag{8}$$

Number average molecular weight can also be determined from boiling point elevation, freezing point depression, and osmotic pressure (Silver , 1987). Using a technique termed osmometry a solution of polymer molecules in solvent is separated by a semipermeable membrane from pure solvent. Only solvent molecules can penetrate the membrane. When the polymer molecules are confined to one side of the chamber, the concentration of the solvent is different in the two components. The thermodynamic drive towards equilibrium results in a different liquid level in the two compartments. The number average molecular weight, $\overline{M_n}$ is calculated from the osmotic pressure Pi, knowing the weight concentration of macromolecule (c) where R and T are the gas constant and absolute temperature, respectively (Figure 6.8).

$$\lim_{c \to 0} \frac{P_i}{c} = \frac{RT}{\overline{M_n}} \tag{9}$$

6.4.2. Weight Average Molecular Weight

Weight average molecular weight, $\overline{M_w}$, can be determined from the intensity of light scattered by a solution of macromolecules with a concentration c. For light scattered at low angles (scattering angles less than 4°) the relationship between $\overline{M_w}$ and c is given by equations (10) and (11) where R_O is the Rayleigh factor of the solution minus the Rayleigh factor of the solvent, B is the second virial coefficient, r is the distance of the sample from the photodetector, I_s and I_o are the intensity of scattered light and incident light, and O is the scattering angle.

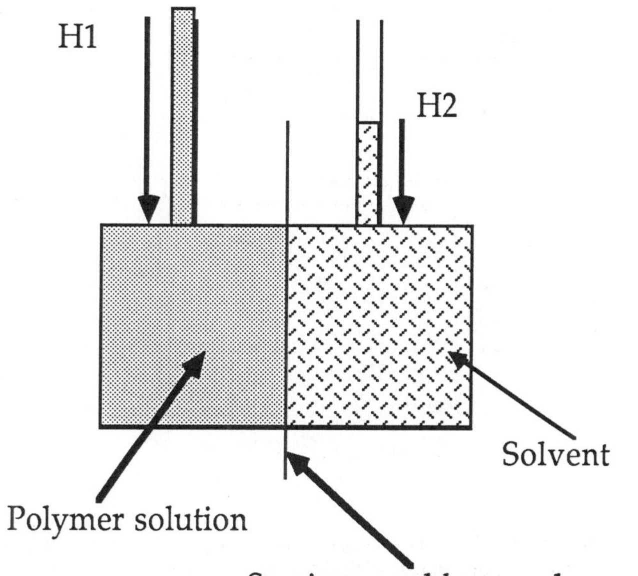

Figure 6.8. Determination of number average molecular weight by osmometry. The number average molecular weight is determined from the height difference between two solutions separated by a semipermeable membrane that allows the diffusion of only low molecular weight species. The solution containing polymer chains causes solvent ions to move from left to right. The number average molecular weight is determined from the product of the height difference.

$$\frac{R_c}{R_0} = \frac{1}{\overline{\overline{M}}_w} + 2Bc \qquad (10)$$

$$R_0 = \frac{r^2 I_s}{I_s\left(1 + \cos^2 B\right)} \qquad (11)$$

A plot of Kc/R_O versus c has an intercept of $1/\overline{M}_W$ (Figure 6.9).

$\overline{M}_W$ can also be determined from equilibrium ultracentrifugation (see Section 6.3) knowing the concentrations (C_1 and C_2) at two points (r_1 and r_2) within the cell, the angular velocity, (w_1), specific volume (v), and density, ρ, using equation (12) (Figure 6.10).

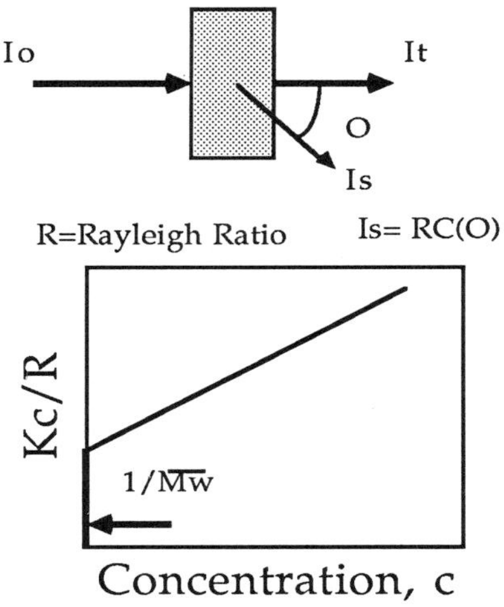

Figure 6.9. Determination of weight average molecular weight by light scattering. When an incident beam of light with intensity I_o impinges on a solution containing a polymer molecule, most of the light intensity is transmitted through the sample (I_t); however, light of intensity, I_s, is scattered at an angle O. The intensity of scattered light is proportional to the Rayleigh ratio, R, times an optical constant, $C(O)$. The weight average molecular weight is obtained from the inverse of the Y-intercept of a plot of a constant K times concentration divided by the Rayleigh factor versus concentration, c.

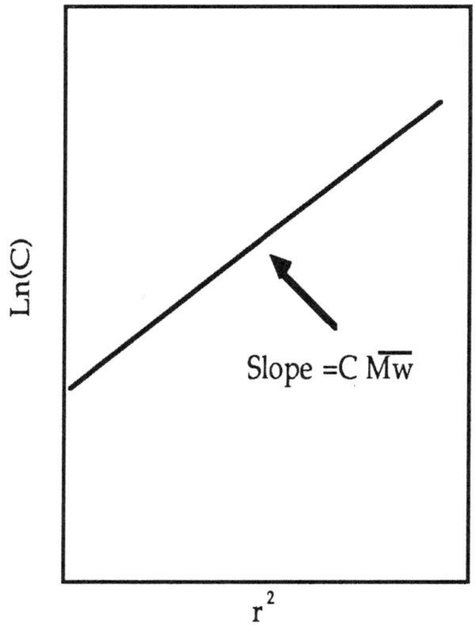

Figure 6.10. Determination of weight average molecular weight from equilibrium ultracentrifugation. The weight average molecular weight is obtained from the slope of a plot of l_n (concentration) versus distance, r that occurs when a polymer molecule moves in a centrifugal force field.

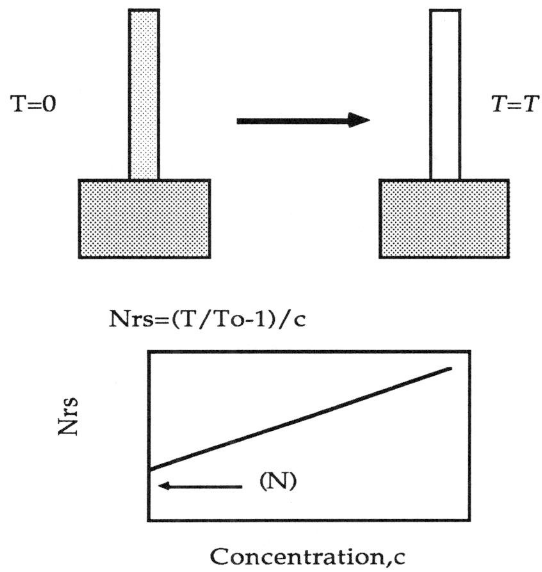

Figure 6.11 . Determination of intrinsic viscosity. The specific relative viscosity (N_{rs}) is obtained from the time required for flow of a polymer solution of known concentration through a capillary of known length. A plot of N_{rs} versus concentration, *c*, yields a Y-intercept equal to the intrinsic viscosity.

$$\overline{M}_w = \frac{2\,RT\ln\left(c_2/c_1\right)}{\left(1 - v\rho\right)\omega^2\left(r_2^2 - r_1^2\right)} \qquad (12)$$

6.4.3. Viscosity Average Molecular Weight

The simplicity of measurement of the viscosity of polymer solutions has led to the development of a relationship between viscosity and viscosity average molecular weight, $\overline{M}_v$. Solution viscosity can be determined from the time (*t*) required for a specified volume of polymer solution to flow through a capillary tube (Figure 6.11). The relative viscosity (N_r) is defined as the ratio of the time required for the solution (*t*) and solvent (t_o) to flow through the capillary. Reduced viscosity (Nrs) is defined as the relative viscosity divided by the concentration (*c*) and intrinsic viscosity *[N]* is defined as the limit as *c* approaches zero of the reduced viscosity (eq. 13).

$$[N] = \lim_{c \to 0}\left[\frac{t - t_0}{t}\right]\frac{1}{c} \qquad (13)$$

Intrinsic viscosity is related to $\overline{M_v}$ by equation (14) where *a* is a constant between 0.5 and 1.0 depending on the solvent. If K1 and a are known

$$[N] = K\,\overline{M_V}^{a} \tag{14}$$

for a polymer then $\overline{M_v}$ can be calculated from experimental values of [*N*]. Values of *K* and *a* are determined from calibration curves of ln [*N*] versus $\overline{M_v}$ similar to Figure 6.12.

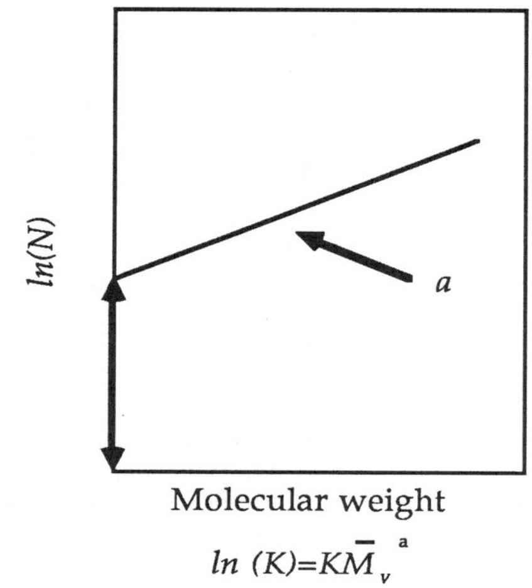

Figure 6.12. Diagram showing relationship between intrinsic viscosity and molecular weight. The molecular weight is obtained from intrinsic viscosity measurements using a calibration curve established using molecular weights determined from light scattering or sedimentation equilibrium. The natural logarithm of the intrinsic viscosity is linearly related to the molecular weight and the shape factor ***a*** is obtained from the slope.

$\overline{M_v}$ usually falls somewhere between the number average ($\overline{M_n}$) and weight average ($\overline{M_w}$) molecular weights. It is used often as an approximation of the average molecular weight since it is easily determined with a minimum investment in equipment and time.

6.5. Determination of Transition Temperatures

Both amorphous and crystalline polymers undergo thermal transitions which characterize changes in structure and physical properties. Amorphous polymers become stiff when cooled below the glass transition temperature and crystalline polymers lose stiffness when heated above the melting point. Therefore determination of the glass transition and melting temperatures are important to characterizing polymer behavior.

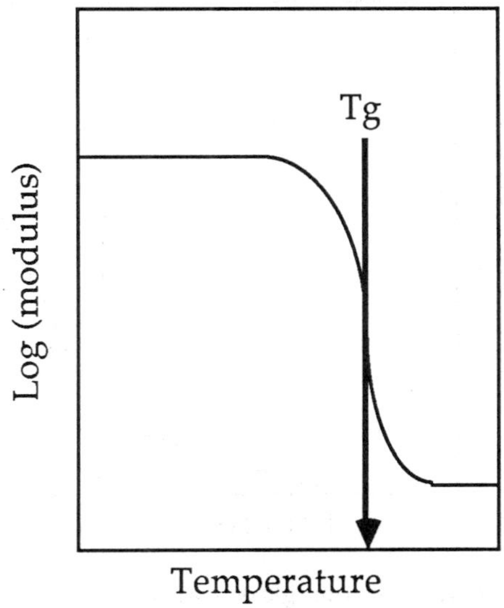

Figure 6.13. Determination of the Glass Transition Temperature From Modulus Measurements. The glass transition temperature is defined as the temperature at which the modulus is half-way between the maximum and minimum values.

6.5.1. Determination of Glass Transition Temperature

The glass transition temperature, T_g, as discussed in Chapter 5, is the temperature at which the flexible backbone of an amorphous polymer chain becomes rigid. It can be determined from measurement of the tensile modulus as a function of temperature and is defined as the temperature at which the stiffness is halfway between that of a glassy polymer (about 10^9 Pa) and a rubber (10^5Pa) as shown in Figure 6.13.

The value of T_g determined from stiffness measurements is dependent on the strain rate, rate of heating of the specimen, molecular weight of the polymer and the volume fraction of polymer in the sample. At high strain rates rubbery polymers behave in a brittle fashion because the polymer chains do not have enough time to unfold during deformation of the sample. Values of tensile strength for rubbers increase almost linearly with increases in log E, where E is the strain rate (Ward, 1983). Changes in T_g are small above a molecular weight of 10,000 as discussed in Section 5.3.2; however, this change must be considered. Polymers containing plasticizers or any additive used for polymerization or processing can lower the value of T_g. Therefore the value of the glass transition temperature depends on the exact content of a polymeric material.

Another approach for determining the glass transition temperature is to measure the volume of a gram of polymer (specific volume) as a function of temperature. At T_g there is a discontinuous increase in the specific volume (Figure 6.14). This technique is termed dilatometry and is also affected by the rate of heating.

Tg can also be estimated as the temperature at which a rubbery material becomes brittle. The brittle temperature is conveniently measured by measuring

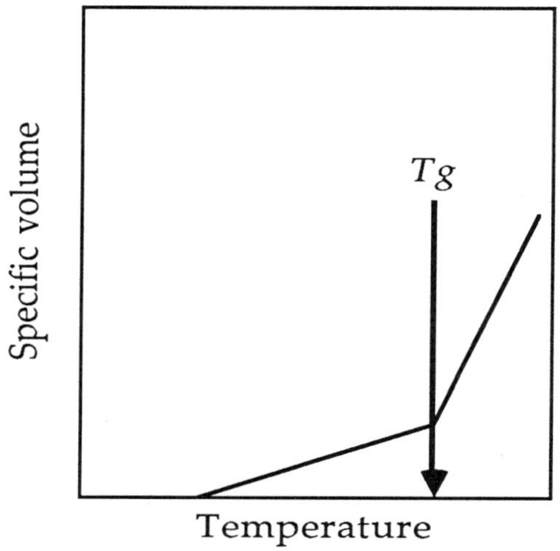

Figure 6.14. Determination of the glass transition temperature from specific volume measurements. The glass transition temperature is the temperature at which a discontinuous change occurs in the volume of a gram of polymer (specific volume).

the temperature at which application of an impact force causes a polymeric material to fracture into many pieces like window glass (Figure 5.14).

6.5.2. Determination of Melting Temperature

Melting temperature (TM) can be determined using mechanical ,volumetric, and optical methods. At the melting temperature, the stiffness decreases discontinuously, the specific volume increases, and the polymer undergoes an optical change from opaque to transparent. Measurement of the temperature dependence of stiffness is done in a similar manner to that reported above for the glass transition. Dilatometry is conducted in a calibrated capillary tube in which small volume changes can be measured. Opacity changes of a crystalline polymer placed in the path of a light beam on a "hot plate" can be observed. The temperature at which the opaque sheet becomes translucent is visually determined.

6.6. Mechanical Property Determination

Polymers that are used in medical devices that are intended to replace structural tissues such as tendon, bone, blood vessels, and skin need to support loads and provide tissue integrity. Mechanical design criteria are intended to mimic the properties of natural tissues.

Static and dynamic tests are used to characterize the mechanical properties of polymers. Static tests refer to the response of a material to a single applied load. Static tests include the change in length of a material under a fixed load (creep) and the change in stress of a sample held at fixed length (stress relaxation). Stress-strain properties are obtained by measuring the force per unit area (stress) as a function of change in length divided by original length (strain) that occurs when a specimen is stretched at a constant rate of extension. The toughness is obtained from the area under the stress-strain curve. Other tests measure tear and abrasion resistance of polymers (Table 6.3 and Figure 6.15).

Table 6.3. Mechanical Tests Used To Characterize Polymers

Test	Property measured
Abrasion	Frictional coefficient
Creep	Characteristic relaxation time
Fatigue	Endurance limit
Sinusoidal loading	Viscoelastic properties
Stress relaxation	Characteristic relaxation time
Tear	Tear resistance
Tensile	Stress-strain behavior, toughness

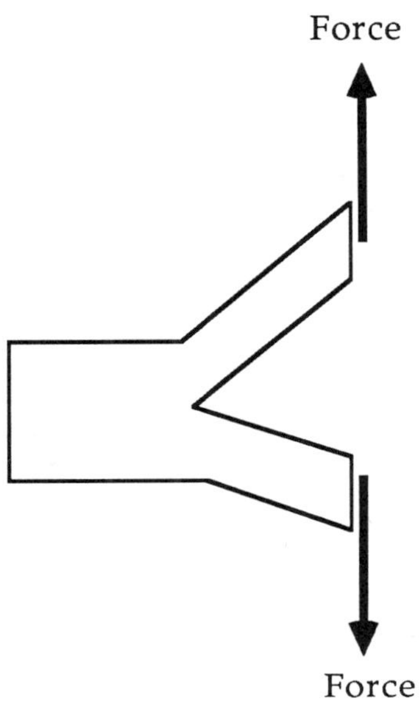

Figure 6.15. Diagram illustrating trouser tear test. The tear resistance of a polymeric material is determined from the force required to pull apart two legs of a trouser made from a polymeric material.

Dynamic tests subject the polymeric material to many repetitions of a load and measure the fatigue life (number of cycles to failure). Fatigue tests are conducted by compressing, stretching, or bending a material repeatedly to a predetermined stress or strain until the specimen breaks or will not support a load. Fatigue as well as other mechanical tests can be conducted in phosphate buffer solution, serum, or a solution containing enzymes to stimulate the effects of tissue fluids. Devices used to replace skeletal tissues must be carefully tested for fatigue in the laboratory before any animal evaluation is conducted.

6.6.1. Static Mechanical Tests

In order to conduct even simple static or dynamic tests several factors need to be considered. The first involves measuring specimen dimensions. Most biomaterials are in the form of thin sheets, tubes, and fibers. Careful measurement of specimen dimensions must be made in order to calculate mechanical parameters. Fiber and thin-sheet thickness measurements can be made on wet specimens using calipers or a microscope with acalibrated eyepiece. Care must be taken not to deform the sample when using calipers or when mounting samples for mechanical testing.

Mechanical testing is usually performed by pulling (tension) or by pressing (compression) on a specimen. Tensile tests are conveniently conducted on sheets, films, and fibers, while compression tests can be conducted on specimens that are not large enough to test in tension. Normally, specimen lengths of 5 cm and widths of at least 1 cm are preferred for tensile testing. Tensile and compressive tests are performed with a device such as an Instron tester.

All polymeric materials are viscoelastic and exhibit a time dependence in their mechanical properties. The time dependence is evaluated by measuring the stress as a function of time at a fixed strain. Knowing the operating load that is encountered in a particular application it is possible to determine whether the deformation observed is acceptable for the desired application. If the allowable strain is limited, such as in the design of an artificial heart valve, then the stress that the material can withstand is obtained from a stress relaxation experiment.

The most popular mechanical test is the stress-strain test. In this test the material is loaded at a constant rate of strain and the resulting stress is recorded (Figure 5.8). The area between the stress and strain axes traced out during the test is proportional to the toughness of the material and the energy required to cause failure (Figure 6.16). Stress relaxation can be performed during a stress-strain experiment by allowing the force to relax to equilibrium after stretching the specimen to a predetermined strain (Figure 6.17). The stiffness of a polymer is determined from the slope of the stress-strain curve. If the slope changes with strain, then the stiffness is estimated as the tangent to the curve at a defined strain (Figure 6.18). The strain at failure indicates the extensional or compressional deformation limits for a polymer to meet a particular end use.

6.6.2. Dynamic Mechanical Tests

Dynamic testing is used to study the response of a material to repeated or cyclic application of a force. Materials subjected to a cyclic stress repeatedly,

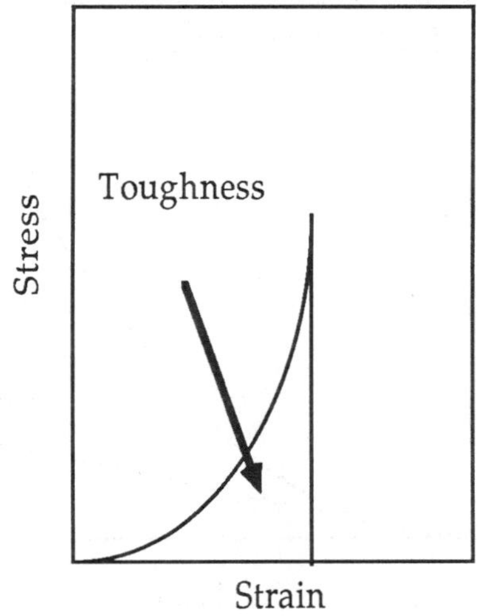

Figure 6.16. Diagram Illustrating Toughness Determination. The toughness of a polymeric material is related to the area under a stress-strain curve.

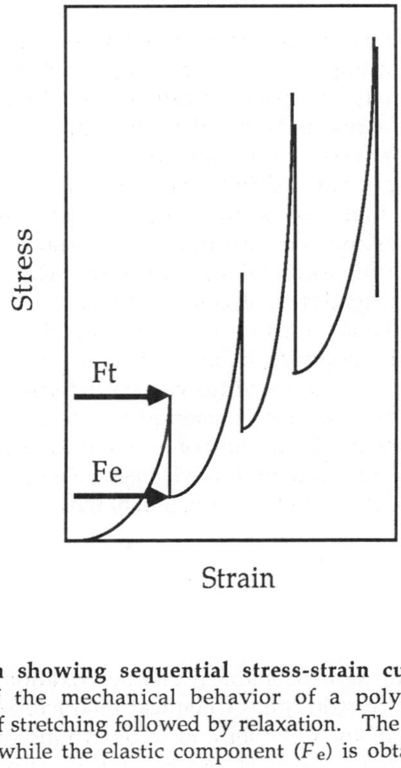

Figure 6.17. Diagram showing sequential stress-strain curve. The elastic and viscous component of the mechanical behavior of a polymeric material can be determined by a cycle of stretching followed by relaxation. The elastic and viscous force components sum to F_t while the elastic component (F_e) is obtained from the force at equilibrium.

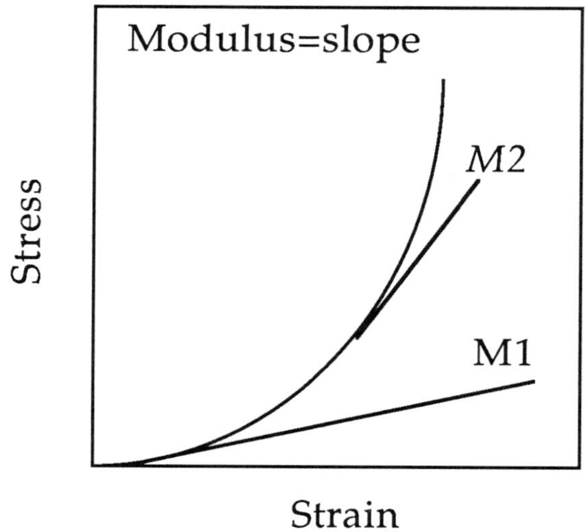

Figure 6.18. Diagram illustrating determination of the tangent modulus. The tangent modulus at a particular strain is determined from the slope of the stress-strain curve as shown for a low (M_1) and moderate strain (M_2). This behavior is unique to viscoelastic materials.

fail at stresses below the failure stress observed after a single deformation. The lower the applied operating stress, the more cycles that are required to cause failure. For some materials, the specimen can be loaded and unloaded an infinite number of times if the stress is below the "endurance limit". Figure 5.15 is a fatigue life plot of stress versus cycles to failure.

Fatique tests can be conducted in tension, compression, and by bending the specimen. The exact mode of testing depends on the loading conditions encountered in the desired application. For instance tendons are cycled repeatedly in tension during locomotion. Therefore cyclic tensile loading is used to test the fatigue life of tendon replacement materials. In comparison, skin is repeatedly cycled in biaxial tension and therefore the fatigue life of a skin substitute is evaluated by applying biaxial tensile loads.

Dynamic measurements of modulus can be determined by imposing a sinusoidal deformation on a polymeric material and determining the stress in phase with the deformation (elastic component) and the stress out of phase with the deformation (viscous component) (Ward, 1983). The elastic (storage) and loss (viscous) moduli can be determined as a function of the deformation and number of cycles.

6.7. Summary

Chemical, physical, and mechanical characterization of polymers is an important aspect of device design. Chemical characterization is required to confirm the repeat unit of the polymer and to determine the level of impurities present. Average molecular weight determination and the distribution of molecular weights play a role in reproducible processing of polymers into devices.

Mechanical testing of polymers is essential to predicting the lifetime of a device in skeletal and nonskeletal applications. Taken together these tests are required to document polymers used in medical devices.

References

Billmeyer, F.W. Textbook of Polymer Science, Wiley Interscience, New York, N.Y., Chapter 8,1984.

Silver, F.H. Biological Materials: Structure, Mechanical Properties, and Modeling of Soft Tissues, NYU Press, New York, Chapters 4 and 7,1987.

Silver, F.H. and Trelstad, R.L. Biopolymers 20,359,1981.

Skoog, D.A. Principles of Instrumental Analysis, Saunders College Publishing, Chapter 18,1985.

Ward, I.M. Mechanical Properties of Solid Polymers, John Wiley and Sons, New York, Chapters. 6 and 12,1983.

Willard, H.H., Merritt, L.L., Dean, J.A. and Settle, F.A. Instrumental Methods of Analysis, 6th edition, Wadsworth Publishing Co., Belmont CA, Chapter 19,1981.

Yphantis, D.A. Biochemistry, 3,297,1964.

Chapter 7

Methods for Evaluating Biocompatibility

7.1. Introduction

Design of medical devices involves fabrication and testing of prototypes until the desired end-use properties are obtained. After initial prototype designs have been fabricated, it is essential to evaluate the biocompatibility of the materials used as well as that of the assembled prototype.

Several tests for evaluating biomaterials and biocompatibility have been developed. A long list of tests can be described, however, we will limit the scope of this chapter to include tests that evaluate the interaction of materials with cells in culture and in tissues. Results of these tests are helpful in screening on a limited basis the possible materials for prototype fabrication. Evaluation of cell and tissue biocompatibility is the primary screening procedure for biomaterials before they can be used in end-use applications. However, as discussed in the report on the Concensus Development Conference on "Clinical Applications of Biomaterials" (Galletti and Boretos, 1983) a balance between risk and benefit must be used in determining how biocompatible a material must be before it can be used clinically.

7.2. Blood Compatibility Evaluation

Interaction of blood components with implants occurs in a wide variety of surgical procedures and is important to the function of surgical implants. Vascular implants should not initiate or deactivate the blood clotting cascade or interfere with platelet aggregation. In contrast, skin implants must prevent excessive bleeding in order to facilitate tissue repair. Therefore the hemostatic activity of an implant varies depending on the exact end-use.

Evaluation of blood compatibility in an animal model or using blood from an animal is complicated by several factors. It is known that blood from different species of animals exhibits differences in clotting properties such as quantitative amounts of blood components, cell activity, fibrinolytic response, and coagulation. In addition, prior to clotting tests, anticoagulants are usually added, which affect measurement of hematologic parameters.

In practice, a set of standard tests are performed on control and uncharacterized surfaces using a defined protocol. Experimental data is compared to that obtained with a control surface such as glass or collagen.

Two major problems to be overcome by a biomaterial are (1) unwanted adhesions that arise from healing of clotted blood, and (2) calcification and stiffening of the material used to repair the cardiovascular system. Biomaterials should not cause platelet adhesion and aggregation, blood coagulation, obstruction of blood flow by deposition of blood cells and molecules, or depletion of electrolytes.

7.2.1. *In Vitro* Methods for Evaluating Hemocompatibility

A variety of *in vitro* methods have been proposed for evaluation of blood compatibility. We will summarize a few of the tests that may be used to evaluate the safety and efficacy of a material. *In vitro* experiments are usually performed as an initial screening test in the early phase of biocompatibility evaluation. Materials that cause clotting and that are incompatible with cells usually do not merit *in vivo* evaluation.

Several hematologic tests are used to evaluate blood compatibility. These place whole blood on a test surface for short times (i.e., 1 sec to 1 min) or until clotting occurs at room temperature. Measurement of clotting time, clot weight, number of adherent platelets, thrombin time, prothrombin time, partial thromboplastin time, platelet factor release, and platelet counts are determined (Table 7.1). Short or excessively prolonged clotting times suggest that the material is either activating or deactivating the clotting system (Figure 7.1) and is evidence of blood incompatibility.

Other tests are used to evaluate the degree of platelet adhesion under controlled conditions. Platelet-rich plasma or suspended platelets are usually used rather than whole blood. The degree of platelet adhesion and the morphology of the attached platelets are also determined. However, the surface protein layer that is built up between the surface of a biomaterial and blood may modify the results obtained with this test (Figure 7.2).

Analysis is also made of the deposition of plasma proteins (see Section 7.6.1) onto the surface of the material such as albumin that may decrease platelet adhesion and activation while adsorption of fibrinogen may have the opposite effect. Protein adsorption can be measured by fluorescence spectroscopy, infrared spectroscopy, immunoassay, radio-immunoassay, chromogen-labeling, radiolabeling, circular dichroism, ellipsometry, contact angle determination, and Fourier-transform infrared spectroscopy. The most commonly used technique involves measurement of I^{125} radiolabeled proteins (albumin, fibrinogen, or gamma globulins) absorbed onto the surface of a biomaterial.

Protein adsorption can also be determined from experiments performed using flowing solutions of either whole blood, platelet-rich plasma, or blood proteins in a specially designed flow chamber. Fourier-transform infrared-attenuated total reflectance spectroscopy (FTIR-ATR) directly measures protein adsorption to a surface and is a most useful method in addition to radiolabeling techniques.

A convenient way to test large surface areas of a material is to coat glass beads with the material and pack the beads into a chromatography column.

Table 7.1. *In vitro* Blood Compatibility Assays

Assay	Reagent	Property measured
Whole blood clotting time	Whole blood	clotting time and weight of blood in contact with a surface
Platelet adhesion and aggregation	Platelet-rich plasma	adhesion and aggregation of platelets in contact with a surface release of radioactive serotonin
Thrombin time	Platelet-poor plasma, bovine thrombin	conversion of fibrinogen to fibrin in blood exposed to a test surface after addition of thrombin
	Platelet poor plasma, bovine prothrombin	conversion of prothrombin to thrombin and fibrinogen to fibrin in blood exposed to a surface

Anticoagulated blood or platelet-enriched plasma is pumped through the column and platelet activity and morphology are compared before and after elution. The scope of this test is limited but it is used to screen and rank blood compatibility of a large number of materials (Figure 7.3).

Recirculation systems consist of inserting the material into a chamber through which blood is continually pumped in a closed system. Under conditions of controlled blood flow, hematologic changes that occur with time include loss of platelets and platelet adhesion to the test surface.

Rheological test chambers using a stirring bar, spinning disc, or other moving surface have also been studied. These tests evaluate the compatibility of blood proteins and platelets with a surface under a variety of controlled shear rate conditions (Figure 7.4).

Cell compatibility is evaluated in cell culture using a standard line of

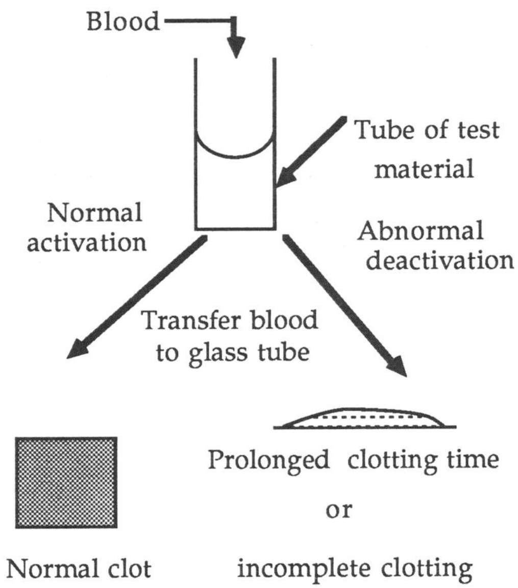

Figure 7.1. Deactivation of blood clotting system. Diagram illustrating transfer of whole blood exposed to a tube constructed out of a test polymer to a control glass tube. If blood exposed to the polymer tube does not clot within 3-5 minutes even after exposure to a glass surface then it is can be concluded that the polymer surface has inactivated or removed blood clotting factors. Polymer surfaces that inactivate the clotting system are not considered blood compatible; polymer surfaces that exhibit prolonged clotting times and do not deactivate the clotting system are deemed blood compatible based on the whole blood clotting time test.

fibroblasts or other cells. Cell attachment, replication, and protein synthesis is evaluated as discussed in Section 7.31.

7.2.2. *In Vivo* Methods for Evaluating Hemocompatibility

Although *in vitro* methods for predicting blood compatibility have made screening of a large number of candidate blood-compatible surfaces possible, *in vivo* tests are still the only totally satisfactory method to evaluate the ability of a material to maintain the fluidity of blood in the vascular system. *In vivo* methods involve implantation of test samples into a number of different test sites and in several different configurations. The surgical procedures used to prepare the animal for these types of procedures requires the use of anticoagulated blood. The material to be tested can be implanted for a predetermined period of time as a vascular graft or it can be tested in a vascular catheter or arterio-venous shunt, and in a cardiac assist device. The material is removed and the amount of clot and platelet aggregates attached are determined. In some cases the animal's platelets are radioactively labeled and the platelet half-life is used to assess the platelet survival time.

For long-term blood-compatibility evaluation, vena cava ring tests are largely used. Rings composed or lined with the material to be tested are inserted

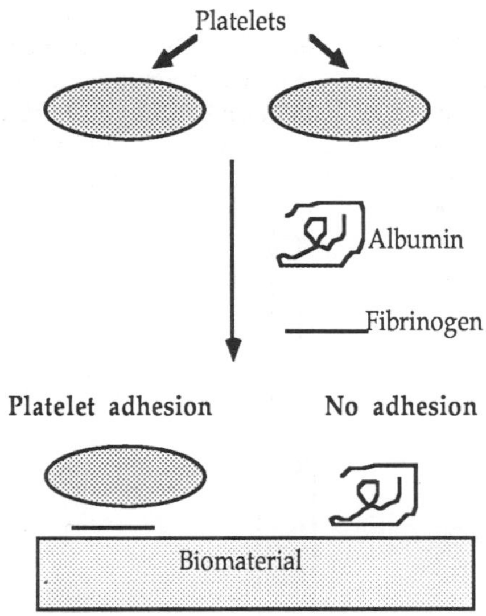

Figure 7.2. Platelet adhesion in the presence of blood proteins. Attachment of blood proteins to a biomaterial surface can activate (fibrinogen) or inactivate (albumin) adhesion of platelets. Diagram shown illustrates the binding of a platelet to a surface to which fibrinogen has previously been adsorbed.

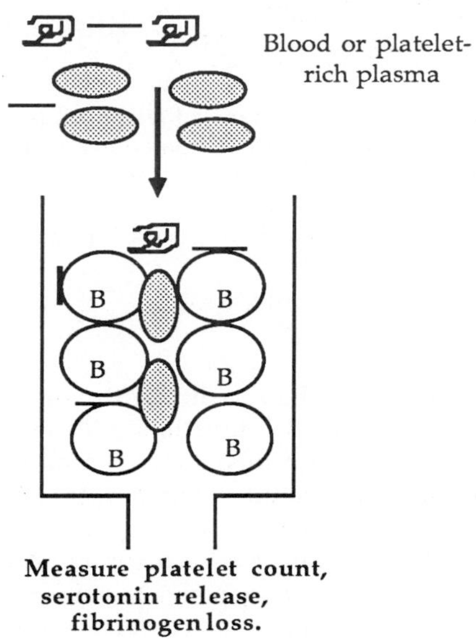

Figure 7.3. Evaluation of protein and platelet adhesion to a biomaterial using column chromatographic techniques. Diagram showing beads of a biomaterial (B) over which blood containing fibrinogen, albumin, and platelets is poured. Adsorption of blood components to the bead surface results in depletion of these components. The fractional depletion of these components after the blood is passed over the column is related to the blood compatibility of a biomaterial.

into the vena cava of a dog and removed after a predetermined period of time (between 2 hr and 2 weeks). The "patency" of the ring is then assessed by weighing the clot and by visually determining the fraction of the vessel that remains open. The degree of embolization can be measured by determining the amount of clot that has been filtered from the blood by the animal's kidney or lungs, when the ring is inserted into the aorta or vena cava, respectively.

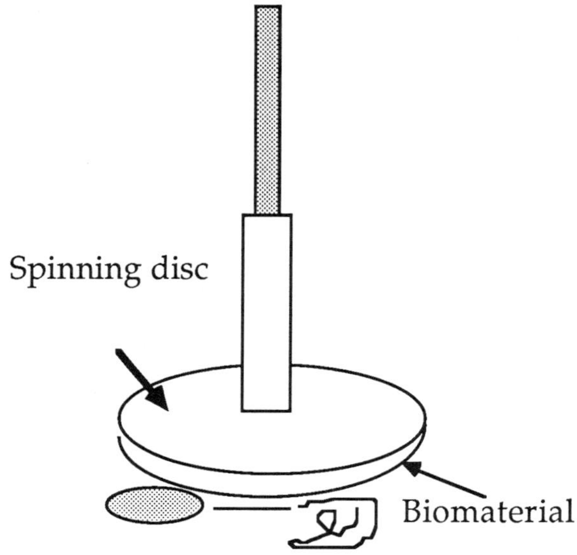

Figure 7.4. Evaluation of blood component adhesion under dynamic conditions. Diagram illustrates the use of a spinning disc composed of a biomaterial to be tested and the adhesion of blood components, including blood proteins and platelets, under shear conditions that simulate flow. Comparisons are made between adhesion to test and control surfaces; adhesion is minimized for materials used in contact with blood.

7.2.3. *Ex Vivo* Methods

The value of *in vitro* tests is the rapidity with which potentially useful materials can be evaluated for blood compatibility. The problem with *in vitro* tests is that they only predict blood compatibility and do not assure it. Animal testing is the only manner that blood compatibility can befully evaluated. However, since the concern for unnecessary animal testing is rising, alternate tests have been developed that are a compromise between *in vitro* and *in vivo* tests and do not expose animals to unnecessary pain and suffering. In *ex vivo* tests the blood of an animal is by passed via a cannulated vessel through a test chamber or section of shunt containing the material to be evaluated and the blood can then be returned to the animal if desired (Figure 7.5). Anticoagulants do not need to be used because of the limited handling of the blood prior to testing. A variety of configurations have been used for the shunt including open tube, floating plate, flow cell, or cells containing a branch or stagnation point. In all of these geometries the interface between the blood and a material is controlled.

Using an open system, a tube is inserted into the test animal's veins and

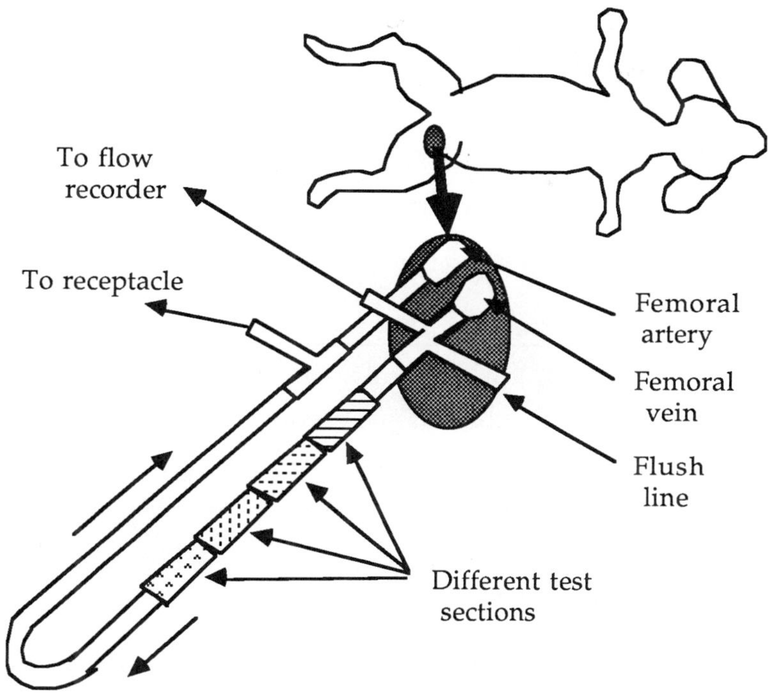

Figure 7.5. *Ex-vivo* shunt used for blood compatibility assessment. Diagram illustrates cannulation site and geometry of an *ex vivo* shunt used to pass blood over a test surface and return it to the host. The shunt is evaluated for protein and blood cell deposition.

blood is allowed to drip over the surface to be tested from the open end of the tube. The time for blood to stop flowing is measured (Dudley clotting time). In a closed system blood is removed from the animal through a shunt and returned to a vein. The flow velocity and surface shear within a shunt containing the material to be tested is controlled. The shunt may pass the blood over a rotating shaft or plate covered with a test material. Arterial-venous and venous-venous shunts that recirculate the blood back to the animal are used in closed systems to evaluate blood compatibility.

7.3. Toxicological Testing

The medical device amendment to the U.S. Food and Drug Act in 1976 requires that medical devices must show safety and efficacy under the intended use of the product. Several organizations publish guidelines for testing biomaterials and devices (American Dental Association and American Society for the Testing of Materials). Normally, short and long-term studies are required since biomaterials can be toxic to cells and tissues directly or via degradation products that only show up after chronic implantation (Figure 7.6).

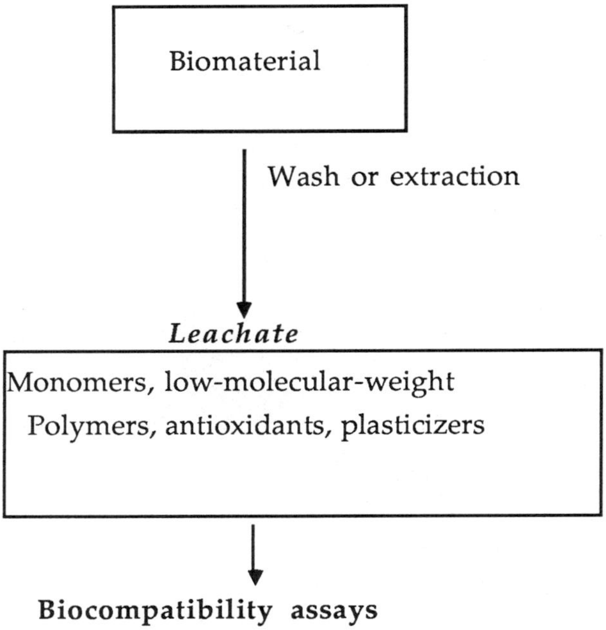

Figure 7.6. Extraction of leachable components from a biomaterial. Diagram illustrating washing and extraction of low- molecular-weight impurities from a biomaterial. Cell and tissue toxicity is normally present in this fraction as opposed to the high-molecular- weight polymeric fraction.

7.3.1. Primary Acute Toxicity Screening

Initial toxicity and biocompatibility screening assays are conducted in cell culture. Materials or extracts of materials in contact with cells under controlled conditions should not result in cell death. Below several specific assays are-presented in detail.

In the agar overlay assay (Figure 7.7) the material to be evaluated is placed in contact with the surface of an agar covered Petri dish. The agar is placed over a monolayer of mouse fibroblasts or chick embryo cells. A dye such as neutral red that is incorporated into living cells, is added into the culture medium. The loss of staining characteristics of the cells 24 hr after the stain is added to the medium indicates that the material or a leachable substance released from the material is toxic to cells. The size of the colorless zone can be measured and recorded as aquantitative indicator of toxicity.

The fluid medium assay (Figure 7.8) is conducted by cementing the material to be tested to the bottom of a plastic petri dish over which a monolayer of cells is grown. After 24 to 72 hr the cells are stained with a dye such as an aqueous crystal violet solution and the degree of destruction of cells is measured from the size of the zone of lysed cells.

Inhibition of cell growth (Figure 7.9) is accomplished by cultivating cells in tissue culture for 72 hr in the presence of biomaterials or their extracts. Inhibition of cell growth is measured either by counting the number of cells

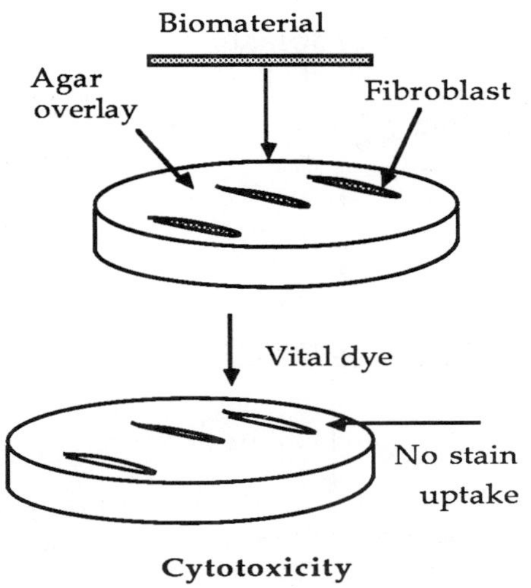

Figure 7.7. Agar overlay assay of cell biocompatibility. Diagram illustrating overlaying an agar gel containg fibroblasts and culture medium with a biomaterial. The cells are exposed to the biomaterial surface for a fixed period of time and then a vital dye is added to determine if cytotoxicity is observed.

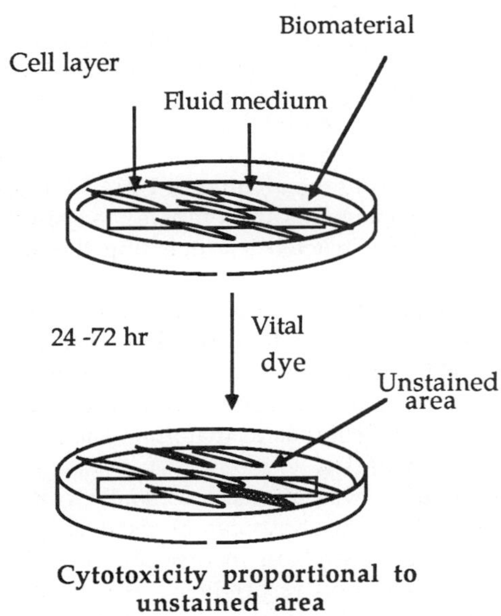

Figure 7.8. Fluid medium assay of cell biocompatibility. Diagram illustrating the growth of fibroblasts in cell culture in a Petri dish containing a biomaterial that is glued to the bottom of the culture dish. Cells are allowed to grow in culture for a predetermined time interval and a vital dye is added. Dead cells are unable to take up the dye and the unstained area is measured and considered to be proportional to the amount of cytotoxicity that is observed.

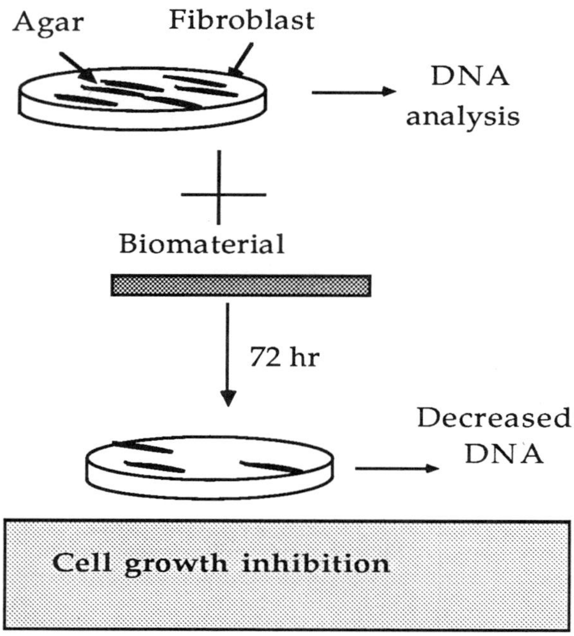

Figure 7.9. Inhibition of cell growth biocompatibilty assay. Diagram illustrating determination of DNA levels before and after exposure of a cell culture to a biomaterial surface. Inhibition of cell growth by a surface results in decreased levels of DNA compared with control tissue culture dishes.

microscopically or by measuring the amount of DNA. The results of cell growth studies in the presence of biomaterials are compared to growth of control cultures. A dose-response curve is constructed by repeating the assay using increased weights of a biomaterial.

Extracts of biomaterials are made for determining their cell growth effects by soaking the material to be tested in distilled water, physiological saline, polyethylene glycol 400, or cotton-seed oil. The extract containing the biomaterial is then autoclaved at 121°C for 1 hour or stored at 37°C for 5 days in sealed tubes. In the case of composite materials each component should be assayed separately for cytotoxicity.

Cell-culture cytotoxicity evaluations are usually run as a screening test. However, modifications of these assays can be used to quantitatively measure DNA and protein synthesis based on incorporation of ^{3}[H] thymidine and ^{14}[C]proline (or an amino acid mix), respectively. A variety of modified versions of the basic cell-culture assay are used for research purposes.

7.3.2. *In Vivo* Toxicity Tests

Once a material has proven compatible with cells the next step is to assay its effects on tissue. The tissue of choice is skin since exposure to a biomaterial can be tested easily. Specifically the material can be brought into contact with the skin or muscle to evaluate potential irritation or it can be injected or implanted under the skin in rats, guinea pigs, or rabbits. The material is implanted for periods of at least 4 days to study the acute inflammation that results. Longer-term studies are used to determine the time required for biodegradation as well as the associated chronic inflammation. Histological evaluation of the tissue reaction is conducted at sacrifice or on a tissue biopsy. In some instances the material is placed in a cage that is implanted beneath the skin. Tissue fluid is then periodically removed using a syringe and the types of cells present observed. Long-term subcutaneous implantation is necessary to evaluate the tumor producing activity of a material. Other models of wounds that are used to test toxicity include intraperitoneal implantation.

7.3.3. Safety Testing:Summary

The purpose of the tests described in this section is to ensure that the product or prototype product is safe, i.e., it does not induce cytotoxicity in culture or cause chronic inflammation or tumor formation when implanted subcutaneously *in vivo*. Although these tests are sufficient to establish the safety of a biomaterial, additional animal tests in end-use applications are necessary to demonstrate proper performance in the intended use. In some cases a biomaterial may have a potentially active leachate that under the intended use has no significant short term effects. For this reason long term animal testing is required to demonstrate complete safety.

7.4. Acute and Chronic Inflammation and Infection Testing

Short- and long-term implantation studies are essential to evaluate the efficacy of an implant. All implants cause some degree of irritation and the success of an implant depends to a large extent on minimizing the inflammatory response and preventing infection. Chronic inflammation results in tissue and/or implant destruction. Therefore it is important to evaluate short- and long term-inflammation induced by an implant.

7.4.1. Inflammatory Response Testing

The inflammatory response to an implanted biomaterial is most effectively evaluated using histological analysis of the tissue response to an implanted biomaterial. The analysis is performed using light and electron microscopic techniques that can identify cells including polymorphonuclear and mononuclear cells. The cells can be easily identified histologically by light microscopy after staining with hematoxylin and eosin (Figure 7.10) and at higher magnification by transmission electron microscopy. Digestion products and the implant/tissue interface can also be analyzed using routine histological procedures although the implant may be difficult to cut during histological processing.

Cell death and ingestion of implanted materials into inflammatory cells can be studied using transmission electron microscopy. Areas of tissue necrosis containing dead macrophages are identified and a grading scale is used to quantitatively analyze the inflammatory response. Specific histochemical

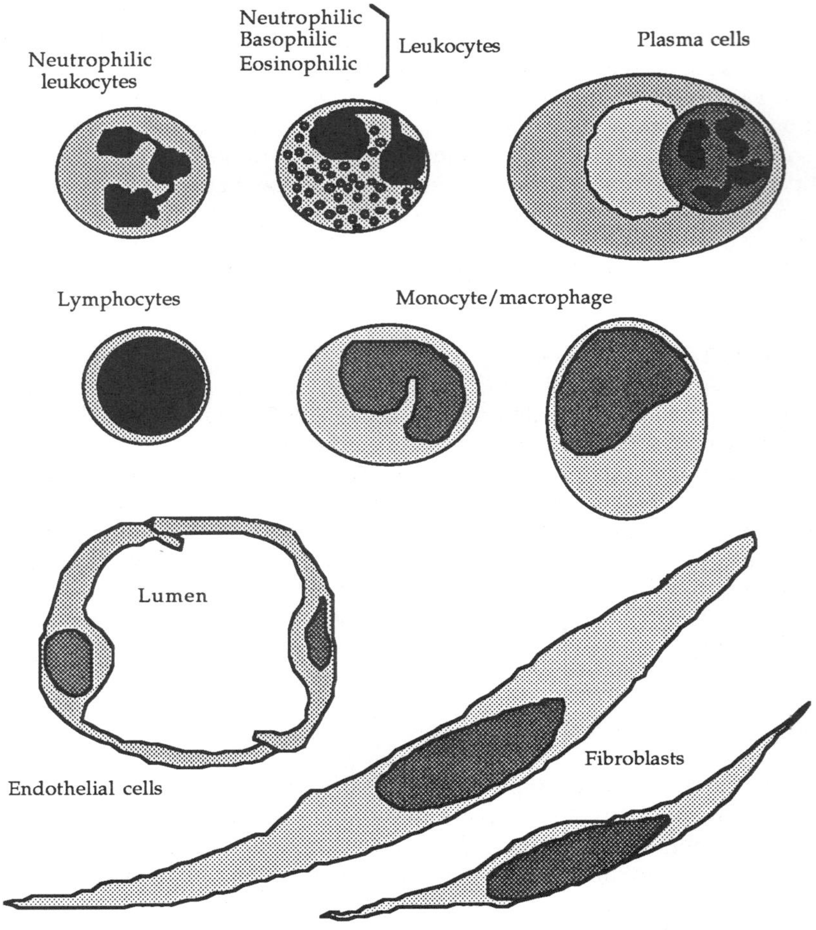

Figure 7.10. Cells observed surrounding implants. Diagram illustrating the types of cells that are seen in a histological section stained with hematoxylin and eosin surrounding an implant in a subcutaneous or intramuscular site. Cells observed include fibroblasts, endothelial cells, leukocytes, plasma cells, lymphocytes, and monocytes/macrophages.

stains are used to visualize the release of enzymes in contact with a biomaterial. Alkaline and acid phosphatase (from lysosomes) are the enzymes most widely analyzed.

Enzyme activity of cells at the implant—tissue interface is quantitatively assessed and used to grade the inflammatory response. Using frozen sections and microscopy it is possible to assess the phagocytic activity by measuring lysosomal hydrolase liberation at the interface. In the case of subcutaneously implanted cages, tissue fluid is used to analyze the enzyme activity.

Signs of chronic inflammation include persistence of inflammatory cells around or inside the implant, the presence of multinucleated giant cells, and persistence of a high degree of vascularization.

The inflammatory phase is generally followed by the formation of a fibrotic capsule containing collagen and connective tissue cells around the implant; the thickness of the capsule following the tissue reaction varies for each material tested. Measurement of the thickness can be used as quantitative data for biocompatibility evaluation. Porous implants are less likely than solid implants to be surrounded by a fibrotic capsule. The fibrotic capsule is a sign of a foreign body reaction.

7.4.2. Infection

Infection is a problem in implants that are sterilized using conventional methods such as autoclaving, gamma irradiation, and exposure to ethylene oxide gas. Although these methods are effective in killing viruses, fungi, spores, and bacteria, these techniques sometimes are ineffective in creating conditions of total sterility.

Infection is normally evaluated histologically based on observation of tissue-implant interfaces. The presence of leukocytes and positive staining for gram-negative organisms (bacteria) are evidence of infection. In some instances the samples are contaminated in a controlled manner with organisms prior to sterilization to evaluate the efficiency of sterilization. In addition, *in vitro* tests are used to evaluate whether contact with a biomaterial inhibits the ability of leukocytes to phagocytize and kill specific bacteria.

7.5. Immunological Testing

Most implants are composed of synthetic and natural polymers that are capable of acting as antigens and stimulating antibody production by plasma cells and activating or deactivating T-cells (see Section 4.25). Immediate hypersensitivity occurs in some animals after repeated exposure to an antigen such as bovine collagen. Circulating antibody-antigen complexes lead to tissue destruction. Decreased helper T-cell activity on exposure to a biomaterial may prevent normal defense against infection. Below we consider several tests that are used to assess the effects of contact of a biomaterial on the functioning humoral (antibody dependent) and cell-mediated (cell dependent) immune reactions.

7.5.1. Humoral Response Testing

Antibodies produced in response to an implant are measured by exposing an animal to either a single dose or several equally spaced intradermal doses. The implant is made into particles that are small enough to inject through a needle. Particles are mixed with phosphate buffered saline or Freund's adjuvant, which boosts the immune response to an antigen, and then are injected subdermally into a rabbit or other animal.

Immediate hypersensitivity is characterized by redness swelling and pain at the site of injection within 24 hr. Delayed hypersensitivity can occur 48 hr post—injection. Antibody production is maximized 14 days after injection. The animal can be reinjected (challenged) with particles of the implant 7 days after the initial exposure to heighten the immune response.

Fourteen days following final injection, blood is drawn from the animal and antibody levels are determined by exposing the antigen to the blood forming antibody—antigen complexes, and measuring the amount of the complex.

Several variations of this test are used. In the guinea pig maximization test the substance is injected with and without Freund's complete adjuvant intramuscularly. On day 7 after injection, the material is applied on a patch to the surface of the skin. A second patch of the material is placed on the skin 21 days after the first patch at a second site and the tissue reaction at the second site is observed after 24 hr. The skin reaction is graded in comparison to controls.

Sensitivity is also determined by placing a patch of the material on the skin and covering it with a water-impermeable dressing. Monomer, extracts of the polymer, degradation products, and the polymer itself can be tested in this manner. The skin reaction to repeated application of the material is graded in comparison to that of controls.

These tests are useful in identifying materials that result in acute hypersensitive reactions and delayed sensitivity. Slight differences in reactions to materials may be due to differences in polymer solubility and a variety of other factors. In practice, these tests are used to screen for materials that generate extreme immunologic reactions.

Insoluble antigens can also be measured by agglutination (Figure 7.11) using red blood cells covered with antibodies to the antigen. Dispersed red blood cells are clumped in the presence of red blood cells covered with antibody-antigen complexes. Antigens can be coupled to a carrier consisting of polystyrene beads. In the presence of antibody the antigen coated beads are agglutinated and form a precipitate. The amount of precipitate is determined as a function of antibody concentration.

Antigen-antibody complexes are also formed using immunosorbent assays in which the antigen is coupled to a substrate and the animal's blood is poured over the substrate. Antibody-antigen complexes are formed on the substrate and the complex is analyzed. Normally the antibody is chemically or radioactively labeled before the complex is formed. The amount of label that is found on the substrate is proportional to the amount of antibody-antigen complex formed (Figure 7.12).

7.5.2. Testing for Cell-Mediated Immunity

These tests are designed to detect the response of lymphocytes to soluble substances that are released from materials during the immune response. These tests usually involve preparing a soluble or colloidal extract of the material to be tested.

The lymphocyte transformation test uses a combined culture of T lymphocytes and monocytes in a medium containing the antigen or an agent that stimulates T-cell proliferation (mitogen) such as Phytohemagglutinin or Concavalin A. The amount of cell division determined using radiolabel techniques is an indicator of the degree of antigen recognition and reflects the memory to that antigen.

Other studies test the normal function of lymphocytes in response to the challenge of a known stimulus. Antigens or toxic leachable materials from

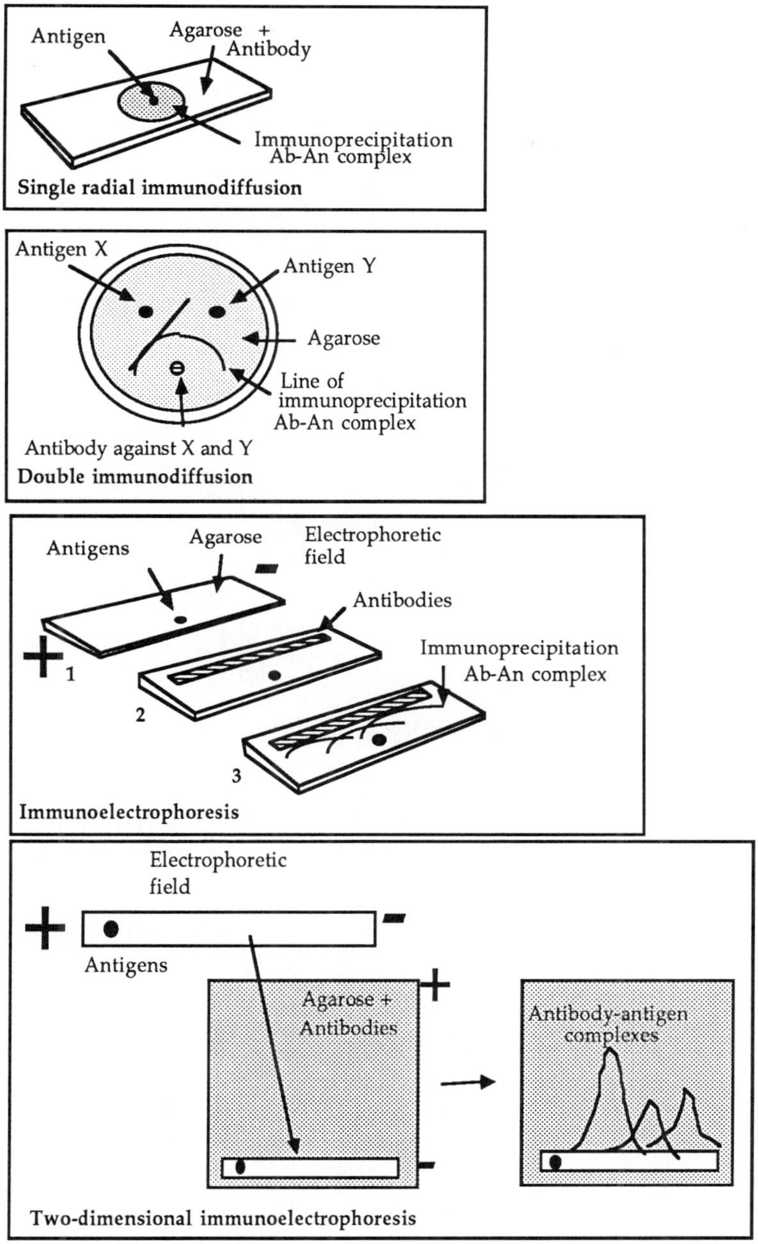

Figure 7.11. Immunological tests used with insoluble antigens.
Diagram illustrating the types of immunological tests that are used to determine the presence of antibodies in the blood of animals that have been immunized with a biomaterial. These tests include single radial immunodiffusion, double diffusion, and immunoelectrophoresis in one and two dimensions.

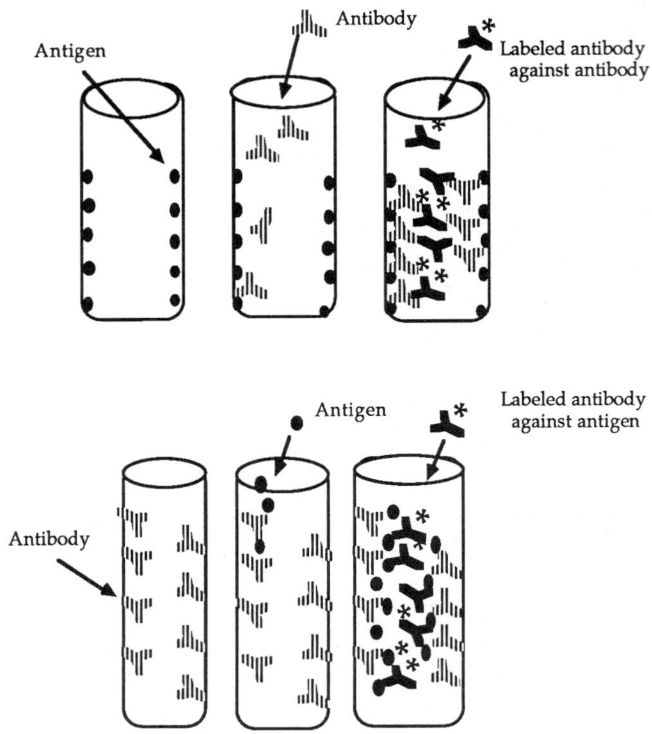

Figure 7.12. Immunoabsorption assay for charaterizing antibodies to an insoluble antigen. Diagram illustrating the binding of radiolabeled antibody to an insoluble antigen that is coated on a surface of a tube. Alternately the antigen is poured over a tube that has been previously labeled with radioactive antibody. Antibodies are labeled with radioactive elements or gold.

implants may inhibit normal lymphocyte function and therefore compromise the inflammatory response. Activated lymphocytes in the presence of an antigen, release factors such as migration inhibition factor (MIF) that inhibits migration of monocytes (Figure 7.13). Lymphocytes and macrophages secrete lymphokines that prevent migration of monocytes or polymorphonuclear cells away from the site of contact. Secretion of these factors is part of the normal inflammatory process(see Section 4.4.).

7.6. Interactions of Materials and Their Degradation Products with Cells and Tissues

Although a battery of tests exist for evaluating the interaction between materials and tissues, care must be taken to evaluate the results of these tests. Sometimes misleading results arise because of surface impurities or the lack of careful cleaning of implant surfaces. The biomaterials scientist must always carefully consider and interpret the results and not use biocompatibility protocols as "black box"procedures.

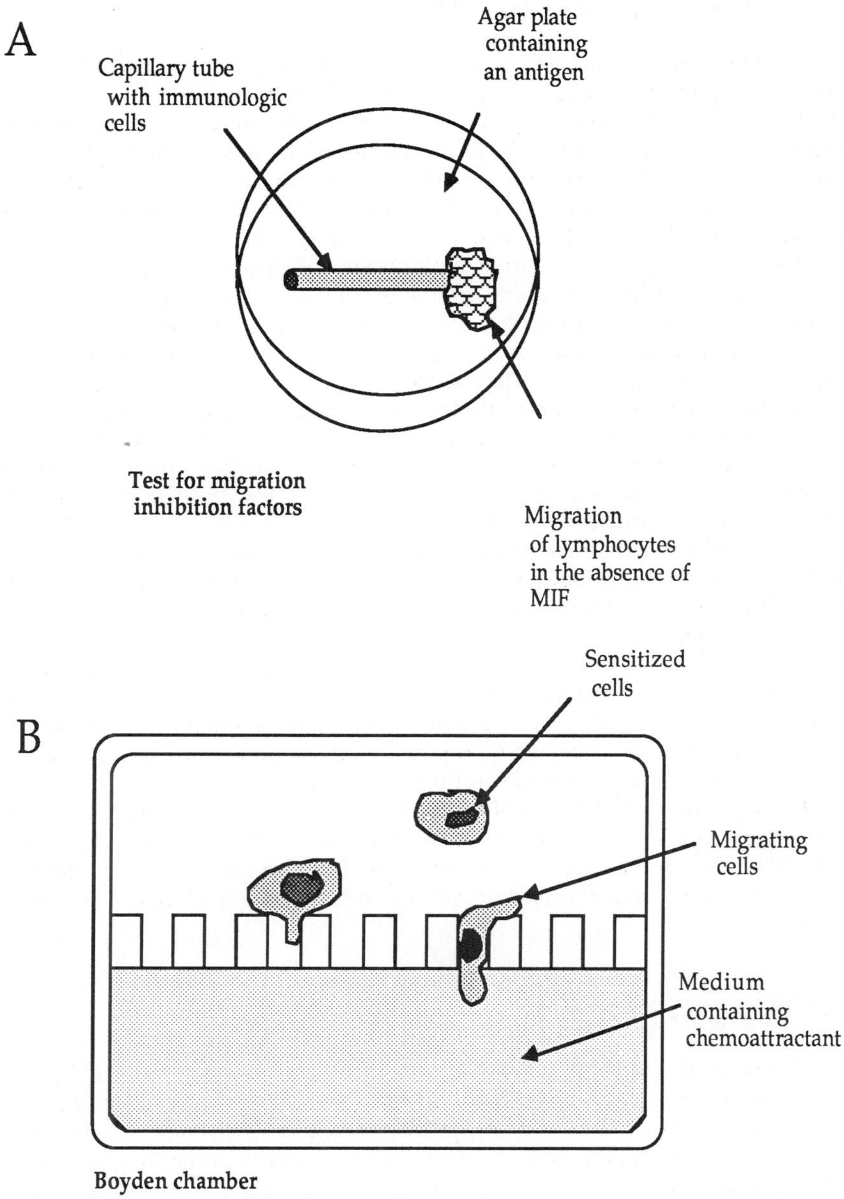

Figure 7.13. Assay for cell-mediated immunogenicity of a polymer. (A) Diagram illustrating the use of a capillary tube filled with immunological cells in a tissue culture plate containing the antigen. Migration of lymphocytes from the tube is observed. Another test (B) involves a Boyden chamber in which migration of sensitized lymphocytes to the biomaterial, extract, or other chemoattractant contained in the lower chamber is observed. If no migration is observed then it is concluded that activated cells are secreting migration inhibition factors.

7.6.1. Protein Adsorption on Biomaterials

Protein adsorption to biomaterials can alter the surface properties and modify compatibility with cells and tissues. Proteins are present in the blood and in the extracellular fluid phase. Protein adsorption to a biomaterial occurs instantaneously on contact with living tissues. Adsorbed protein modifies cell behavior by causing a conformational change, by inducing enzymatic activity and by activation of complement or coagulation factors that can lead to tissue destruction and clotting or be toxic to surrounding tissues.

Protein-adsorption measurements consists of 3 stages: (1) the exposure of a material to a solution containing the ^{125}I-radiolabeled form of the protein of interest (2) removal of unadsorbed protein by rinsing, and (3) measurement of the remaining radioactivity retained by the surface. Protein adsorption is a function of the surface chemistry and surface morphology.

Surface chemistry and morphology is evaluated using scanning electron microscopy, nuclear magnetic resonance, contact angle measurements, spectroscopy and differential scanning calorimetry. Protein adsorption measurements are sensitive to the presence of surface films, methods of handling of the surface and the source of the radioactive protein.

The surface of the material to be tested should be cleaned and prepared carefully including degassing of the solvents. The protein preparation should be of a high purity and stability and can be radiolabeled in the laboratory using a standard protocol.

Biomaterials in the form of films and tubes can be tested for static adsorption and adsorption at a fixed flow rate. Adsorption is studied at increasing concentrations of a single protein for a fixed time (adsorption isotherm) or with two proteins (competitive adsorption from binary mixtures). Whole blood or purified protein solutions are studied.

In situ radioiodination involves radiolabeling of proteins after adsorption onto a surface. The adsorbed proteins are then eluted and analyzed by gel electrophoresis. Similar techniques can be used *in vivo* where anticoagulants are used to prevent clotting and biomaterials can be ranked according to the degree of protein adsorption.

7.6.2. Biodegradation Products

Polymers degrade in vivo due to hydrolysis reactions, attack by free radicals, and enzyme-mediated breakdown. Degradation may occur rapidly in condensation polymers and synthetic polymers containing peptide bonds or it may require periods of months to years. In the case of polymers that degrade slowly, it is necessary to develop methods that accelerate chain cleavage in order to evaluate the effects of degradation products on cells and tissues.

Radioactively labeled polymers can be incubated with tissue homogenates or cells and tissues in culture. The rate of release of radioactivity into the medium is assayed and compared to standards. Release of labeled products into the medium can be used to analyze quantitatively the percentage of the implant that is solubilized.

Degradation products are tested in cell culture by measuring their effects on cell replication, cloning efficiency, cell viability, cell morphology, and accumulation of intracellular degradation products. Studies are performed using phase-contrast microscopy, labeling techniques, and scanning and transmission electron microscopy.

Material characterization includes molecular weight (before and after implantation), chemical composition, and mechanical properties.

7.7. Testing for Carcinogenic Potential

The potential to induce cell injury that ultimately may lead to cell transformation and tumor formation is associated with the long term use of any implant. A variety of tests are used to detect mutations in the genetic material of a cell induced by a biomaterial.

Mutation potential can be studied by exposing a test surface or a leachate to mammalian, bacterial, fungal, or yeast cells in culture. Exposure to the test surface results in inhibition of cell replication if a zone of inhibition is observed. The amount of inhibition can be quantitatively determined; however, a chromosomal map must be constructed to identify the mutation.

Mammalian cell transformation can be tested by exposing a specific cell line to a material and injecting passaged cells into an animal. The animal is checked periodically for signs of tumors. Critical to this evaluation is the choice of an animal (normally rats are used), site of implantation (subcutaneous, pulmonary tissue, abdominal cavity), the geometry and porosity of the implant, and the duration of exposure (usually at least 2 yr). Animals are sacrificed at predetermined time intervals or when they become morbid, and gross pathology and histology are performed on all tissues.

7.8. Summary

It is implicit in the discussion of biocompatibility that the material be well characterized physically and chemically from the outset. Characterization includes chemical composition, melting point, glass transition temperature, water and lipid solubility, molecular weight, degree of crosslinking, and physical structure. Techniques such as scanning microscopy, nuclear magnetic resonance, infrared spectroscopy, differential scanning calorimetry, thermogravimetric analysis, thermomechanical analysis, and electron spectroscopy are used in analysis of polymers.

Biocompatibility tests involve material contact with cells in culture, implantation in animals, and skin patch testing. Of primary concern in these tests are the screening of materials to establish safety. Safety is defined as the lack of cytotoxicity in cell culture, absence of circulating antibodies in immunized animals as well as the lack of hypersensitivity or prolonged inflammation. Long-term studies of tumor-generating potential are normally conducted after the implant has been shown to have efficacy in an end-use application.

The efficacy of a biomaterial or device is usually evaluated in the end-use application. Several end-use applications are discussed in the following four chapters.

Bibliography

Anderson, J.M. and Miller, K.M. Biomaterials, 5,5,1984.

Adsolom, D.R., Zingg, W. and Newmann, A.W. J. Biomed. Mat. Res. 21,161,1987.

Benson, W.H. and Stockhouse, R.A. Drug Chem. Toxicol. 9,275,1986.

Galletti, P.M. and Boretos, J.N. J. Biomed. Mat. Res. 17,539,1983.

Hench, L.L. and Wilson, J. Science, 226,630,1984.

Johnson, H.J., Northup, S.J., Seagraves, P.A., Garvin, P.J. and Wallin, R.F. J. Biomed. Mat. Res. 17,571,1983.

Marchant, R., Hiltner, A., Hamlin, C., Radinoitch, A., Slobodkin, R. and Anderson, J.M. J. Biomed. Mat. Res. 17,301,1983.

Pizzoferrato, A., Vespucci, A., Giapetti, G. and Stea, S. Biomaterials 6,346,1985.

Williams, D.F. Techniques of Biocompatibility Testing, CRC Press, Boca Raton, FL, 1986.

Part II. Biomaterial End–Use Applications

Chapter 8

Wound Dressings and Artificial Skin

8.1. Introduction

Thermal, chemical and mechanical loss of skin results in functional and cosmetic changes that if left uncorrected affect the lives of many each year.

The role of skin implants (sometimes referred to as skin substitutes, artificial skin or wound dressings) is (1) to replace the function of lost skin, (2) to protect wounds from protein and fluid losses, prevent bacterial invasion, and dissipate mechanical stress (external), and finally (3) to improve and stimulate wound healing. As mentioned in Chapter 4, the presence of dead tissues delays wound healing and increases the risk of infection. Initial wound management is the first procedure to be performed before implanting a skin substitute and consists of wound debridement, disinfection by topical antimicrobial agents, and excision of necrotic tissues (more commonly used in burn wounds). The exact treatment protocol depends on the severity of the wound (Figures 8.1 and 8.2).

The properties required of skin implants are summarized in Table 8.1. Adherence to wound tissue is one of the main properties required of a skin substitute because this prevents bacterial growth and completely protects the wound. Adherence depends on the conformability, hydrophilicity, and morphological structure of the implant. Complete incorporation of the implant within the tissue is useful if the material is biodegradable. Proper moisture balance is also required to prevent either excess fluid build-up or dessication of repair tissue. The transmissibility of gas and water vapor should be characterized. Other characteristics described in Table 8.1 are also necessary.

A variety of approaches have been used for repair of the form and function of damaged skin. Perhaps the simplest device to accelerate the repair of a skin wound is a bandaid. The bandaid consists of an absorbant material to remove pus and other materials from the wound and an adhesive polymer film that maintains contact of the absorbant material with the wound. The polymer film also prevents moisture loss from the wound and in this manner aids in repair of

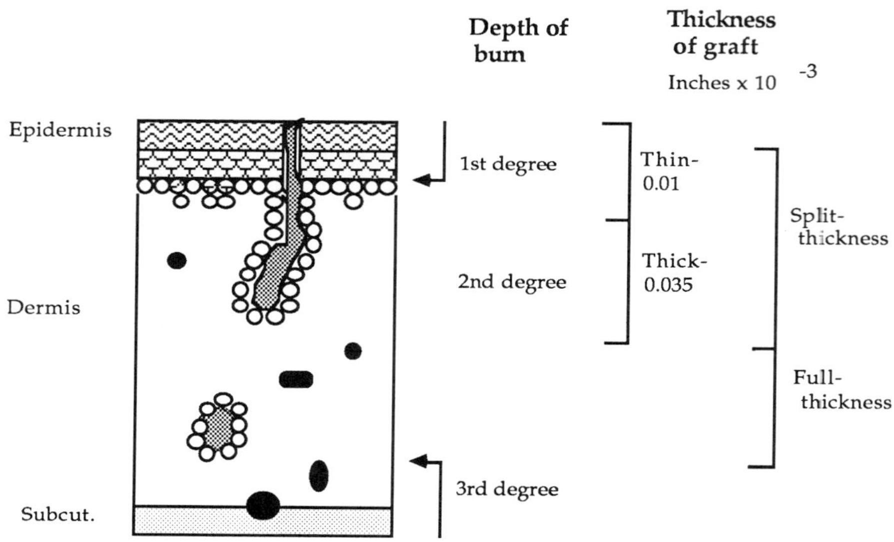

Figure 8.1. Classification of burn injury. Diagram illustrating depth of burn, graft thickness, and location within the skin. Split-thickness grafts used to cover burn wounds include epidermis and a portion of the dermis while full thickness grafts include epidermis and most of the dermis

shallow wounds that involve loss of the epidermis.

Extension of the bandaid approach involves covering large open wounds with a synthetic polymer film that prevents heat and moisture loss and allows the wound to heal. This approach, which is discussed further below, is a temporary solution that "buys time" for the surgeon. Ultimately this approach requires skin grafting of the patient's own skin over the area denuded of epidermal cells.

Numerous studies have been reviewed describing the use of synthetic polymers for skin implants (see Quinn *et al.*, 1985; Wise, 1985). These materials are useful for repair of shallow wounds of small surface area. They also can be used to cover wounds of large surface areas for periods of several weeks while granulation tissue forms at the base of the wounds. Granulation tissue serves as a good base over which a skin graft can be placed. These materials are passive wound dressings since they are not incorporated into the wound.

A modification of the bandaid approach involves production of a composite of a biological polymer and the polymeric membrane. The biological polymer is prepared in a form that encourages ingrowth of dermal cells while the polymeric coating maintains heat and moisture in the wound. Again the ultimate success of this procedure requires successful autografting of the patient's own skin. The benefit of this approach is that repair tissue is promoted, upon which a thin-thickness autograft can be used to cover the denuded areas of skin.

Another approach involves culturing epidermal cells from the patient in tissue culture for periods of three to four weeks or until a continuous sheet is formed. This sheet is then used to cover areas of skin that had been previously covered with a passive wound dressing.

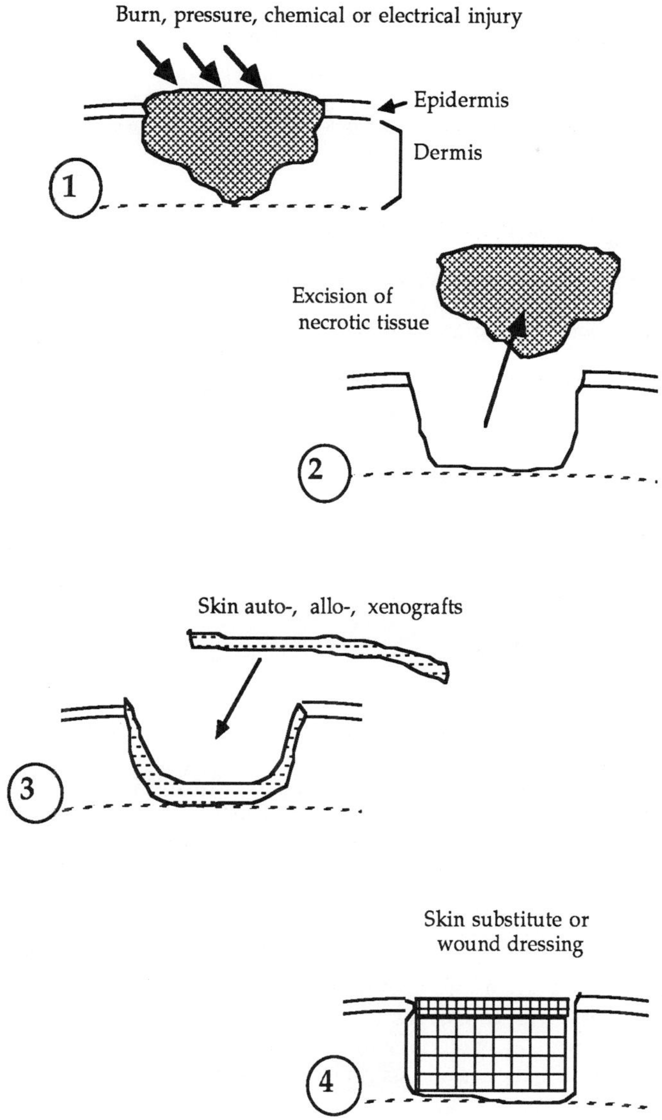

Figure 8.2. Steps in treatment of burns. Burn treatment involves a series of steps including loss of epidermis and dermis (1), excision of necrotic tissue (2), protection of shallow wounds with a skin graft (3), or temporary wound coverage with a wound dressing or skin substitute (4).

Table 8.1. Properties of an Ideal Wound Dressing

Protection from microbial invasion from the environment
Optimal water permeability to prevent either dessication of the wound or fluid accumulation under the covering
Capability to adhere well to the wound and to be readily removable without causing any damage to the tissue beneath the cover
Prevention of excessive formation of granulation tissue
Optimal elasticity to facilitate an intimate cover of the wound
Readily available material, in any size and thickness
Nonantigenic
Nontoxic
Capable of being sterilized
Stability during storage
Drug delivery reservior for any medication (antibiotics)
Protects against mechanical insults

A modification of this approach is to co-culture the patient's dermal cells (fibroblasts) and epidermal cells on a biological support such as a collagen sponge and then transplantion of the combination onto the patient. The benefit of this approach is that the patient has functional dermal and epidermal components of skin. The disadvantage of this approach is that the skin lacks normal appendages such as hair follicles, pigments, sweat glands, and a complement of immunologically active cells found in normal skin.

Skin autografts are the method of choice for patients with body surfaces that are largely covered by healthy skin. The disadvantages of skin grafting are (1) that it requires a surgical procedure to harvest pieces of skin, (2) donor sites are themselves open wounds and may become scarred, and (3) skin grafts sometimes die.

The remainder of this chapter is devoted to reviewing the types of wound dressings that are used as coverings for skin. These include materials classified as biological tissues, biological tissue derivatives and wound coverings.

8.2. Biological Tissues as Dressings

A variety of biological tissues have been used as skin replacements and wound dressings. These include skin autografts, allografts, xenografts, amniotic membranes, and other tissues. Skin autografts have the benefit that they provide a permanent epidermal covering. Below, each of these materials are discussed in detail.

8.2.1. Skin Autografts

Skin autografts (Hunt *et al.*, 1979; Tavis *et al.*, 1977; Alexander *et al.*, 1981) consist of either epidermis and a thin layer of dermis (thin-thickness graft) or epidermis and dermis (thick-thickness graft). A skin autograft can be transplanted onto a clean wound bed, but most of the time the grafting is performed after the wound bed is scraped clean of dead tissue and pus. Adherence to the wound bed as well as revascularization of the skin autograft are increased when healing is allowed to initiate over a 3- to 6- week period. Healing occurs in the presence of synthetic or semisynthetic dressings used to provide a temporary covering for the wound bed. These dressings aid healing by maintaining heat and humidity at the wound site.

Skin autografts can be expanded (1.5 - 9 times their original size) into a mesh (Alexander *et al.*, 1981) using a device termed a tissue expander. The advantage of using meshed skin is that bacteria, serum, and blood are easily washed out through the holes in the mesh. Different skin-graft thicknesses are used for specific surgical applications.

Adherence of a skin autograft to the underlying recipient bed is normally accomplished 3-4 days postimplantation. Skin autografts can be stored at 4°C in culture medium for periods up to 2 weeks prior to transplantation. The biology of tissue transplantation is discussed in Chapter 10.

Skin autografts are rapidly invaded after implantation by capillaries from the wound bed. These capillaries are necessary for supply of nutrients and survival of the graft.

8.2.2. Skin Allografts

When skin autografts are unavailable and a temporary dressing is needed to cover the wound, skin allografts can be used fresh (from living donor) or, more commonly, after thawing (frozen cadaver-skin allografts). Frozen cadaver skin (May and DeClement, 1981) is prepared using cryoprotectants such as dimethyl sulfoxide or glycerol that allow storage in liquid nitrogen at -160°C for periods of 6 months or longer. Skin is then rapidly thawed to body temperature and washed with saline solution the day of transplantation.

The harvesting procedure for skin banking requires extra surgery. Decreased numbers of donors and the difficulty involved in collecting skin limits the use of this biological tissue. In addition, the rejection reaction associated with cadaver transplants limits the duration of their use to 2-4 days. Bacterial contamination is another problem that is associated with prolonged use of cadaver skin.

Implantation of a skin allograft results in its vascularization; however, if the allograft is used to cover the wound permanently, rejection occurs leaving behind only the collagen of the dermal portion. Allografts are removed and replaced every 2 or 3 days because of the rejection problem and can be expanded as described for skin autografts.

Allografts are used by many burn surgeons as a temporary covering to keep the wound clean and to allow formation of granulation tissue upon which skin autografts can be placed.

8.2.3. Skin Xenografts

Frozen, irradiated, and dried skin grafts from animals are the most useful xenografts; lyophylized porcine skin can be stored in a dehydrated state at room temperature for long periods of time (Koff *et al.*, 1978; Morris *et al.*, 1979). Due to the rejection reaction these grafts are used in contact with the wound for periods no longer than 2 days. They are replaced at regular intervals with fresh xenograft material. Adherence of a xenograft is similar to that observed with an allograft; however, the graft is more sensitive to enzymatic degradation by enzymes secreted from wound tissues and bacteria compared to allografts.

Variable results have been reported for burn wounds covered with xenografts. In some cases the xenograft is chemically treated after removal of the epidermis and skin appendages to prevent the rejection process. Dermodress is such a product treated with dialdehydes to modify the antigenic determinants that give rise to the rejection process.

8.2.4. Amniotic Membrane

Amnion is a biological membrane that surrounds the embyro during development and has been used as a burn dressing (Robson *et al.*, 1974; Walker *et al.*, 1977). This membrane can be obtained after delivery of the baby and is washed with a chemical preservative such as sodium hypochloride (Thompson and Parks,1984). It is 0.02 - 0.50 mm thick and contains an epithelial cell layer, connective tissue, and fibroblasts. Because of durability problems the amnion is sometimes difficult to handle. In addition, amnion will elicit a rejection response. Normally, amnion is removed three days after grafting because of the problem associated with rejection.

8.3. Dermal Repair Materials

The rejection problem associated with allografts, xenografts, and amnion resulted in the recognition that temporary wound dressings were needed to protect tissues prior to skin grafting. Noncellular biological materials do not stimulate the rejection process and can be incorporated into wound repair tissue. In contrast, synthetic polymers can be manufactured using continuous processes by cost effective means. Synthetic and noncellular dressings will be considered below in detail.

8.3.1. Synthetic Polymeric Dressings

Synthetic polymers derived from petroleum products can be easily manufactured using conventional technology into films, fibers, sheets, and sponges. For this reason these materials have received attention as potential wound dressings for deep wounds.

Synthetic polymers have several advantages such as ability to adhere to the wound edges, ability to drape to the wound contour, and ease of use. The major

Table 8.2. Synthetic Polymeric Wound Dressings

Traditional dressings
Gauze
Lint
Gauze swab
Nonwoven viscose swab

Paraffin gauze dressings
Open mesh nylon fabric
Nonwoven pad coated with aluminum

Foam
Laminated polyurethane foam (covered with a polyurethane film or a microporous Teflon)
Nonlaminate, non-adherent foam (polyurethane)
Closed-cell polyurethane foam
Silastic foam
Poly(ε-caprolactone) foam (covered with a polyurethane film)

Film dressings
Polyurethane
Copolymer of dimethylaminoethyl methacrylate and acetonitrile
Hydroxy ethylmethacrylate (HEMA) and silastic film
Poly (ε-caprolactone) film
Poly(ethoxyethylmethacrylate)film
Silicone polymer membrane
Poly(hydroxyethyl methacrylate)(HEMA) and polyethylene glycol

Granular or past dressing
Dextran polymer
Activated charcoal cloth actisorb

Hydrocolloids and hydrogels
Natural hydrocolloid and polyurethane foam layer
Hydrogel:Polyacrylamide and Agar
Sodium carboxymethylcellulose and polyurethane film
Gelable polysaccharide (agarose crosslinked with a polyacrylamide)
Graft T starch copolymer
Polyethylene oxide copolymer - poly (oxypropylene)
Poly (acrylamide) and polyurethane film

Other
Biobrane;silicone rubber membrane; knitted nylon fabric + porcine collagen peptides

disadvantage is the lack of biological properties such as enhancing wound healing via attraction of cells involved in the healing process.

These dressings are used as coverings for deep (full-thickness) burns and skin ulcers. In these applications synthetic polymeric dressings create an inert environment that controls water and heat passage from the wound while preventing bacterial infiltration (Table 8.2). The most widely used dermal dressings are in the form of films, foams or hydrogels (Walder *et al.*, 1969; Winter, 1975; Eaglstein and Mertz, 1978;Wood *et al.*, 1977; O'Neill, 1973; Tavis *et al.*, 1979).

Foams are synthetic polymer chains that are highly spaced by introduction of a gas (foaming agent). The foam is prepared under pressure or conditions that allow for the expansion of the constituent polymer chains. Under the appropriate conditions the gas is released, resulting in polymer chains that are highly separated by gas molecules (Figure 8.3).

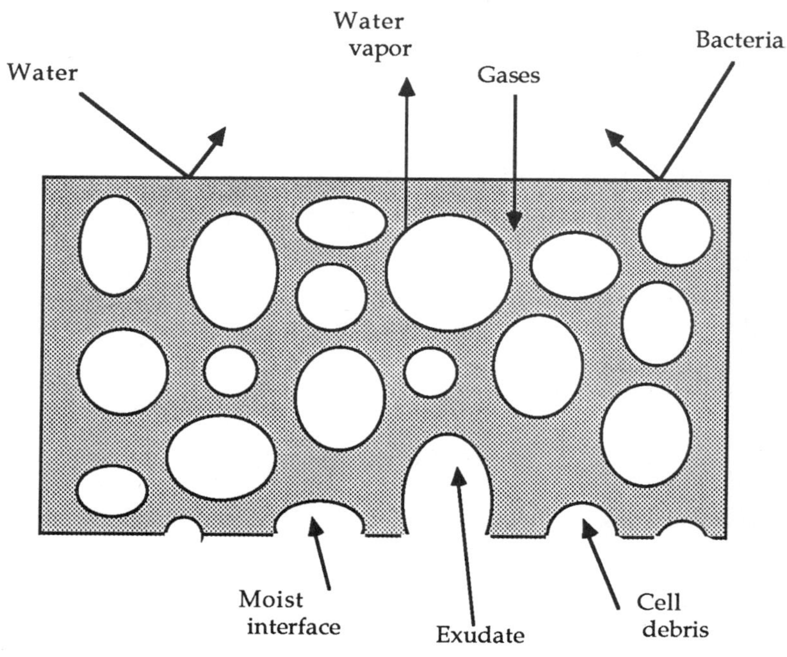

Figure 8.3. Structure of a polymer foam. Polymer foams are formed by release of a gas during processing of a polymer matrix leading to the formation of pores in the structure into which cell debris and tissue exudates are adsorbed. The foam is designed to control the rate of water vapor loss from the wound and prevent bacterial invasion.

Hydrogels are similar to foams except that the spaces between the polymer chains is filled by water molecules (Ratner and Hoffman, 1975; Dressler *et al.*, 1980). The high degree of swelling provides spaces between polymer chains and results in a material that contains mostly water. Hydrogels with high degrees of swelling have physical properties similar to those of soft contact lenses.

Hydrogels are sometimes mechanically supported by a mesh fabric. They are available in dry sheets or in the form of granules that can absorb wound fluids.

Hydrogels absorb more fluid than hydrocolloids. They are impermeable to bacteria, but permeable to oxygen and water vapor but do not adhere to the wound bed. Antimicrobial agents can be incorporated into hydrocolloids and hydrogels.

Colloids consist of a suspension of finely divided particles in a continuous medium. The particles are the dispersed phase or the colloid. If the dispersing medium is water, the colloid is termed a hydrocolloid (Figure 8.4).

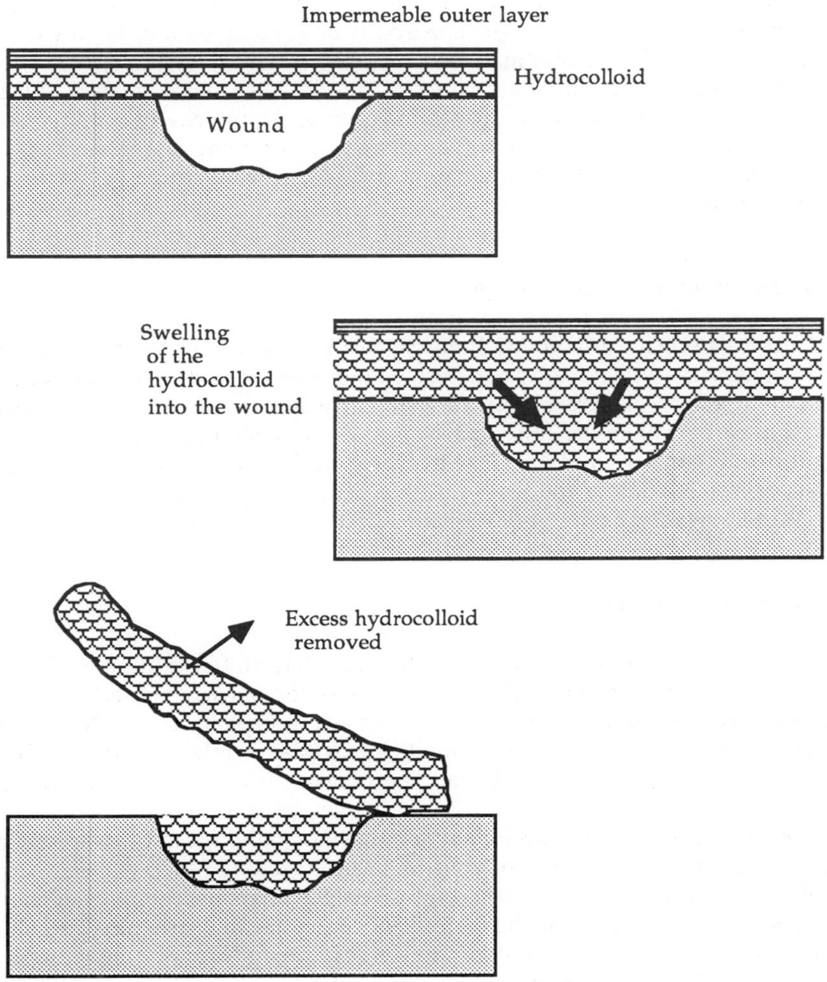

Figure 8.4. Use of a hydrocolloid gel. Shallow wounds can be covered with a hydrocolloid gel with an impermeable outer layer. Once in contact with the wound the hydrocolloid swells and excess hydrocolloid is removed.

8.3.1.1. Foam Dressings

The main properties of foams are their conformability, porosity, and insulation and increased adherence to wounds. Polymers such as polyurethane are used to make foam that is stabilized after contact with the wound. These materials are used to fill the wound cavity and when placed in contact with the wound they present a smooth, hydrophilic porous surface. Excess water (exudate composed of wound fluids) and cell debris are absorbed and retained inside the foam. In addition, foams protect the wound from excess pressure (as observed in skin ulcers or pressure sores). The external surface of foams can be treated to increase hydrophobicity and to protect against bacterial proliferation. Water vapor and gas transmission can be adjusted by varying the volume fraction of polymer and crosslink density. The one disadvantage of a foam is the low tensile strength and lack of structural integrity.

Other foams are made of a mixture of silicone base and a catalyst. These foams expand and swell to fill the wound cavity. As the cavity diminishes in size, the foam is replaced. Such foams are not adherent to the wound and consequently do not cause pain when removed. Foams made of polyvinyl alcohol (Ivalon) can also be used.

8.3.1.1. Hydrocolloids and Hydrogels

These materials consist of an impermeable outer layer made of a polymeric film such as polyurethane and an adhesive inner surface. Natural hydrocolloids are made in a form of films or particles, similarly synthetic hydrocolloids (made of sodium carboxymethylcellulose particles) are available in sheet, powder, and paste forms (Figures 8.4 and 8.5). The hydrocolloid is placed onto the wound and begins absorbing exudate and wound fluids. The final product is a soft gel that remains in the wound cavity until it is rinsed out with saline solution.

8.3.2. Biological Polymers

A variety of biological polymers including collagen, fibrin, fibronectin, and hyaluronic acid have been studied as dressings for dermal wounds. Unlike synthetic polymers, which at the very best act as inert coverings for the wound, biological polymers have unique properties that play a role in normal wound healing. Of the polymers listed above, type I collagen has been evaluated most extensively.

Collagen is the major noncellular component of skin. It can be purified and reconstituted into a wide variety of physical forms including films, sponges, tubes, and fibers (Chvapil *et al.*,1973). The high strength of reconstituted collagenous materials is a consequence of the rigidity of the molecular structure, crosslink density, and parallel alignment of molecules. In addition, the biological properties that make it suitable as an implant include low antigenicity, controllable biodegradation rate, and enhanced wound-healing properties (Pachence *et al.*,1987).

Type I collagen can be extracted from animal and human skin and tendons in quantities large enough for commercial production of wound dressings. Safety and efficacy studies of type I collagen as a hemostatic agent, injectable tissue filler, and other applications support further biomaterial development.

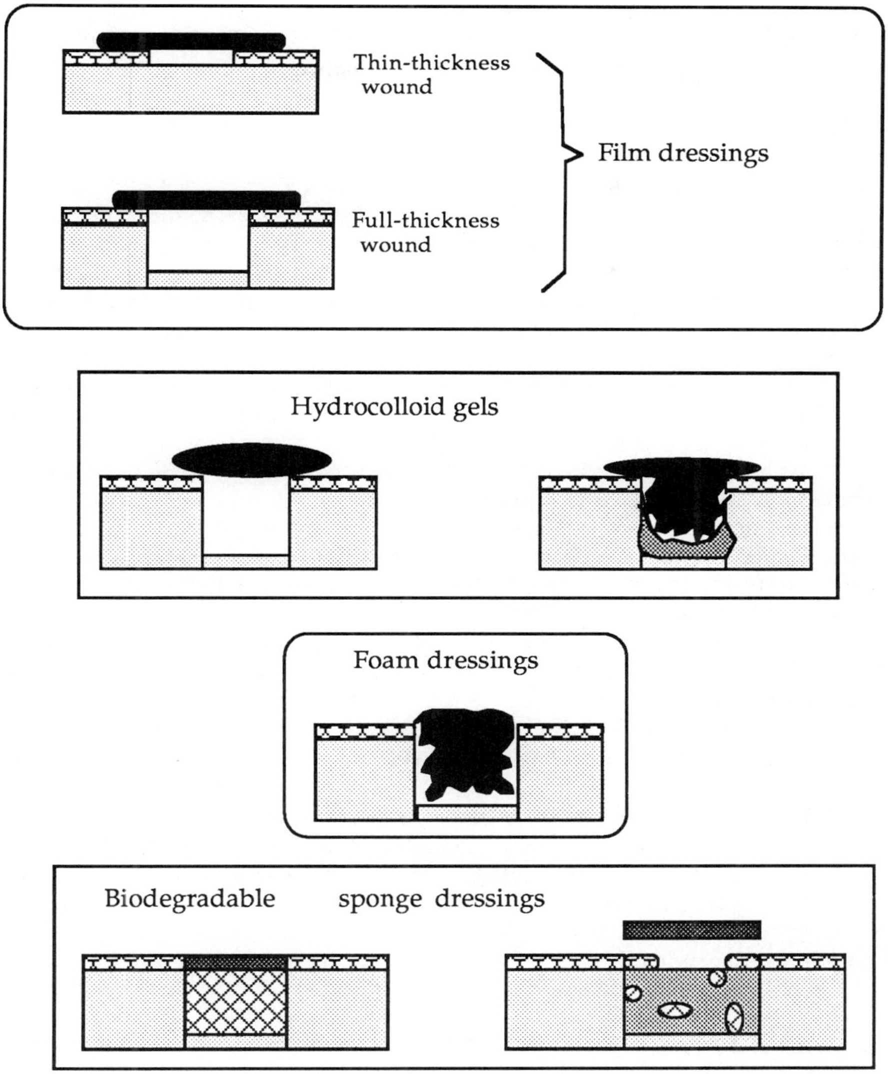

Figure 8.5. Types of wound dressings. Dressings include films used to cover thin and full-thickness wounds, hydrocolloid gels, foam dressings, and biodegrable sponge dressings. Biodegradable sponge dressings are the only classification that does not need to be removed from the wound.

8.3.2.1. Collagenous Materials

A number of research groups have developed porous collagenous materials that aid in wound debridement and as awound dressing (Chvapil, 1977; Yannas and Burke, 1980; Doillon *et al.*, 1984, 1986). Each of these materials employs a three-dimensional porous structure (sponge) in the design of a dermal dressing. Some of these materials are intended to accelerate repair of the dermis (Doillon *et al.*, 1986). A three-dimensional pore structure allows tissue ingrowth into the material and with time the wound tissue and implant cannot be separated. The implant is ultimately degraded and is replaced by normal scar tissue.

Collagen sponge formation involves freeze-drying a type I collagen acidic dispersion. The freeze-dried sponge is then crosslinked using different physical treatments such as severe dehydration or irradiation and/or chemical treatments including exposure to carbodiimide, glutaraldehyde, formaldehyde, or diisocyanates (Doillon and Silver, 1986; Yannas *et al.*, 1980; Speer *et al.*, 1980). However, the presence of unreacted chemical crosslinking agents or chemicals released by biodegradation at long term into the wound can slow down wound healing.

The surface and interior of the sponge can be modified by varying the condition of manufacture. Sponges can be coated with growth factors or antimicrobial agents to facilitate healing.

It has been found that sponges with pore sizes between 50 and 300 μm in diameter and interchannel communications between the surface and interior of the sponge significantly increase tissue ingrowth and consequently enhance wound healing (Doillon *et al.*, 1986; Dagalakis *et al.*, 1980). Channel structures result in orientation of fibroblasts and newly deposited collagen parallel to the fibers on the surface of the sponge (Doillon *et al.*, 1984). The incorporation or coprecipitation of molecules that stimulate wound healing (hyaluronic acid, glycosaminoglycans, and glycoproteins) improves tissue ingrowth and also wound healing (Doillon and Silver, 1986).

Other uses of a collagen sponge include debridement of the wounds (Chvapil, 1977). In this instance the collagen by virtue of its hydrophilic property can adsorb wound fluids and cellular debris. Daily replacement of a collagen debriding agent can efficiently cleanse and debride the wound without introducing trauma and at long term stimulates granulation tissue (Doillon *et al.*, 1988a). The advantage of using collagen instead of a synthetic polymer is that residual collagen remaining in the wound will not interfere with healing. In contrast residual synthetic polymer remaining in the wound causes inflammation and prolonged healing times.

8.3.3. Artificial Dermis

Wound dressings are designed as temporary coverings or biodegradable substrates that are invaded and then replaced by host tissue. In comparison, the function of the dermis can only be replaced by cell-populated implants. As a result of advances in cell culture techniques, several approaches have been developed to make cell-populated matrices. These matrices consist of fibroblasts that are cultivated *in vitro* on a support (Figure 8.6).

Fibroblast cultures *in vitro* on a biological material such as a type I collagen can be used as an "artificial dermis". *In vitro* collagenous matrices are successfully used as supports for fibroblast growth. Initially, fibroblasts populate the matrix and then synthesize and deposit new connective tissue. At long term these fibroblasts reconstitute their own matrix. By growing such materials in culture, large surfaces of artificial dermis can be prepared from a small tissue

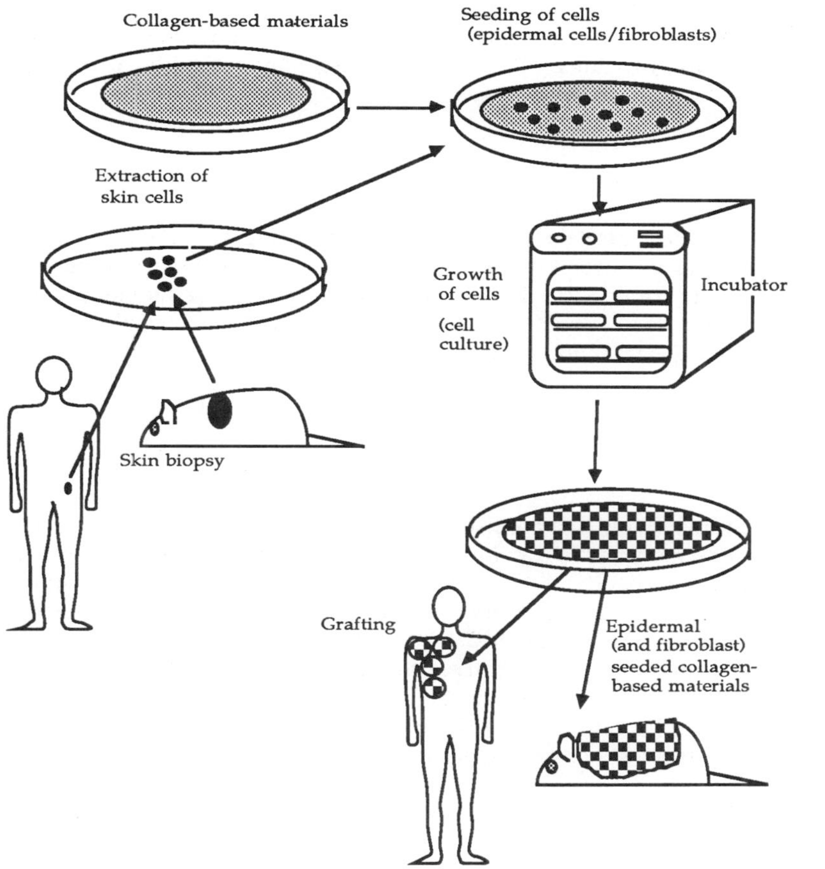

Figure 8.6. Cell expansion procedures. Epidermal and dermal cells from the skin are obtained from humans and animals and expanded in tissue culture. These cells are grown on plastic, a feeder layer of irradiated cells, a collagen matrix or skin (cadaver or animal). The expanded tissue cultured skin is then grafted on animals or humans.

biopsy. The use of allogenic cells (cells from skin excision) allows for immediate availability of cultured dermis.

The collagen matrix can be in the form of a sponge, gel, or fibers. Collagen solutions in dilute acid are mixed with fibroblasts *in vitro* forming a gel after neutralization with a base. The resulting collagen gel is contracted in the presence of fibroblasts in cell culture.

8.4. Epidermal Replacement

Dermal dressings and cell-seeded dermal replacements are a necessary part of the skin repair problem. Ultimately the wound must be covered with an epithelial cell layer. In some instances the epithelium will spontaneously migrate over the dermis. This occurs with shallow wounds or wounds that have been filled in with new dermal tissue. Epidermal migration is often accelerated by preventing excessive moisture loss from the wound. This is accomplished with film and membrane dressings. Below we discuss polymeric film dressings employing synthetic and natural polymers.

8.4.1. Synthetic Polymers

Several synthetic materials are effective in protecting superficial wounds from excessive water permeability and bacterial invasion, which are properties of normal epidermis (Table 8.2). Successful applications of these synthetic films include dressings for superficial tangential wounds (dermabrasion), split-thickness skin graft donor sites and superficial burns (first- and second degree burns).

8.4.1.1. Polymeric films

Polymeric films (Table 8.2) consist of thin layer of polymer made of either silicone, polyurethane, polyvinyl chloride, or polyethylene. Films are flexible and impermeable to bacteria. However, they have a low water-vapor transmission and lead to pooling of exudate under the dressing; for this reason they do not adhere to the wound bed. Most of these films have a synthetic adhesive as a coating on the inner surface that adheres well to dry skin at the wound margins but does not adhere to the wound site. The fact that the film adheres well to the edges of normal skin creates an enclosed environment that protects the wound against bacterial invasion. Increased adherence to the wound is achieved by addition of a second layer containing synthetic (gauze or gels) or biological (collagenous) materials. Tissue adhesion occurs via fluid adsorption or cellular migration into the lower layer. In addition, antibacterial agents (such as iodine) can be incorporated into the polymeric membrane.

8.4.1.2. Other Membranelike Dressings

Another approach that has been used involves spreading a monomer or polymer powder onto the wound; the monomer polymerizes and forms a sheet after a solvent containing a crosslinking agent is sprayed on the powder. One example involves spreading polyhydroxyethylmethacrylate (PHEMA) monomer powder onto the wound which is then sprayed with a small amount of polyethyleneglycol 400 (PEG). The PEG solubilizes the powder forming a uniform homogenous sheet that conforms to the wound contour. PHEMA and PEG film is adherent to wound tissue because it is crosslinked in the wound bed. However, these films are excessively permeable to water and can result in excessive fluid loss. The film is elastic, nonantigenic, and inert and provides an effective antimicrobial barrier. However, the film tends to crack or peel off. Antimicrobial agents such as silver sulphadiazine can be combined with the film.

Films are also produced by mixing gelatin and glycerol in hot water and then stirring the mixture into an aqueous emulsion of acrylic copolymers, the mixture is then evaporated. The film obtained is durable, stable, reproducible, controls the water-vapor transmission, and has a high water permeability.

8.4.2. Artificial epidermis

In applications where epidermal migration over the wound is slow, skin grafting or application of a tissue-culture-expanded epidermis facilitates healing of large wounds.

The combination of epidermal cells grown in culture with a synthetic or biological substrate is termed an artificial epidermis. However, in most cases, actual application of epidermal cells is performed by inverting the layer of epidermal cells on a support so that only the epidermal layer is in contact with the wound surface.

Several different techniques are used to grow epidermal cells on plastic. One technique uses a feeder layer of fibroblasts on which epidermal cells are grown (Green *et al.*,1979; O'Connor *et al.*, 1981). Epidermal cells grown in this manner show better differentiation and tissue-cultured epidermis grown has been used successfully on patients. This technique uses lethally irradiated 3T3 fibroblasts as the feeder layer on which epidermal cells can grow and differentiate. From a small piece of epidermis (1 cm^2), a layer of 6,000 cm^2 of epidermal cells can be grown in culture. Epidermal cells can be removed from the plastic using enzymes without destroying the layer of cells. They are then supported by means of a gauze or a collagen film (Eisinger *et al.*, 1979) that is used as a vehicle to aid the transport of the implant. The layer of cells tends to retract during such manipulation.

Other techniques use collagenous materials as a support for growth of cells; epidermal cells grown on collagen films and sponges show good differentiation (Doillon *et al.*, 1988b). The use of collagen sponge for epidermal cell growth allows for the possibility of growth of skin appendages in the sponge pores. In this manner skin appendages can be grown in culture and isolated for reintroduction into an "artificial epidermis".

Implantation of epidermal cells on a collagen sponge onto wounds allows for ingrowth of wound tissue and increases the adherence of such grafts.

8.5. Total Skin Replacement

8.5.1. Composite Synthetic and Semisynthetic Wound Dressings

Most skin implants have been made by combining two or more materials as described above. The outer layer is used as an epidermis and is designed for durability, elasticity, and eventual suturability to wound edges. The inner layer is designed for maximum adherence and elasticity and replaces the function of the dermis. Composite materials usually consist of a fabric mesh (cotton gauze, dacron flocking) bonded to a synthetic polymer film.

Biobrane (Tavis *et al.*, 1979; Frank *et al.*, 1981) is a composite of a silicone rubber layer attached to a knitted nylon fabric that also contains porcine collagen-derived peptides. The biological component is believed to increase adhesion to the wound and attract new connective tissue cells.

Epigard consists of an inner layer of reticulated polyurethane foam laminated to an outer sheet of microporous polypropylene film. It adheres firmly to the wound and is stripped from the wound when it is replaced.

Other composite skin implants use polytetrafluoroethylene or polyurethane films as the outer layer, and pectin or methylcellulose as inner adhesive layer. Methylcellulose is more adherent to the wound and impermeable to bacteria and water.

Polymeric wound dressings are used only on small shallow wounds that can heal by epithelial cell movement across the wound from the wound edges.

8.5.2. Composite Collagenous Materials as Wound Dressings

One of the earliest artificial skins was developed by Yannas and coworkers at Massachusetts Institute of Technology (Yannas and Burke, 1980) and consisted of a bilayer containing a silicone membrane attached to an inner layer of collagen sponge. The collagen sponge was formed in the presence of chondroitin-6-sulfate and crosslinked using glutaraldehyde. The silicone layer prevents excessive water-vapor transmission and bacterial invasion and can be sutured onto the wound edges. Such skin implants are incorporated into body tissue and are characterized by a controlled rate of biodegradation. Tissue ingrowth appears within the porous structure of the sponge. The collagen is degraded at long term (more than one month) at which time the silicone layer spontaneously peels as epithelialization occurs (about 15 days).

This composite acts as a temporary wound dressing by inducing dermal tissue formation; however, a skin graft is later required to cover large areas of wound with an epidermis. Skin grafting usually occurs 21 to 28 days after the composite is placed on wounds.

Adherence of the composite to the wound bed occurs by incorporation of wound tissue. The risk of infection is decreased as healing proceeds. However, at the time of implantation, air, blood, and liquid accumulation below the implant should be minimized to decrease the chance of infection. This is of particular importance in mobile areas such as on the surface of joints.

Similar composites have also been developed using diffent glycosaminoglycans (such as hyaluronic acid), glycoproteins (such as fibronectin) and other agents known to increase attachment, mobility, and cell proliferation.

8.5.3. Complete "Artificial Skin"

The complete artificial skin consists of a cellular dermis and epidermis and a connective tissue matrix containing blood vessels and glands. Although in principle the techniques are available to produce such a material, no such composite exists today. However, fibroblast- and epidermal cell-seeded matrices are currently available (Bell, *et al.*, 1981; Bell, 1981; Doillon *et al.*, 1988b) (Figure 8.7).

A fibroblast- populated collagen-based material upon which epidermal cells are cultured is an ideal skin substitute. The details of successful clinical use of tissue-cultured materials have been partially worked out. The time required for growth of epidermal cells in tissue culture is about 21 days during which time

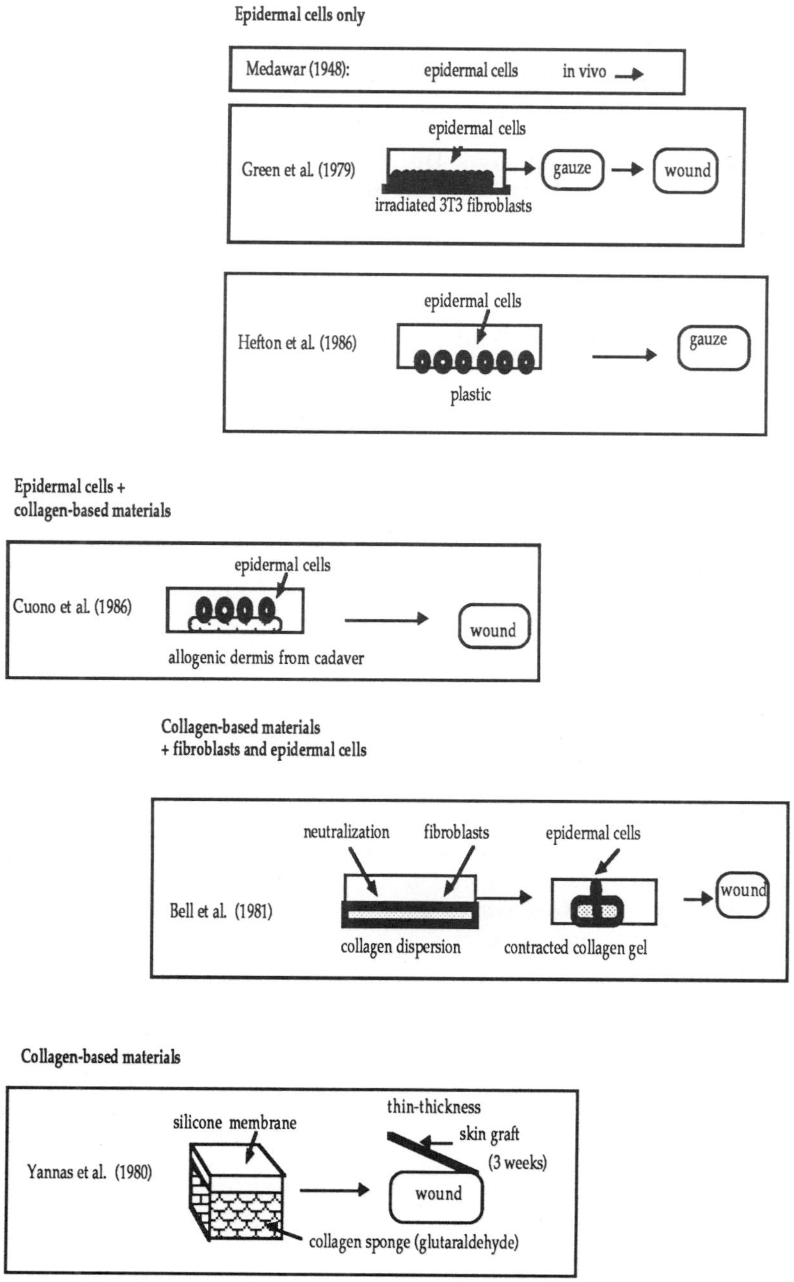

Figure 8.7. Summary of artificial skin studies. Artificial skin composed of tissue culture expanded epidermal cells, epidermal cells in the presence of a collagen matrix, fibroblasts and epidermal cells in a collagen gel, and collagen-based materials have been developed. References to important literature are cited and a summary of methods is given.

one of the temporary dressings described above can be used. Unless the wound is very deep, in which case a tissue-cultured dermis is used, a tissue-cultured epidermis is then applied to the wound. However, the use of a dermal implant is sometimes useful to prevent the formation of scar tissue. Some patients receiving only epidermal grafts have been observed to have tight skin that exhibits scarring at the graft periphery. Ultimately, it is envisioned that the time required for cell- culture expansion can be reduced via use of allogenic cells in combination with biologic or biodegradable supports. Further investigation is required but scientific research is directed at testing the use of living allogenic skin implants that will be available immediately after injury to cover exposed areas of skin.

Bell and coworkers report preliminary results using a contractile collagen lattice that is populated by fibroblasts and seeded with epidermal cells (Bell *et al.*, 1981). This lattice is perforated on one side and small skin biopsies are implanted into these holes. The results of *in vitro* experiments show epidermal cell differentiation and growth over the superficial surface of the collagen lattice.

The use of a collagen sponge allows feeding of cells in culture within the matrix through the sponge pores. The collagen sponge is a useful substrate for growth of epidermal cells since it is a vehicle that can modulate adhesion to the wound bed by infiltration of repair tissue and allow for better graft acceptance by the host.

8.6. Summary

A variety of natural and synthetic polymers have been used as wound dressings and artificial dermal and epidermal components. Synthetic polymers are generally used to create a barrier to bacteria and prevent heat and fluid loss from a wound. They are temporary wound coverings that are removed prior to skin grafting (large surface area wounds). Biological polymers such as collagen can be used as biodegradable wound dressings and are incorporated into wound tissue. They also promote wound healing and act as substrates for epidermal cells and fibroblasts grown in cell culture.

References

Alexander, J.W., MacMillian, B.G., Law, E. and Kittur, D.S. J. Trauma 21,433,1981.

Bell, E., Buttle, D.J., Ehrlich, H.P. and Nakatsuji, T. Science 211,1052,1981.

Bell, E. Plast. Reconstruct. Surg. 67,386,1981.

Chvapil, M., Kronenthal, R. and van Winkle, W., Jr. Int. Rev. Connect. Tissue Res. 6,1,1973.

Chvapil, M. J. Biomed. Mater. Res. 11,721,1977.

Cuono, C., Langdon, R. and McGuire, J. Lancet 1123, May, 1986.

Dagalakis, N., Flink, J., Stasikelis, P., Burke, J.F. and Yannas, I.V. J. Biomed. Mat. Res. 14,511,1980.

Doillon, C.J., Whyne, C.F., Berg, R.A., Olson, R.M. and Silver, F.H. Scanning Electron Microscopy, III,1313,1984.

Doillon, C.J., Whyne, C.F., Brandwein, S. and Silver, F.H. J. Biomed. Mat. Res. 20,1219,1986.

Doillon, C.J. and Silver, F.H. Biomaterials 7,3,1986.

Doillon, C.J., Silver, F.H., Olson, R.M., Kamath, C.Y. and Berg, R.A. Scanning Microscopy International 2,985,1988a.

Doillon, C.J., Wasserman, A.J., Berg, R.A. and Silver, F.H. Biomaterials 9,91,1988b.

Dressler, D.P., Barbee, W.K. and Sprenger, R. J. Trauma, 20,1024,1980.

Eaglstein, W.H. and Mertz, P.M. J. Invest. Dermatol. 71,382,1978.

Eisinger, M., JiSoo, L., Hefton, J.M., Darzynkiewicz, Z., Chiao, J.W. and de Harven, E. Proc. Natl. Acad. Sci. USA. 76,5340,1979.

Frank, D.H., Wachtel, T.L., Frank, H.A. and Sanders, R. Surg. Forum 31,552,1981.

Green, H., Kehinde, O. and Thomas, J. Proc. Natl. Acad. Sci. USA. 76,5665,1979.

Hefton, J.M., Caldwell, D., Biozes, D.G., Balin, A.K. and Carter, D.M. J. Am. Acad. Dermatol. 14,399,1986.

Hunt, J.L., Sato, R. and Baxter, C. Ann. Surg. 189,147,1979.

Koff, I., Kaufman, L., Lahav, M. and Schechter, I. Eur. Surg. Res. 11,114,1978.

May, S.R. and DeClement, F.A. Cryobiology 17,617,1981.

Medawar, P.B. Quart. J. Microscop. Sci. 89,187,1948.

Morris, D.M., Hall, G.M. and Elias, E.G. Surg. Gyn. Obstet. 149,893,1979.

O'Connor, N.E., Mulliken, J.B., Banks—Schlegel, S., Kehinde, O. and Green, H. Lancet 1,75,1981.

O'Neill, J.A., Jr. J. Ped. Surg. 8,705,1973.

Pachence, J., Berg, R.A. and Silver, F.H. Medical Device and Diagnostic Industry, January, p. 49,1987.

Quinn, K.J., Courtney, J.M., Courtney, J.M., Evans, J.H., Gaylor, J.D.S. and Reid, W.H. Biomaterials 6,369,1985.

Ratner, B.D., Hoffman, A.S. and Andrade, J.D. (eds.) Symposium Divn. Polymer Chemistry, Inc., 170th Meet. ACS, Chicago, Ill. August p. 27,1975.

Robson, M.C., Krizek, T.H., Koss, N. and Samburg, J.L. Surg. Gyn. Obstet. 136,904,1974.

Speer, D.P., Chvapil, M., Eskelson, C.D. and Ulreich, J. J Biomed. Mat. Res. 14,753,1980.

Tavis, M.J., Thornton, J.W., Harney, J.H., Danet, R.T., Woodroof, E.A. and Bartlett, R.H. Surg. Forum 28,522,1977.

Tavis, M.J., Thornton, J.W., Bartlett, R.H, Roth, J.C. and Woodroof, E.A. Burns 7,123,1979

Thompson, P.D. and Parks, D.H. In: Burn Wound Coverings, edited by D.L. Wise, Boca Raton, Florida, Vol. 1, p. 47,1984.

Walder, A.I., May, P.D., Bingham, C.P. and Wright, J.R. Trans. Am. Soc. Artif. Int. Organs 15,29,1969.

Walker, A.B., Cooney, D.R. and Allen, J.E. J. Pediatr. Surg. 12,391,1977.

Winter, G.D. Plast. Reconstruct. Surg. 56,531,1975.

Wise, D.L. Burn Wound Coverings. CRC Press, Florida, Volumes I & II, 1985.

Wood, R.A.B., Williams, R.H.P. and Hughes, L.E. Br. J. Surg. 64,554,1977.

Yannas, I.V. and Burke, J.F. J. Biomed. Mat. Res. 14,65,1980.

Yannas, I.V., Burke, J.F., Gordon, P.L., Huang, C. and Rubenstein, R.H. J. Biomed. Mat. Res. 14,107,1980.

Chapter 9

Replacement of Skeletal Soft-Tissue Structures

9.1. Introduction

Mechanical trauma to knee and facial structures is a problem faced by military personnel, laborers and factory workers, athletes competing in contact and noncontact sports, and all individuals during daily life. Development of protective devices as well as restraint systems for passenger cars has helped reduce the number of serious injuries. In cases where injury occurs, the use of medical devices to repair or replace the function of damaged soft tissues has led to restoration of cosmetic and functional properties. This chapter deals with materials used to replace the function of tendon, ligaments, and cartilage.

9.2. Tendon, Ligament and Cartilage Replacement

As discussed in Chapter 2 (Table 2.1) all mammals are able to convert chemical energy into mechanical forces. This ability requires the transmission of forces generated by muscle to the skeletal components involved in locomotion including tendon, ligament, cartilage, and bone. Tendons transmit forces between muscle and bone while ligaments provide the attachment between bones in joints. Opposing muscle forces are converted into translation and rotation of joint components that are required for movement. Cartilage that lines the surface of long bones acts as a shock absorber to buffer the effects of impact loads. Below a brief review of the anatomy and physiology of the knee joint and the hand are presented as well as the design problems faced by the biomaterials scientist in replacing soft tissue joint structures.

9.2.1. Physiology of Joints in Knee and Hand

The knee and hand are complex biologic structures that are cycled through as many as 10^8 cycles during a lifetime. Design of man-made structures to cycle this many times without failure is a sophisticated task that requires an understanding of how normal joints work in the knee as well as the hand.

Although many soft-tissue orthopaedic injuries involve trauma, such as sports- and job- related injuries, as the life expectancy of our population increases the number of "wear and tear" diseases increases. "Wear and tear" diseases arise from repeated cycling of skeletal tissues over many cycles and result in hands, shoulders, and knees that no longer move without pain. The pain is associated with loss of soft-tissue structures such as cartilage that normally prevent bone-to-bone contact. In addition, in concert with wear and tear, inflammatory diseases such as rheumatoid arthritis contribute to the loss of function of joints in the hand and knee.

The goal of the biomaterials scientist is to understand the structure and function of normal joint and skeletal tissues as well as the pathological conditions that limit the useful lifetime of biological structures. This information provides the basis for design of soft-tissue implants to replace ligaments, tendons, and cartilage.

9.2.1.1. Anatomy of Knee Joint

The anatomy of a knee joint is rather complex. The components involved in rotation include the femur, tibia, fibula, cruciate ligaments, and collateral ligaments (see Muller,1983).

The weight of an individual when standing upright on both legs, is concentrated through the center of gravity that is situated between the legs in the torso. This weight is transmitted to the long bones of the knee including the femur, tibia, and fibula (Figure 9.1). Articular cartilage lining the surface of the tibia and femur acts as a shock absorber to buffer vertical motion and prevent the tibia and femur from touching. During vertical motion of the knee the collateral ligaments that are part of the iliotibial tract as well as the cruciate ligaments prevent hyperseparation of the tibia and femur. The cruciate ligaments are also important in controlling rotational motion of the tibia with respect to the femur.

As internal rotation occurs in the knee the cruciate ligaments are twisted around each other, causing the tibia to become pressed against the femur preventing further rotation. The collateral ligaments and the iliotibial tract aid in control and limit rotation of the knee (Figure 9.2) as a result of the application of external forces to the knee (Muller, 1983). Injuries to these ligaments commonly occur in athletes and as a result of mechanical impact to the knee.

In addition to the collateral and cruciate ligaments, the patellar tendon and ligament stabilizes the knee (Figure 9.1) by connecting the gastrocnemius muscle (which is attached to the femur) to the patellar bone and then to the surface of the tibia. Patellar bone protects the rotational surface of the tibia in the front of the knee.

A variety of other structural components are also present to stabilize the knee joint (Muller, 1983); however, a detailed biomechanical analysis of the knee is beyond the scope of this book.

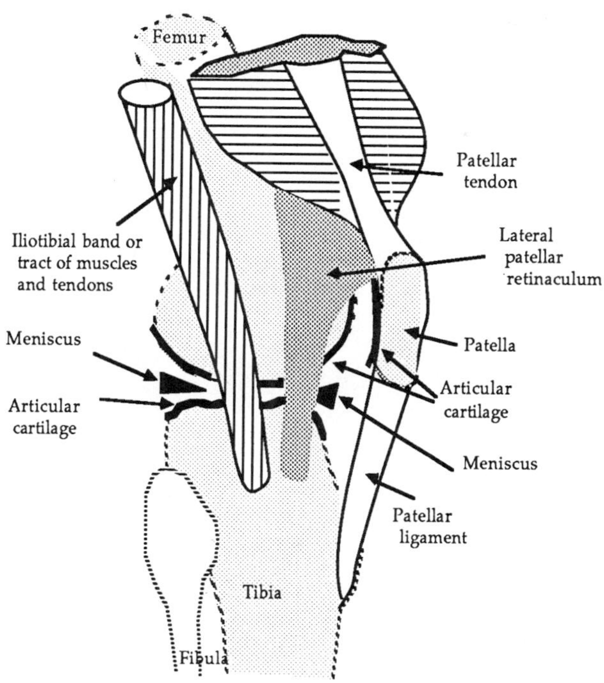

Figure 9.1. Anatomic diagram of the knee. Diagram showing some of the bones, tendons, ligaments, and muscles of the knee including the patella, tibia, femur, iliotibial band, patellar ligament, and tendon.

In summary, the knee contains several cablelike connective tissue structures that limit rotation and extension of the long bones that act to transmit force. Absence or failure of the anterior cruciate ligament associated with a sports injury limits "cutting" ability but does not prevent joint motion. Failure or injury to the collateral ligaments causes joint instability and, if left uncorrected, premature joint wear. Therefore biomaterials are used to reconstruct these structures in the knee.

9.2.1.2. Anatomy of Joints in Hand

Movement of the fingers in the hand is controlled by an unusual cablelike connective tissue in the form of long flexor tendons (Figure 9.3). These tendons are unusual since they can be greater than a foot in length and connect the muscles in the arm to movement in the finger through a system of biological "pulleys". In order for these tendons to transmit force to the joints in the fingers, each of the flexor tendons in the hand is covered by a nonadhering sheath. The sheath provides protection and blood supply to the fibroblasts and other tendon cells. Figure 9.3 is a diagram of the hand showing the flexor tendons that run through synovial sheaths. Mechanical injury to the tendons and sheaths of the hand as a result of industrial or other accidents and mechanical trauma associated with

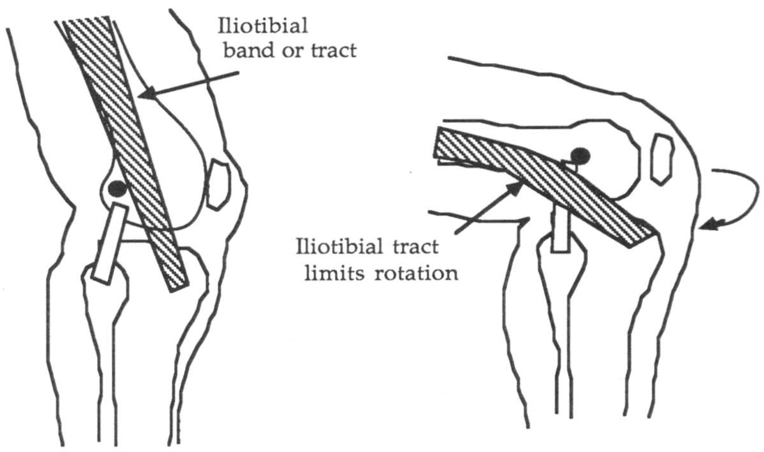

Figure 9.2. Stabilization of the knee by the iliotibial tract.
Diagram of stabilization of knee by the iliotibial tract in bent and unbent joint conformations.

impact loads can lead to impairment of the gliding function and loss of mobility.

9.2.2. Anatomy of Cartilage Structures in The Nose

As a result of traumatic injury to the face or congenital malformation, reconstruction of the nose is a common clinical practice. Reconstruction of the nose requires the use of materials to recontour the surfaces and replace biological-tissues that cannot be surgically repaired. In this discussion we will focus on the structure of the nose since the contour of the nose reflects the shape of the cartilagenous framework.

Anatomically, the nose is composed of the outer and inner nose. The outer nose is of triangular form and projects downward. The base contains the nostrils separated from each other by a septum (Figure 9.4). Internally the nose is covered by a mucous membrane that is continous with the external skin. A cartilagenous framework consisting of five cartilages rests on the superior maxillary and nasal bones. These cartilages include two upper and lower cartilages and the septum cartilage that runs through the center of the nose.

Contour and other defects of the nose are largely altered by changing the cartilagenous structures in the nose. Addition or removal of cartilage to either the base of the nostrils, the lateral cartilage, or the top of the septum are common procedures in nasal surgery.

9.3. Design of Artificial Tendons and Ligaments

Artificial tendons are needed to replace the gliding function of the long tendons in the hand that are damaged in farm and industrial accidents. It is not uncommon for the plastic surgeon to reconstruct tendons in the hand that have

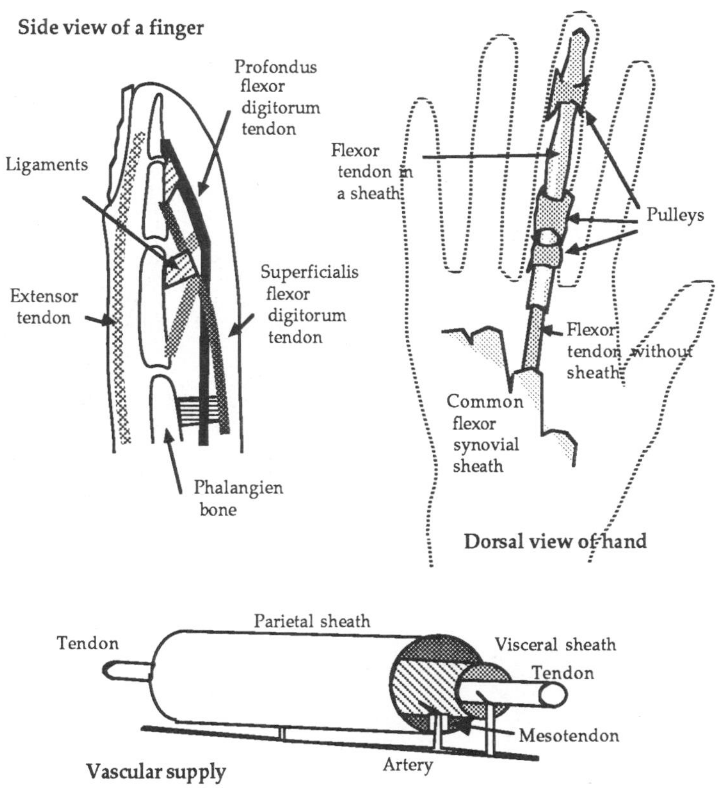

Figure 9.3. Anatomy of flexor tendons. Diagram illustrating side and back views of flexor tendons in the hand and the synovial sheath. Flexor tendons are surrounded by a synovial sheath and run through pulleys. The tendon sheath has its own vascular supply that provides nutrients.

been severed as a result of home and automobile accidents. Proper gliding function occurs only under optimum conditions when the tendon ends can be sutured together and healing of the tendon ends occurs without adhesions forming between the tendon and the tendon sheath. The design criteria for tendon replacement in the hand are listed in Table 9.1.

In contrast to the hand where tissue ingrowth is likely to cause adhesions between the synovial sheath and tendon and is therefore undesirable, in the joint space vascular ingrowth is necessary for long-term survival of tendon and ligament replacements and is therefore desirable. In addition, the design criteria for ligaments exposed to synovial fluid (intraarticular implants) are more stringent compared to extraarticular implants since blood supply and tissue ingrowth are more difficult to achieve within the joint space.

Replacement of tendons and ligaments in a location outside the synovial space is less rigorous than replacement of ligaments within the joint. Extraarticular tendon and ligaments are surrounded by an environment that can supply blood vessels and connective tissue cells that rapidly infiltrate into an implant providing attachment and ultimately supporting mechanical loads.

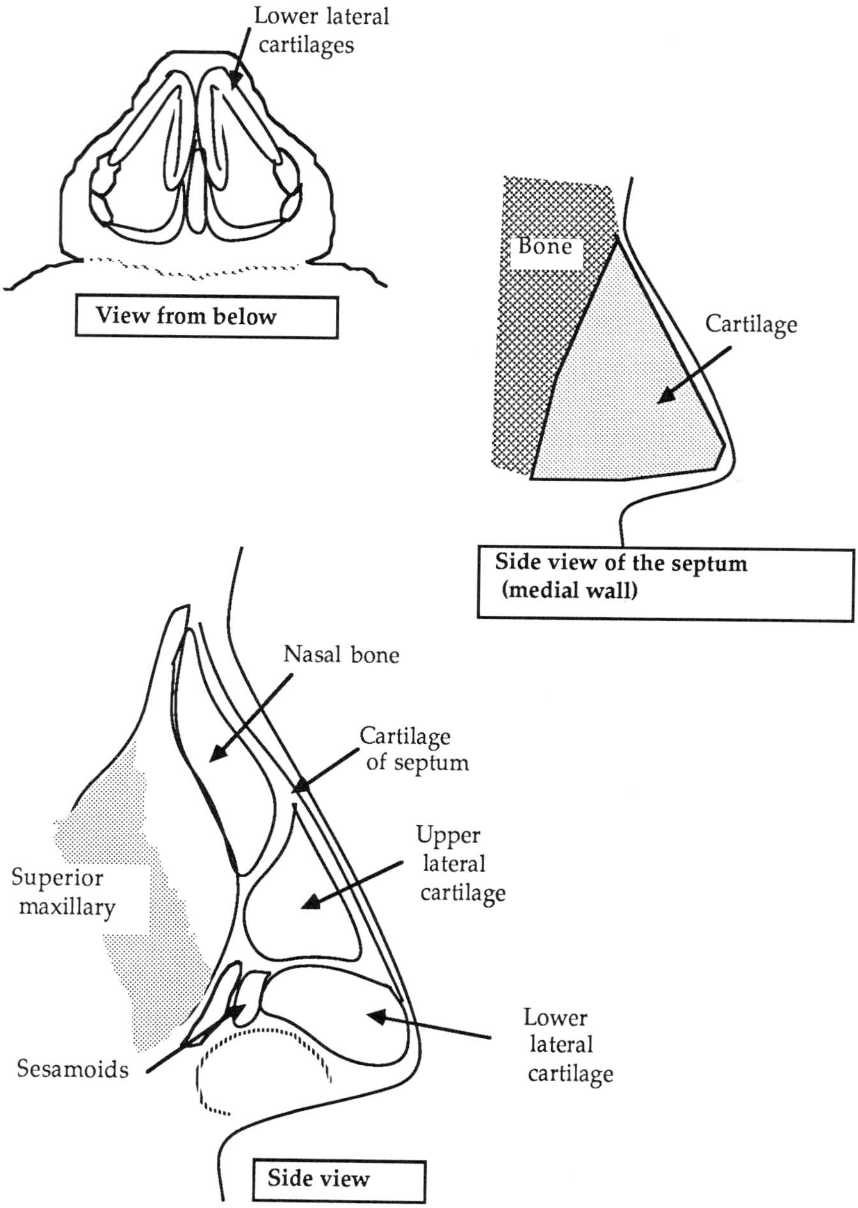

Figure 9.4. Anatomy of the nose. Diagram illustrating side and bottom views of the nose and the cartilages that give shape and contour. These cartilages include the nasal septum and upper and lower cartilages.

Table 9.1. Design Criteria For Gliding Tendon Implants[a]

Tendon	Criteria	Implant materials
Long tendons of hand	High strength load transmission element (tendon)	Tendon - Dacron
	nonadherent synovial sheath	Sheath - Silicone rubber
	breaking strength of 10N	
	High level of biocompatibility	
	low degree of inflammation	

[a] based on Bader and Curtin (1968), Hunter and Salisbury (1971), and Salisbury *et al.*, (1974).

Implants within the joint space are exposed to the degradative enzymes present in synovial fluid; tissue and vascular ingrowth can occur only from extra articular sources. Therefore, the design criteria for tendons and ligaments reflect the environments of these implants.

Several approaches have been used to repair or replace tendons and ligaments. We will focus on the methods and materials used to repair or replace the flexor and Achilles tendons and the anterior cruciate ligament (ACL) in the sections below.

9.3.1. Repair of Flexor Tendon

Flexor tendons that are disrupted by trauma and disease are normally repaired by suturing. Despite improved suturing techniques (Ketchum *et al.*, 1977; Kessler and Nissim, 1968) clinical results are often disappointing largely due to adhesions between the tendon and surrounding tissues (Hunter and Jaeger,1977). Pring *et al.* (1968) indicate that an artificial tendon should have a tensile strength of 1500 N (Newtons) and an elongation at break of 13%.

Flexor-tendon prosthetics date back to the use of silicone rods and partial function tendon replacement (Hunter, 1965) and use of silicone rubber underlays or pulleys (Bader *et al.*, 1967, 1968). Bader and Curtin (1968) developed a tendon prosthesis using silicone rubber sheeting impregnated with Dacron suture material that contained polyester mesh end plates. This prosthesis was used to replace the foreleg extensor tendon in canines for periods as long as eight months.

Studies by Salisbury *et al.* (1974) investigated the use of a Dacron tape covered with a Silastic sheath. The Dacron tape was used as an analog of the synovial sheath in the form of nonporous braided woven and knitted tapes. A wire impregnated knitted Dacron tendon was produced. Prototypes were implanted into the legs of chickens for four weeks.

Murray and Semple (1977) reviewed a variety of tendon prostheses. Most of these prostheses made use of polymeric materials in the form of fibers, velours, tubes, and tapes.

9.3.2. Repair of Tendons and Ligaments of the Knee

A variety of biological structures and synthetic polymers have been used to replace tendons and ligaments of the knee. Biological structures such as patellar tendon have been successfully used to replace the function of the anterior cruciate ligament. This tendon acts as a scaffold for new tissue ingrowth as well as for vascular ingrowth. Synthetic polymer-based implants also have been used as a scaffold for tissue ingrowth and are eventually replaced with a new tissue.

Biological transplants including patellar tendon, fascia lata, gracilis tendon, semitendineous tendon, and meniscus (Figures 9.1 and 9.5) have been used to replace the ACL (Friedman *et al.*, 1985 for a review). Except for fascia lata the above structures are already present in the knee. Friedman *et al.* (1985) reported a success rate of 75% to 85% in the three year follow up to studies using these transplants.

Physical and chemical treatment of biological grafts has been reported as a means to make allograft and xenograft material antigenically acceptable. Shino *et al.* (1986) reported the use of allogenic tendon grafts stored at -80°C. Arnoczky *et al.* (1986) studied rejection of fresh and frozen patellar tendon allografts in dogs. They demonstrated that while fresh grafts incited a marked inflammatory and rejection response, deep-frozen allografts appeared benign. Frank *et al.* (1988) showed that deep frozen rabbit medial collateral ligaments stored at -70°C did not totally prevent collagen synthesis by cells in the substance of these ligaments even after being frozen for up to six weeks. Other studies report the use of glutaraldehyde-treated xenograft anterior cruciate ligament replacements (McMaster,1985) as well as glutaraldehyde-treated pericardium (Chvapil *et al.*, 1987). Van Stensel *et al.* (1987) report the observation of severe inflammation of the synovial lining (synovitis) associated with use of glutaraldehyde treated bovine xenografts.

Synthetic polymers that have been used in tendon and ligament prostheses include polyurethane (Thompson and Sezgin,1963), silicone rubber (Bowen and Dyer, 1962; Chong *et al.*,1972), polyester (Amstultz *et al.*, 1976; Park *et al.*, 1985), polypropylene (McPherson *et al.*, 1985), poly (dioxanone) (Claes *et al.*, 1987), poly (acrylamide) (Leandri *et al.*, 1987) and poly (tetrafluoroethylene) (Bolton and Bruchman, 1985). Carbon fibers (Jenkins *et al.*, 1977; Jenkins, 1978) derived from synthetic polymeric fibers have also been used in tendon and ligament replacement. Devices involving synthetic polymers that are synthesized by

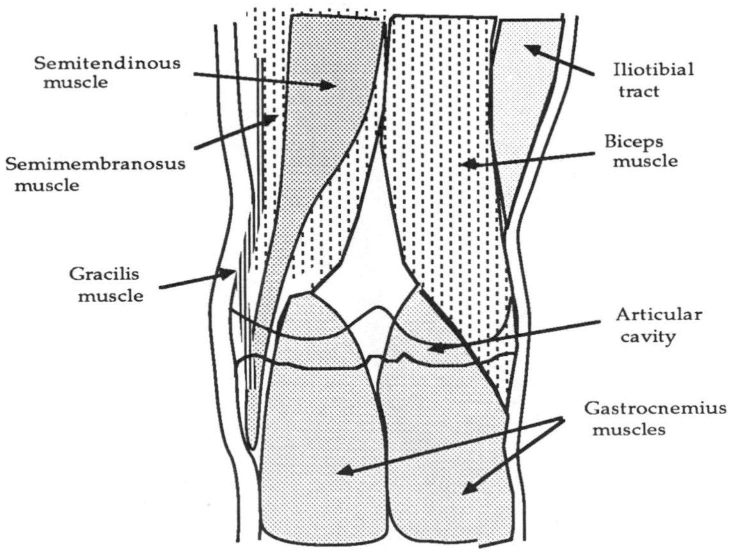

Figure 9.5. Biological structures from the knee used in tendon/ligament replacement. Replacement of the anterior cruciate ligament is accomplished using semitendinous, semimembranosus and gracilis tendons as well as with part of the iliotibial tract.

condensation polymerization including, polyamides and polyesters, are hydrolytically degraded by tissue fluids and eventually lose their strength characteristics. In addition, the degradation products of these polymers may cause tissue inflammation as well as cellular changes (Goodship and Cooke,1985). Silicone rubber- and teflon polymer-based devices appear to be more stable in the presence of tissue fluids and do not exhibit a similar decrease in tensile strength. Carbon fiber implants appear to retain much of their initial strength after implantation; however, some fiber fragments are found in synovial fluid, regional lymph nodes and the reticular endothelial system. Alexander *et al.* (1981) have coated carbon fibers with polylactic acid (PLA) which biodegrades over a period of weeks as new collagenous tissue is synthesized and deposited. These workers have observed little or no migration of carbon fibers from the immediate area of the implant. Due to the early problem of carbon fiber migration, interest in carbon fiber implants continues on a limited basis.

In the case of carbon fiber implants as well as fibrous polymer-based implants, a neotendon (new tendon) is formed as a result of deposition of collagenous repair tissue. The fiber orientation of the prosthesis serves as a scaffold to direct the alignment of newly synthesized and deposited collagen fibers. As the repair process progresses, tensile loads are borne by both the ligament and tendon prosthesis as well as the new tendon. Some authors believe that the fiber stiffness of the prosthesis should be matched closely with that of normal tendon to ensure that the collagen fibers of the new tendon have a crimped pattern similar to that of normal tendons and ligaments (Goodship *et al.*,1985). Kennedy *et al.* (1976) estimate that the maximum load that the ACL can support is between 48 and 64 kg while the strain is between 30% and 36% depending on

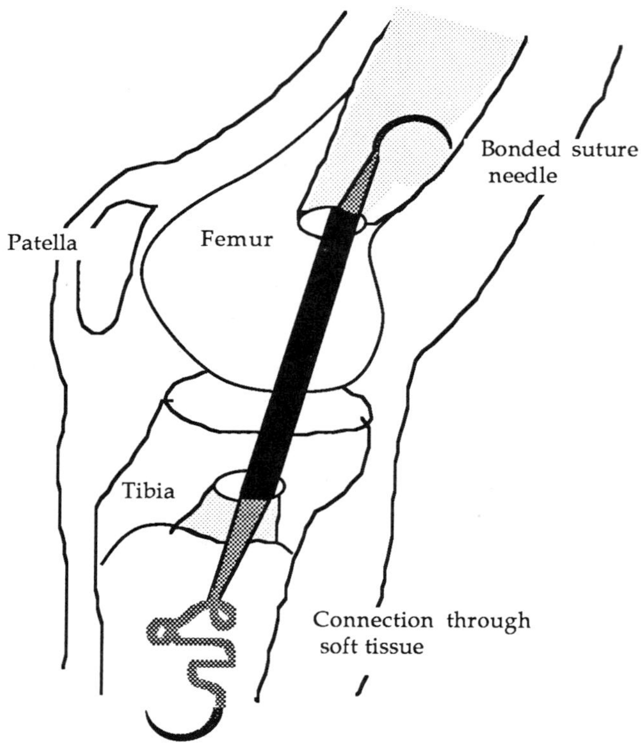

Figure 9.6. Replacement of medial collateral ligament. Diagram illustrating replacement of the medial collateral ligament using a prothesis with a suture bonded to each end. The suture needle is used to stitch the ends of the prosthesis to the stumps of the residual tendon. In some cases the prosthesis is sutured to the outside of the tendon and acts as a stent.

the strain rate.

Surgical repair of ligaments and tendons is a complex process. Repair of a collateral ligament is somewhat straight forward involving the weaving of a prosthesis through tendon stumps if they are present (Figure 9.6). ACL replacement can be achieved by (Figure 9.7) drilling holes through the tibia and femur and pulling the prosthesis through these holes. The prosthesis ends are then stapled or wrapped around a screw attached to the bone.

9.3.3. Biocompatibility of Tendon and Ligament Implants

Biocompatibility of ligament and tendon devices is very similar to biocompatibility of devices used to replace skin and cardiovascular tissue. The surface charge of the fibrous component of the device may play a role in the attraction of cells and the adsorption of cell attachment factors such as fibronectin. It has been suggested that a negatively charged surface prevents cell attraction and blood clotting. High energy surfaces may be associated with cell

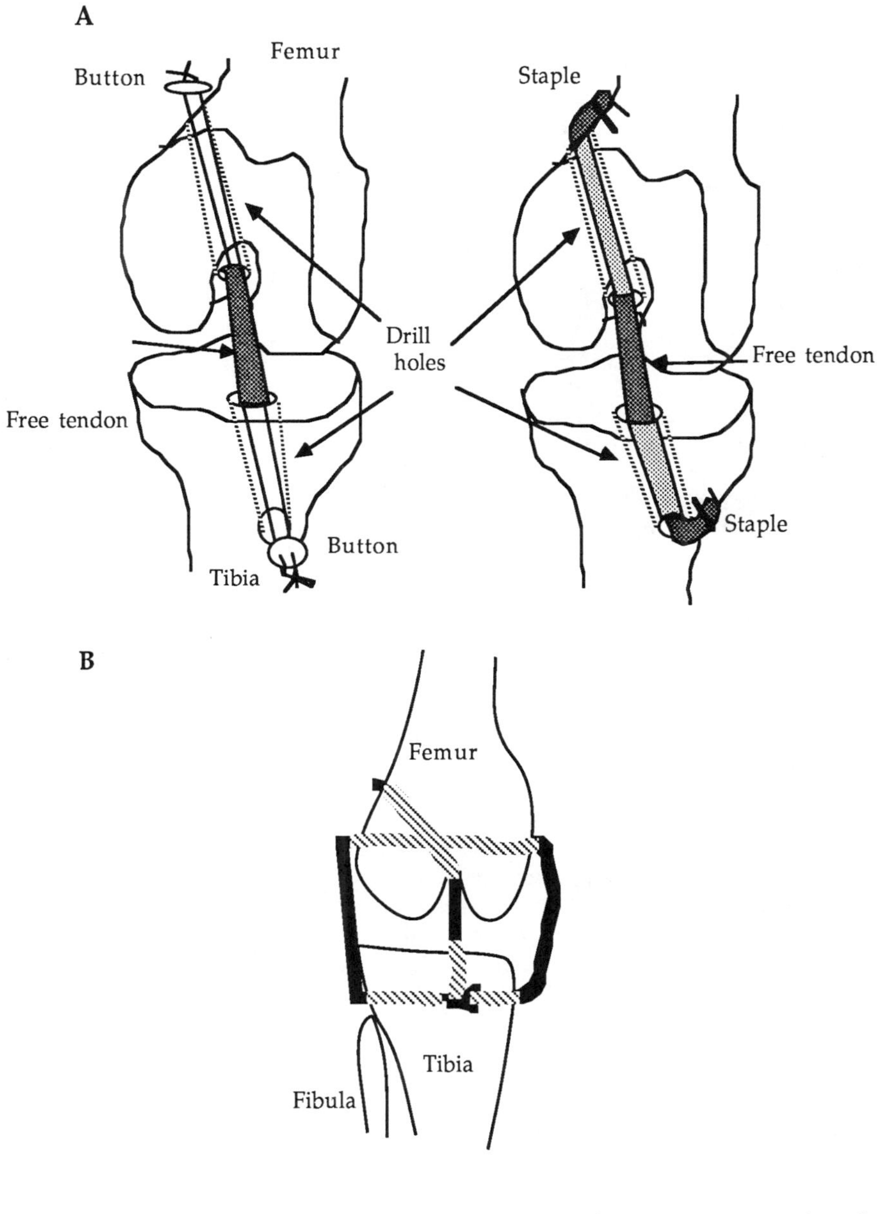

Figure 9.7. Reconstruction of the anterior cruciate ligament (ACL) using the tunnel technique. (A). Diagram illustrating reconstruction of the anterior cruciate ligament by drilling holes through the tibia and femur and pulling the prosthesis through the bone tunnels. The prosthesis is anchored to the bone with staples or by wrapping the prosthesis around a screw inserted into the bone. (B). A modification of the tunnel technique involves creating additional tunnels through the bone to fix the implant in place. The prosthesis tested in this model replaces the function of the ACL and collateral ligaments.

attachment and spreading.

The extent and rate of tissue ingrowth is dictated by the site at which the prosthesis is implanted and the physical form of the implant. Replacement or augmentation of the collateral and Achilles tendon (extra-articular replacement) is often accompanied by rapid infiltration of the implant with fibroblasts and connective tissue components because of the accessible blood supply and close proximity to undifferentiated mesenchymal cells.

Replacement of the ACL (intra-articular replacement) is somewhat more difficult since the blood supply to this site is limited. The implant is often wrapped with a layer of living tissue to accelerate cellular ingrowth. It has also been observed that synovial fluid found in the joint inhibits cell division. In addition, fibroblasts from ACL do not proliferate as readily as fibroblasts from extraarticular tendons and ligaments.

Cell infiltration into the implant is directly related to the degree of porosity of the implant similar to that reported for skin and cardiovascular prostheses. Tumor formation is also decreased by increasing the porosity of the implant. In general, the biocompatibility of both natural and synthetic polymer implants is improved by the presence of pores in excess of 50 um in diameter that allow rapid cellular infiltration. Fibroblast ingrowth into porous implants occurs best when the fiber diameter is close to cellular dimensions.

9.4. Cartilage Implants

Degenerative loss of articular cartilage from the surface of long bones is normally an irreversible process. However, in some instances autologous transplants (Bernum, 1974; Matzen and Matzen, 1974; Skoog and Johansson, 1976) result in formation of new articular cartilage.

Materials such as fascia, poly (vinyl alcohol) sponges, ceramics, and collagen have been used in the repair of cartilage (Breck, 1967; Chiroff *et al.*, 1975; Cobey, 1967; Cucin *et al.*,1972; Henefer *et al.*, 1968; Hulbert *et al.*, 1974, Krajicek *et al.*, 1964; Mitchell and Shepard, 1976; Speer et al., 1979).

Human cartilage transplants are widely used in facial reconstructive surgery (Donald, 1986). Cartilage grafts are frequently used to recontour and augment the nose via projecting and lifting the tip (Peck, 1984). Cartilage is inserted through-an incision in the mucous membrane to project the tip of the nose (Glasgold, 1984). As diagrammed in Figure 9.8, other grafts are used or removed from the dorsum, tip, and the wall of the nostril (alar graft) (Sheen, 1978; Anderson and Reis, 1986).

Nasal septum autografts are the preferred graft for reconstruction of the nose (Glasgold, 1984). Results of experimental studies by Peer (1946, 1948) indicate that live autografts grow in size when implanted into children over a period of 1 1/2 years. Meyer and Walia (1983) reported that cartilage autografts remained intact but showed no growth in animals for 2 years.

Live cartilage autografts are not always available for surgical reconstruction. For this reason chemically or physically treated homografts have been used. Several storage media have been used for homografts including Merthiolate®, alcohol, Cialit®, and formalin (Donald, 1986).

Hagerty *et al.* (1960) showed that within a few days of storage in a mixture of Merthiolate® and saline, cells, (chondrocytes) were killed. However, Muhlbauer *et al.* (1971) reported long-term success using Merthiolate® treated

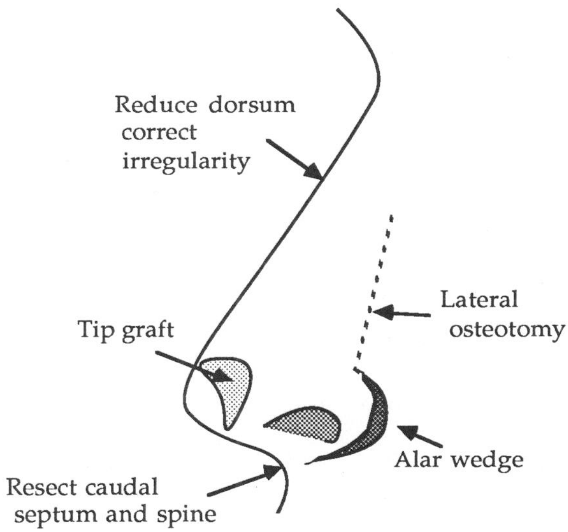

Figure 9.8. Cartilage grafts for nasal reconstruction. Diagram illustrating the use of cartilage grafts to correct defects in the dorsum, tip, and alar areas of the nose.

homografts. Donald and Col (1982) confirmed the long-term survival of Merthiolate® treated as well as irradiated cartilage homografts.

Dingman and Crabb (1960) and Schuller *et al.* (1977) described the use of irradiated cartilage homografts to repair facial defects. In follow-up studies that lasted up to three years there was no evidence of resorption. Recently Donald (1986) showed a high rate of resorption of irradiated cartilage grafts in animals.

Another source of cartilage for tip projection is from the ear. Conchal cartilage autografts from the ear are used as a source for this grafting procedure because of a similarity in shape and substance to normal nasal tip cartilage (Peck, 1984). The posterior aspect of the ear is the preferred donor site.

9.5. Summary

A variety of biological and synthetic polymeric structures are used to replace tendons and ligaments of the knee. The poor fatigue lifetime of synthetic polymers limits their use in ACL repair and therefore biological structures such as live patellar tendon are commonly used.

Replacement of gliding tendons in the hand requires the use of a sheath as well as pulleys. Silicone rubber is commonly used for these applications because of limited adhesion to the surrounding tissue.

Biological cartilage grafts are frequently used in the nose. These grafts are used to contour or augment tissue that is congenitally disfigured or traumatized by injury.

References

Alexander, H., Parsons, J.R., Straughder, I.D., Cocoran, S.F., Mayott, C.W. and Weiss, A.B. Orthop. Rev. 10,41,1981.

Amstultz, H.C., Coulston, W.F. and David, E. J. Biomed. Res. 10,47,1976.

Anderson, J.R. and Ries, W.R. Rhinoplasty: Emphasizing The External Approach, Thieme Med. Pubs. Inc., New York, 1986.

Arnoczky, S.P., Warren, R.F. and Ashlock, M.A. J. Bone J. Surg. 68A 376,1986.

Bader, K.F., Sethi, G. and Schneewind, J. Proc. Inst. Med. Chicago 26,311,1967.

Bader, K.F. and Curtin, J.W. Arch. Surg. 97,406,1968.

Bader, K.F., Sethi, G. and Curtin, J.W. Plast. Reconstr. Surg. 41,157,1968.

Bernum, P. Acta Orthop. Scand (Suppl.) 176,1,1974.

Bolton, C.W. and Bruchman, W.C. Clin. Orthop. Rel. Res.196,202,1985.

Bowen, S.F. and Dyer, J.A. Invest. Opthalmol. 1,4,1962.

Breck, L. Clin. Orthop. 54,133,1967.

Chiroff, R.T., White, E.W., Weber, J.N. and Roy, D.N. J. Biomed. Mat. Res. Symp. 6,29,1975.

Chong, J.K., Cramer, L.M. and Culf, N.K. J. Trauma 12,104,1972.

Chvapil, M., Gibeault, D. and Wang, T—F. J. Biomed. Mat. Res. 21,1383,1987.

Claes, L., Durselen, L., Kiefer, H. and Mohr, W. J. Biomed. Mat. Res. 21,319,1987.

Cobey, M.C. Clin. Orthop. Rel. Res. 54,139,1967.

Cucin, R.L., Goulian, D., Stenzel, K.H. and Rubin, A.L. J. Surg. Res. 12,318,1972.

Dingman, R.O. and Crabb, W.C. Plastic and Reconstruct. Surg. 28,562,1960.

Donald, P.J. and Col, A. Otolaryngology Head Neck Surg. 90,85,1982.

Donald, P.J. Laryngoscope 96,786,1986.

Frank, C., Edwards, P., McDonald, D., Bodie, D. and Sabiston, P. J. Orthopaedic Res. 6,95,1988.

Friedman, M.J., Sherman, O.H., Fox, J.M., DelPizzo, W., Snyder, J. and Ferkel, R.J. Clin. Orthop. Rel. Res. 196,9,1985.

Glasgold, A.I. Amer. J. Cosmetic Surg. 1,41,1984.

Goodship, A.E., Wilcock, S.A. and Shah, J.S. Clin. Orthop. Rel. Res. 196,61,1985.

Goodship, A.E. and Cooke, P. CRC Critical Reviews in Biocompatibility 2,303,1985.

Hagerty, R.F., Calhoon, T.B., Lee, W.J., Jr. and Cuttino, J.T. Surg. Gynecol. Obstet. 110,229,1960.

Henefer, E.P., McFall, T.A. and Hauschild, D.C. J. Hand Surg. 10B,577,1968.

Hulbert, S.F., Cooke, F.W., Klawitter, J.J. and Leonard, R.B. Mat. Sci. Res. 5,417,1974.

Hunter, J.M. and Salisbury, R.E. J. Bone Jt. Surg. 53—A,829,1971.

Hunter, J.M. Amer. J. Surg. 109,325,1965.

Hunter, J.M. and Jaeger, S.H. Orthopaed. Clin N. Amer. 8,473,1977.

Jenkins, D.H.R., Forster I.W., McKibbin, B. and Rales, Z.A. J. Bone Jt. Surg. 59-B,53,1977.

Jenkins, D.H.R. J. Bone Jt. Surg. 60—B,520,1978.

Kennedy, J.C., Hawkins, R.J., Willis, R.B. and Danylchuk, K.D. J. Bone Jt. Surg. 58-B,350,1976.

Kessler, I. and Nissim, F. Acta Orthop. Scand. 40,587,1968.

Ketchum, L.D., Martin, N.L. and Kappel, D.A. Plas. Reconstr. Surg. 59,708,1977.

Krajicek, M., Zastava, V. and Chvapil, M. J. Surg. Res. 4,290,1964.

Leandri, J., Dahhan, Ph., Rey, P., Gerger, D., Tarragano, O. and Cerol, M. IEEE/Ninth Annual Conference Eng. Med. Biology p. 1071, 1987.

Matzen, P.F. and Matzen, P.F., Jr. In: Coarthrosis Surgical and Conservative Treatment, edited by A. Rutt, Springer—Verlag,1974.

McMaster, W.C. Clin. Orthop. Rel. Res.196,196,1985.

McPherson, G.K., Mendenhall, H.V., Gibbons, D.F., Plenk, H., Rottman, W., Sanford, J.B., Kennedy , J.H. Clin. Orthop. Rel. Res.196,186,1985.

Meyer, R. and Walia, I.S. In: Biomaterials in Reconstructive Surgery, edited by L.R. Rubin, C.V. Mosby Co., St. Louis 32,509,1983.

Mitchell, N. and Shepard, N. J. Bone Jt. Surg. 58—A,230,1976.

Muhlbauer, W.D., Schmidt-Tintermann, U. and Glaser, M. Brit. J. Plast. Surg. 24,325,1971.

Muller, W. The Knee, Springer-Verlag, Berlin, New York, 1983.

Murray, G.A.W. and Semple, J.C. J. Biomed. Eng. 1,177,1977.

Park, J.P., Grana, W.A. and Chitwood, J.S. Chem. Orthop. 196,175,1985.

Peck, G.C. Techniques in Aesthetic Rhinoplasty, Thieme Med. Pubs. Inc., New York, 1984.

Peer, L.A. Plast. Reconst. Surg. 1,108,1946.

Peer, L.A. Otolaryngology 42,384,1948.

Pring, D.J., Amis, A.A. and Hauschild, D.C. J. Oral Surg. 26,577,1968.

Salisbury, R.E., Mason, A.D., Levine, N., Pruitt, B.A. and Wade, C.W.R. J. Trauma 14,580,1974.

Schuller, D.E., Bardach, J. and Krause, C.J. Arch. Otolaryngol. 103,12,1977.

Sheen, J.H. Aesthetic Rhinoplasty, C.V. Mosby Col., St. Louis, 1978.

Shino, K., Kumiera, F., Inoue, M. and Ono, K. J. Bone Jt. Surg. 68—B,739,1986.

Skoog, T. and Johansson, S.H. Plast. Reconst. Surg. 57,1,1976.

Speer, D.P., Chvapil, M., Volz, R.G. and Holmes, M.D. Clin. Orthop. Rel. Res. 144,326,1979.

Thompson, F.R. and Sezgin, M.A. J. Bone Jt. Surg. 44—A,1605,1963.

Van Steensel, C.J., Schreuder, O., Van Den Bosch, B.F., VanPaassen, H.C., Menke, H.E., Voorhorst, G. and Gratama, S. J. Bone Jt. Surg. 69—A,860,1987.

Chapter 10

Cardiovascular Implants

10.1. Introduction

In 1968 the death rate from cardiovascular disease was 54%. Since 1968 the rate has steadily declined but continues to be about 20% overall of persons between the ages of 36 and 74. Many deaths from cardiovascular disease are attributed to atherosclerosis, a pathological process affecting the large- and medium-sized muscular arteries especially the aorta, coronary and cerebral arteries (Robbins and Kumar, 1987). The basic lesion, the atheroma or fibrofatty plaque, has a core of lipid and is covered by a fibrous cap that partially occludes the lumen of the artery. Plaque formation leads to defects in the wall of the aorta, heart attack, and/or stroke. In the case of an aortic defect the wall can be replaced with a vascular graft consisting of a synthetic polymer. Correction of coronary artery disease is normally achieved using a vein from thepatient's leg which is used to bypass occluded areas of the artery (Figure 10.1). Small diameter synthetic polymer tubes that do not occlude from clotting are under development.

Once an area of heart muscle dies from deficiency of oxygen and blood flow (heart attack) it must repair itself. For short periods of time a heart assist device such as an intraaortic balloon pump or an artificial heart can be used. In cases of extensive tissue damage a heart transplant is required.

This chapter will deal with the polymeric materials and transplants used in cardiovascular applications.

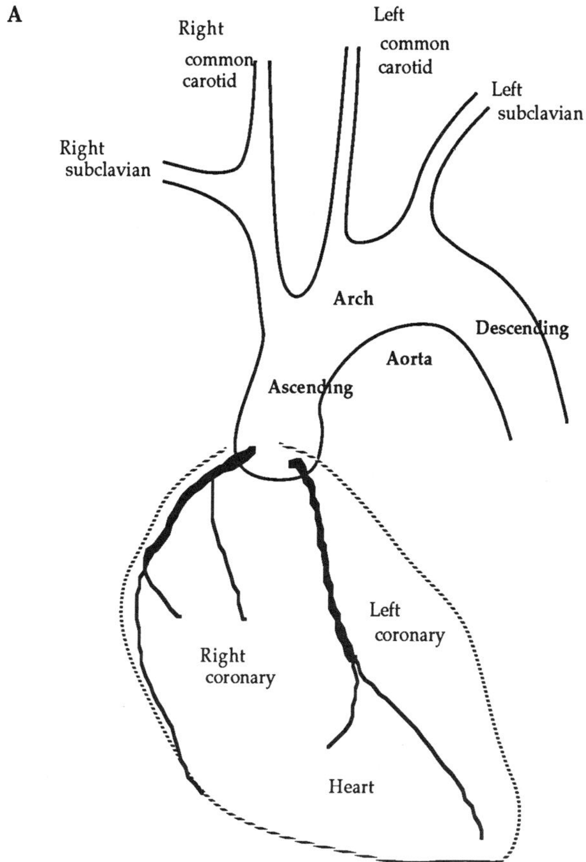

Figure 10.1. Aortic structure and bypass techniques. Diagram illustrating the aorta and some of its branches as well as the coronary arteries (A).

10.2. Classification of Vascular Protheses

Vascular prostheses for replacement of small and large arteries can be classified based on their material composition and according to their diameter. The distinction between large- and small-diameter vessel replacements is arbitrary, and is primarily related to the degree of blood compatibility. Large-vessel replacement (aorta) is well established using a variety of synthetic polymers and is not critically limited by polymer blood compatibility. In contrast replacement of small vessels is limited by blood compatibility considerations. The most commonly replaced vessels are arteries and is the focus of the discussion below (Figure 10.1c).

10.3. Synthetic Materials for Vascular Replacement: Introduction

Vascular grafts made from synthetic fibers are processed into woven and

B

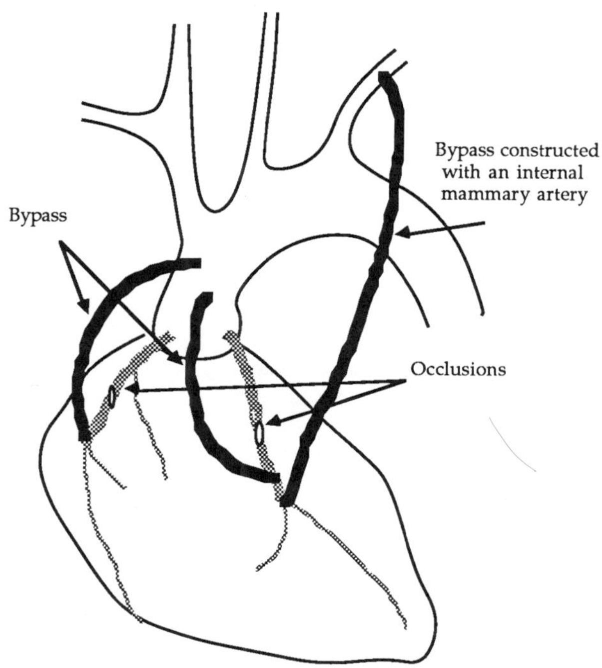

Figure 10.1. (cont.) Normal coronary procedures use veins or arteries to bypass coronary artery (B).

non—woven fabrics using standard textile manufacturing procedures (Figure 10.2). On the macroscopic level, the porosity of these fabrics to fluid is dependent on the diameter and the number of fibers used. On the microscopic level the pores between the fibers in vascular grafts promote tissue ingrowth from the tissue-implant interface which leads to adhesion to the surrounding tissue. Fabrics with pores of 10-45 μm promote optimal tissue ingrowth and rapid wound healing. The liquid permeability of vascular grafts (porosity) is an index-expressed in ml/cm^2 per min relating the rate of water movement through a fabric at a pressure of 120 mm Hg. Porosity is typically measured by pressurizing tubes of the material to be tested. Most prostheses display a porosity of 2300 to 5300 ml/cm^2 per min. Blood loss is minimized as the porosity decreases; however, tissue ingrowth occurs more rapidly as the porosity is increased. Therefore, a balance between blood loss and tissue ingrowth is obtained for prostheses with porosities in this range.

Other parameters of importance in addition to porosity are the biocompatibility, blood compatibility of the surface, stiffness, fatigue lifetime, and handling characteristics of the prostheses. These properties depend on the type of fiber used and the geometry of the finished textile product.

The properties of a textile product reflect the properties of the yarn used to make fabric. A yarn consists of individual fibers that are twisted together in a bundle. Continuous individual fiber is commonly referred to as monofilament.

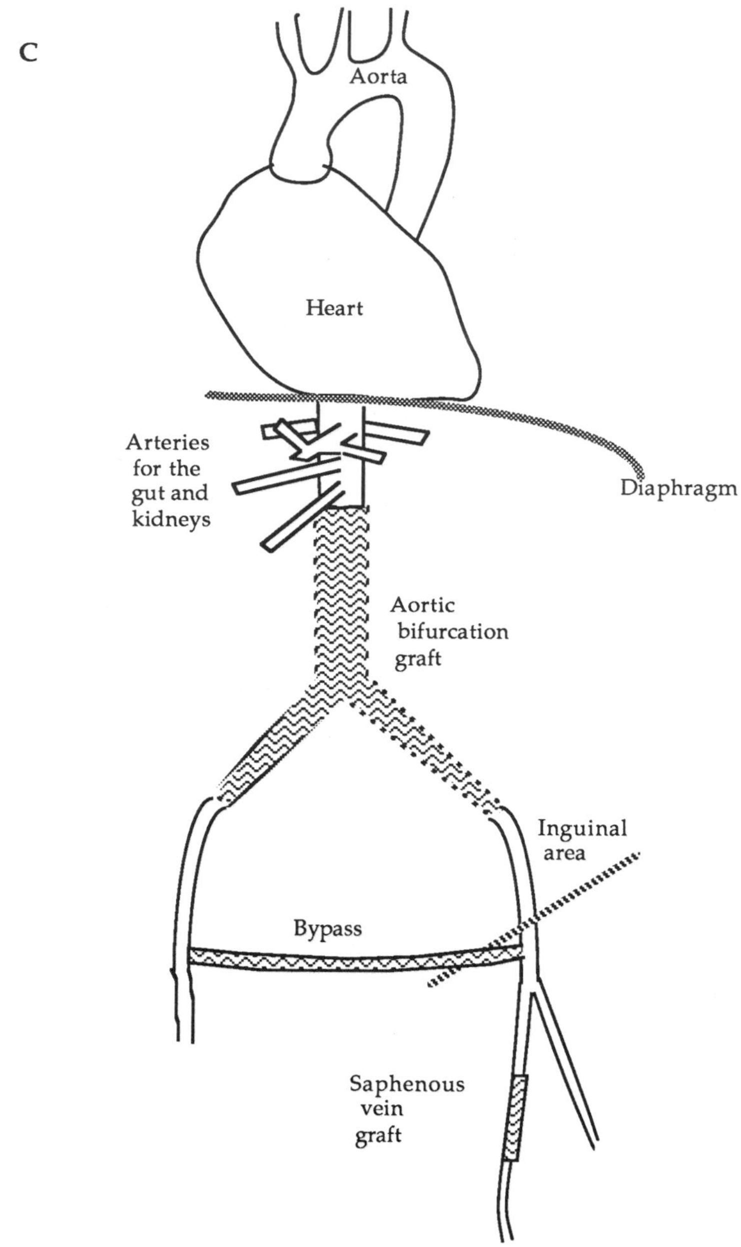

Figure 10.1. (cont.) Grafts are used to replace sections of the aorta and to connect branches in the abdominal region (C).

The degree of twist (angle of twist) determines the surface texture of the yarn. A highly twisted yarn resists extension more than a yarn that is loosely twisted.

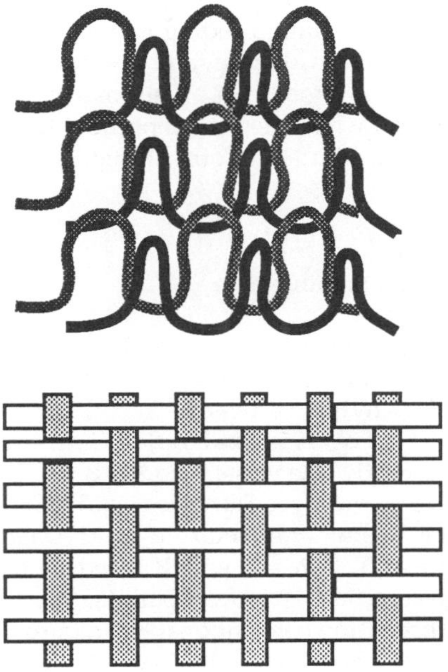

Figure 10.2. Configuration of knitted and woven fabrics. Diagram comparing knitted and woven fabrics.

Yarn is then woven or knitted into fabrics.

The direction of the axis of the yarn in the prosthesis is either warp, weft, or bias when the yarn in a woven structure is parallel, circumferential, or at an angle to the longitudinal direction of the prosthesis, respectively. When the yarn is running in the bias direction as compared to the warp direction the prosthesis can be extended further along its axis before failure.

The diameter of a yarn is given in a unit termed denier. Woven structures are made by crisscrossing layers of fibers that are orthogonal to each other. Because the yarns are passed over and under each other, interfiber friction makes the structure rigid. This results in a high stiffness and strength and low extensibility before failure. The ability of a fabric to "drape"or lie smoothly over a contoured surface is related to the denier of the yarn and the fiber stiffness. Woven fabrics are made using conventional loom technology.

Knitted structures are prepared by looping the yarn with a moving needle. The loops of yarn (Figure 10.2) are interconnected forming a continuous network of loops. Knitted fabrics are highly porous and have lower ultimate tensile strengths than woven structures. The tightness of the knit is expressed in needles/inch.

A fabric that has loops of yarn that extend perpendicular to its surface is termed a velour. These loops of fabric provide anchorage to the surrounding tissue and increase the effective surface area for tissue ingrowth.

Some vascular grafts are crimped into a zig—zag configuration to increase the flexibility and decrease the stiffness of the prosthesis along its axis. Crimping may result in uneven tissue ingrowth on the inner surface of the graft which can interfere with blood flow.

During surgery some vascular grafts are preclotted with the patient's blood to decrease the porosity of the graft and limit bleeding into extravascular tissues. During preclotting, blood proteins such as albumin and fibrinogen are integrated with blood cells into the wall of the prosthesis. This material is eventually replaced by connective tissue to form what is termed a pseudointina. Nonporous prostheses are sometimes coated with albumin or other benign blood components to inhibit thrombus formation.

Synthetic Polymers Used in Vascular Prostheses

The use of grafts made from polymeric materials dates back to the 1950s when a variety of materials including Vinyon N (Voorhees *et al.*, 1952; Blakemore and Voorhees, 1954), nylon (Edwards and Tapp, 1955), Orlon (Hufnagel, 1955), Dacron (DeBakey *et al.*, 1958) and Teflon (Edwards, 1959) were used in vascular surgery. Currently the most widely used large-diameter prostheses are made of Dacron (for review see Sauvage *et al.*,1986).

Poly (ethylene terephthalate) commonly referred to as Dacron and poly (tetrafluoroethylene) referred to as Teflon are widely used to make vascular prostheses. Dacron is a polyester that is biocompatible, inert, and can be sterilized without deleterious effects using conventional techniques and is used as a replacement for large diameter vessels (> 1 cm). It is a condensation polymer of ethylene glycol and terephthalic acid or ethylene glycol and dimethylterephthalate with a catalyst. A linear polymer containing the repeat unit listed in Table 5.1 is formed into a woven or knitted fabric.

Results of clinical studies conducted since the 1950s suggest that fibroblast ingrowth into a Dacron porous prosthesis occurs from the surrounding tissue. At long term the "pseudointima" is composed of a compacted fibrin network of varying thickness but lacks an endothelial cell lining.

Dacron prostheses have been divided into three categories, i.e., tightly woven/crimped, knitted /crimped /nonsupported, and knitted /noncrimped / nonsupported. The first category of prostheses (tightly woven/crimped) is used without preclotting in large-diameter, high-flow arteries because of their lack of porosity (Sauvage *et al.*, 1986). The handling characteristics including suturability of these prostheses is currently being improved. Knitted/crimped Dacron prostheses are used because of their excellent handling characteristics and acceptable performance.

Knitted Dacron grafts introduced in the mid—1950s were weft knit using heavy yarns at a low density (Crawford *et al.*, 1958). By the 1960s a filamentous texture was imparted to the surface of the prosthesis by addition of a velour component (Hall *et al.*,1967; Noon and Debakey, 1977; Sauvage *et al.*, 1971; Bennett *et al.*, 1977). In addition, warp knitting technologies have resulted in the commercial production of Microvel by Meadox Medical Inc. with filamentous inner and outer surfaces and Bionit produced by Bard with a filamentous external surface. The velour surface decreases the time and increases the efficiency of preclotting while tissue ingrowth is increased in the presence of a velour outer surface. Knitted dacron aortic bifurcation grafts show patency and satisfactory performance in about 95% of the cases studied for 10 years. The patency rate is 75% to 80% for Dacron grafts as a replacement for the aorta and femoral artery (Noon and Debakey, 1977; Sauvage *et al.*, 1982). Table 10.1 from Sauvage *et al.*

Table 10.1. Patency Rate For Aortic Grafts [a]

Graft location	Graft type	Knitting method	Number	Patency at five years, %
Aortic bifurcation	Original external velour	Weft	373 graft limbs	99.6
	Bionit	Warp	156 graft limbs	---
Aortofemoral	Original external velour	Weft	236 graft limbs	100
	Bionit	Warp	70 graft limbs	100
Femorofemoral	Original external velour	Weft	71 grafts	80
	Bionit	Warp	23 grafts	93
Axillofemoral	Original external velour	Weft	92 grafts	67
	EXS	Weft	52 grafts	97
Femoropopliteal, above-knee	Original external	Weft	114 grafts	53
	EXS	Weft	60 grafts	84

[a] From Sauvage *et al.* (1986).

(1986) lists the patency rate for different aortic grafts.

In contrast to the high patency observed for knitted/crimped/unsupported Dacron grafts used to replace the abdominal aorta and iliac arteries, the patencies for illiofemoral and above-the-knee femoropopliteal bypass, below-the-knee femoropopliteal bypass, and below-the-knee femorotibial bypass are much lower. This led to the design of a knitted/noncrimped/externally supported Dacron graft (EXS, Table 10.1) that was resistant to kinking (Kenney *et al.*, 1982). These flexible grafts show improved patency as demonstrated by the data reported in Table 10.1.

Teflon is a linear polymer that is formed by addition polymerization and is used as a replacement for small-diameter vessels (< 1 cm). This polymer has the advantage of inertness and chemical stability. Some progress has been made in modifying the pore diameter of expanded teflon to improve tissue ingrowth. Vascular grafts of GORE-TEX are reinforced externally to improve the properties of the implant. Neointima formation is observed after implantation.

Complications to surgical procedures involving large-diameter Dacron prostheses occur primarily at long term. Bending, expansion, deposition of blood clots on the inner wall, fragmentation, embolization of the neointima, proliferation of the neointima into the vessel lumen, deposition of fibrous tissue at the suture lines, and fatigue fraying of the graft material are most commonly observed.

Replacement of small-diameter arteries is normally achieved using a vein graft from the patient (see Section 10.5 below). When the appropriate vein is either unavailable because of prior surgery or is found to be inadequate because of its small diameter, short length, or pathological structure (Veith *et al.*,1979) a synthetic polymer in tube form is used (Veith *et al.*,1986).

Use of expanded polytetrafluoroethylene (PTFE) (see Boyce,1982; Graham and Bergan, 1982 for reviews) vascular grafts were reported in the 1970s by Matsumoto et al. (1973) followed by several clinical reports (Hearn and Charlesworth, 1979; Stansel *et al.*, 1979; Burnham *et al.*, 1978; Campbell *et al.*, 1979; Veith *et al.*, 1978; Haimov *et al.*, 1979; Gupta and Veith, 1980). A recent study (Veith *et al.*, 1986) suggests that the 6-year cumulative life patency rate of femoropopliteal reconstructions with PTFE is 55%. Femorofemoral and axillofemoral bypasses have a 7 1/2 year cumulative life patency rate of 83% and 75%, respectively. The 5 year patency rates for axillopopliteal PTFE bypasses is 45%.

A variety of extruded polymeric tubes of diameters less than 1.0 cm are currently under development for replacement of small-caliber arteries and veins. Poly (tetrafluoroethylene) and polyurethane (Table 5.1) are the most widely used polymers and can be manufactured into a tube with pores as small as 30 μm in diameter yet has little porosity to blood. Urethane copolymers, graphite-impregnated poly(tetrafluoroethylene), carbon—graphite fiber reinforced poly (methylmethacrylate), and pyrolytic carbon—coated prostheses are being explored (Wright *et al.*, 1986; Wildeviur *et al.*, 1987; Chignier et al., 1984). The limiting factor to date has been the blood compatibility of the surface in contact with blood. Since most of these materials cause thrombosis (blood clotting and platelet aggregation) they must be covered with a nonthrombogenic coating such as albumin, glycosaminoglycans such as heparin, anticoagulants or other materials. However, these coatings are passively released into body fluids as the surface is eroded during blood flow. Heparin coating is achieved via electrostatic interaction with a polymer surface, physical entanglement in a water swollen polymer gel, or attachment using free radicals (see Section 11.5.4.1). Heparinlike polysaccharides such as chitin, other glycosaminoglycans, and cationic or anionic polysaccharides have potential in the design of small-diameter vascular prostheses.

In an effort to improve the biocompatibility of vascular grafts containing synthetic polymers such as Dacron and poly (tetrafluoroethylene) an effort has been made to line the lumen of vascular grafts with endothelium.

One method for obtaining endothelial cells involves mechanical shearing (Herring *et al.*, 1978). Cells derived in this fashion were mixed with blood used to preclot a Dacron prosthesis.

Another approach is to enzymatically remove endothelial cells, followed by 2 weeks of growth in cell culture, and preclot seeding of a velour knitted Dacron graft (Graham *et al.*,1980). A variation of this approach is to seed the endothelial cells into Dacron and expanded poly (tetrafluoroethylene) prosthesis immediately after enzymatic isolation (Burkel *et al.*,1981, 1982; Graham *et al.*, 1980, 1982).

Using either of these techniques a minimum number of cells is required for effective seeding. Using Dacron grafts the number is reported to be 8000 cells/cm^2

(Stanley *et al.*, 1986). Future work will involve defining better graft substrates forseeding.

10.4. Biocompatibility of Synthetic Polymers Used In Vascular Prostheses

The tissue response to a vascular prosthesis is very similar to the general wound healing response that is initiated by trauma or surgery. The outer surface of the prosthesis is surrounded by connective tissue and therefore heals in a fashion similar to a subcutaneous wound while the inner surface is in contact with large volumes of blood. The prosthesis is surrounded on its outer surface by a capsule containing fibroblasts, collagen, and other connective tissue components. The interstices in the wall of the prosthesis are infiltrated by fibrous connective tissue similar to that found in the capsule and on the inner surface a pseudointima forms that is in contact with the flowing blood.

The pseudointima consists of fibrin and platelets which is eventually replaced by fibroblasts and collagen. It is rarely seen that host endothelial cells grow down the tubular prosthesis and form a neointima a process that is observed in normal vessels.

During the preclotting process, blood cells and blood proteins become trapped in the interstices in the prosthesis wall and form a fibrin network. The repair tissue on the outside of the graft is very cellular and vascular. Blood cells are present in the wall of the graft such as lymphocytes and macrophages. A foreign-body response is observed in a graft made of poly (ethylene terephthalate) throughout the life of the graft; however, no foreign-body reaction is observed in prostheses containing poly (tetrafluoroethylene).

The surgical procedure required for reconstruction of the vascular system is complicated by infection related to the prosthesis. Infection occurs infrequently but can be fatal if the graft is not removed. Mortality rates from infected grafts range from 25% to 75% and depend on the site of infection (Christensen and Eklof, 1984). Incorporation of a broad spectrum antibiotic into a vascular graft has been tested (Henry *et al.*,1981) as a method to reduce graft infection.

10.5. Biologically Derived Vascular Grafts

An arterial autograft is the ideal material for replacing a damaged vessel. However, use of arterial autografts is possible in only a few situations. Cells in arterial autografts retain their ability to divide and synthesize connective tissue components and can therefore grow in proportion to the normal bodily growth rate when used in children. The flexibility of an autograft matches that of the host tissue and when it is sutured in place healing occurs rapidly.

Host vein grafts are the prostheses of choice for replacement of small-diameter arteries. The saphenous vein is the major external vein in the lower leg and is widely used to replace coronary arteries in bypass operations. It is surgically accessible and the ends can be tied off without having deleterious effects. Implantation of this vein into the arterial system has several long-term disadvantages including proliferation of fibroblasts into the intima, lumenal narrowing by deposition of connective tissue, arteriosclerotic buildup in the lumen, degenerative ballooning out of the wall (aneurysm), and arteriosclerotic aneurysm formation. Storage of saphenous veins is commonly done in heparinized blood products and mechanical distension is used to increase the diameter.

The internal mammary artery arises from the subclavian artery and gives rise to the intercostal arteries. It is also used in coronary bypass surgery.

Coronary bypass surgery is accomplished by harvesting the saphenous vein or the internal mammary artery using a surgical procedure. Internal mammary artery grafts are used to bypass left anterior descending coronary artery blockages, whereas saphenous vein grafts are used to bypass most other coronary arteries (Leather and Karmody, 1982; Kantrowitz, 1986). The bypass is constructed by suturing one end of the vein into the aorta and the other end into the coronary artery beyond the point of occlusion (Figure 10.1b). About 150,000 to 200,000 of these procedures are done yearly (Kantrowitz, 1986).

A second procedure uses the mammary artery rather than a vein in lengths long enough to bypass the left anterior descending coronary artery. Kantrowitz (1986) reviews the results of several studies published on the rate of occlusion of saphenous vein and mammary artery grafts in coronary surgery and concludes that saphenous vein graft occlusions were two to eight times higher than those of the internal mammary artery.

Other vascular tissues that have been used as artery grafts include the common internal and external and the superficial femoral arteries (Ehrenfield *et al.*, 1982). Other vein grafts include the cephalic and jugular.

Human umbilical cord vein grafts can be used for replacement of small-diameter vessels. Glutaraldehyde treated human umbilical vein graft supported with a polyester mesh, commercially known as Biograft® (Dardik, 1986), is an acceptable alternative to use of the saphenous vein. Complications include thrombosis, infection, aneurysm, and foreign-body reactions. At long term fibrous connective tissue is deposited around the implant in the form of a capsule that results in stiffening of the graft.

Human umbilical cord vessels are an alternative to vein and artery grafts for lower limb revascularization (Dardik, 1986). Improved mechanical properties and biocompatibility were achieved after chemical crosslinking (Dardik and Dardik, 1975; Dardik *et al.*, 1975). The grafts are covered with a reinforcing net of polyester mesh or other synthetic polymer. The limb salvage rate using human umbilical cord grafts to replace popliteal and tibial arteries is higher than 50% according to Dardik (1986).

Allografts (sometimes referred to as xenografts) are vascular tissues from other animals that are chemically treated before use as a vascular prosthesis. Chemical treatment is required to prevent rejection due to the presence of antigenic material. Antigenic material induces plaque and aneurysm formation and ingrowth of connective tissue. Glutaraldehyde is most commonly used to chemically treat allografts. Glutaraldehyde-treated allografts have greater durability; however, these grafts can become calcified and brittle over time.

The use of arterial allograft transplantation for replacement of the aorta or reconstruction of femoral vessels is questioned based on the results of long-term studies that indicate that severe atheromatous degeneration and aneurysm formation led to 5-year patency rates of 56% for aortic replacements and 26% for femoral grafts (Meade *et al.*, 1966). These grafts failed as a result of loss of normal structural components including elastin (Halpert *et al.*, 1960). Perloff (1986) describes the use of vascular allografts as well as the cumulative patency rate.

Glutaraldehyde-treated bovine vascular grafts have been reinforced on the outside with a Dacron mesh. Other approaches include use of collagen tubes (see Sawyer *et al.*, 1986 for a review), and collagen gels containing smooth muscle cells. Successful use of the later materials in animals has not been demonstrated.

Another approach involves subcutaneous implantation of a silicone mandril covered by two layers of Dacron mesh. After 6-8 weeks the silicone tube is removed leaving a thin tube composed of fibrous connective tissue that is used as a coronary bypass. However, this fibrous connective tissue tube is stiff and fails at long term in part due to proliferation o ffibroblasts in the intima and deposition of fibrous connective tissue narrowing the lumen.

10.6. Heart Valve Prostheses

Cardiac valves are involved in a variety of disease sresulting from congenital deformity, normal calcification, rheumatic heart disease, bacterial infection, syphilis, rheumatoid arthritis, and genetic disorders of connective tissues (see Robbins and Kumar, 1987). These conditions lead to inadequate valve closure, leakage of blood due to calcification, and deposition of fibrin and other blood elements upon the leaflets. Improper valve functioning results in impaired blood flow to the heart and other organs and results in limited tolerance to exercise.

Patients with severe valvular heart disease who undergo cardiac valve replacement have increased survival, enhanced cardiovascular function, and improved quality of life (Rahimtoola, 1983; Cohn, 1984). The total world-wide prosthetic valve market is estimated at $147 million (Starr, 1986).

Deaths within 30 days following valvular surgery are related to the poor condition of the heart or to surgical complications (Schoen *et al.*, 1983). However, deaths that occur 1 month post-surgery are a result of complications in about 50% of cases studied (Schoen *et al.*, 1983).

Synthetic and biologically derived materials (allografts) have received much attention as valve substitutes (see Schoen,1987a for a review). Mechanical valves consist of a base which is sutured in place and provides a structural framework for the body of the valve, a flow occluder that prevents blood leakage, and guides that control the motion of the occluder (Figure 10.3). Biological tissues used for valve replacement include allografts (human cadaver valves), heterografts (porcine aortic valves, bovine pericardial tissue), and autografts of valves or connective tissue. Most tissue valves are glutaraldehyde-preserved and undergo degenerative processes that limit their useful lifetime (Schoen and Levy, 1984).

The most widely used heart-valve substitutes have been ball valves (The Starr-Edwards), tilting disc valves (Bjork-Shiley), and the biological tissue derived from porcine aortic valves and bovine pericardium.

Biologically derived heart valves have several advantages including their natural form and biocompatibility; however, there are several disadvantages including durability, valve-to-valve variation, *in vivo* calcification and the need for long term anticoagulant therapy.

Porcine aortic valves and bovine pericardium are used after treatment with

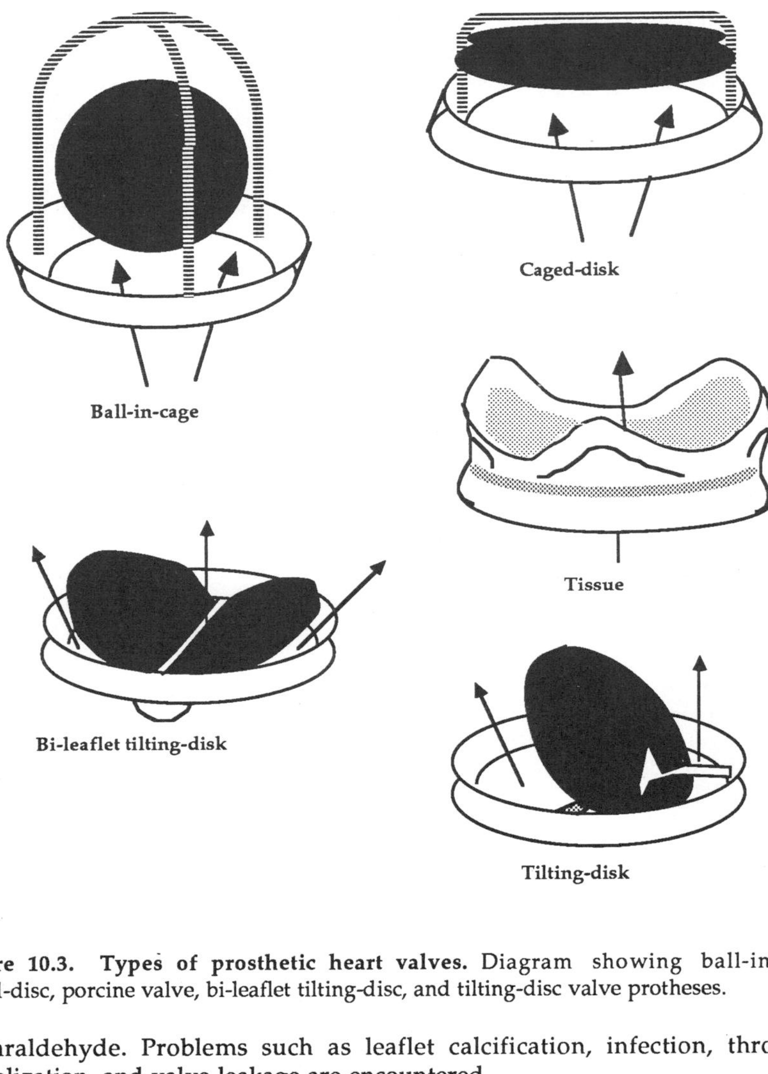

Figure 10.3. Types of prosthetic heart valves. Diagram showing ball-in-cage, caged-disc, porcine valve, bi-leaflet tilting-disc, and tilting-disc valve protheses.

glutaraldehyde. Problems such as leaflet calcification, infection, thrombo-embolization, and valve leakage are encountered.

Mechanical valves can be manufactured to precise specifications and have long-term durability. Designs include ball-in-cage, disc, and tilting-disc valves (Figure 10.3). The results of implantation tests depend not only on the type of valve but also on the location. The valve is attached to host tissue using a cloth sewing ring that is typically made from poly (tetrafluoroethylene) or poly (ethylene terephthalate).

In caged-ball valves the ball is usually made of silicone rubber but because of progressive increases in ball size after implantation due to adsorption of blood lipids the material is not currently used. The cage can be made of titanium or othe ralloy.

The main problem associated with the use of the ball and cage valve and the other mechanical valves is the need for continued anticoagulant therapy (Schoen, 1987b). With the disc prosthesis low cardiac output and arythmias are observed.

Early designs for ball-in-cage valves deformed as a result of lipid deposition (Chin *et al.*, 1971; Hylen *et al.*,1970), however, this problem has been overcome by changing the fabrication process for the silicone ball (Hylen *et al.*, 1972).

Cloth-covered caged-ball valves with metal or polymeric supports that limit the motion of the ball suffer cloth abrasion and disintegration or fibrous growth that leads to leakage (Shah *et al.*, 1978; Schoen *et al.*, 1984). Tilting-disk valves and caged-disk valves with polymeric disks show disk wear that may ultimately lead to decreased performance (Roberts *et al.*, 1975; Silver and Wilson, 1977). Use of pyrolytic carbon as the occluder, and supporting covering material has decreased abrasive wear of disks and metallic cage components (Schoen *et al.*, 1982; Silver, 1980; Silver *et al.*, 1985).

A complication of the use of glutaraldehyde-treated porcine aortic valve bioprostheses is calcification leading to tearing and valve leakage (Schoen, 1987a). Calcium deposits are seen as early as 3 years post-operatively while some valves show no deposits after 10 years (Schoen and Hobson, 1985).

10.7. Cardiac Replacement

About 88% of the deaths related to cardiac diseases are due to ischemic heart disease also known as coronary heart disease. Many of these patients benefit from coronary bypass procedures previously described or other treatments. However, a small fraction of the remaining 12% of deaths due to cardiac disease are the result of other diseases that primarily affect the heart including rheumatic and congenital heart disease and infective endocarditis (bacterial infection of the heart) (Buja, 1987). These conditions lead to cardiac failure and the need for replacement of the heart.

No permanent replacement for the heart exists except for a transplant, although a variety of temporary heart devices are available. Temporary heart devices include assist and implantable artificial heart devices. Results of recent studies on artificial hearts have been reported throughout the news media providing hope that someday a totally artificial heart will be possible.

The history of experimental heart transplantation dates back to the early 1900s (see Cooper, 1984 for a review). Experiments reported by Mann *et al*. (1933) involved the series coupling of a donor heart with that of the recipient by connecting the recipient common carotid artery to the donor aorta and the pulmonary artery of the donor was connected to the internal jugular vein of the recipient. In this manner the donor coronary system was perfused with the recipient's blood. After 4 days the donor heart showed signs of irregular pulse and histologically was filled with lymphocytes, mononuclear cells, and neutrophils suggestive of the rejection process.

Another animal model involved implantation of a second heart into the chest cavity. Demikhov (see Cooper, 1984) transplanted a second heart with and without lungs into the chest cavity of a recipient dog (Figure 10.4).

In 1967 Barnard reported the transplantation of a human donor heart to a 57-year-old man; however the patient died of pneumonia and an autopsy of the heart showed features of acute rejection. Improved patient selection, use of immunosuppressive drugs, prevention of infection, and early diagnosis of acute rejection has led to further improvement in results obtained using this procedure

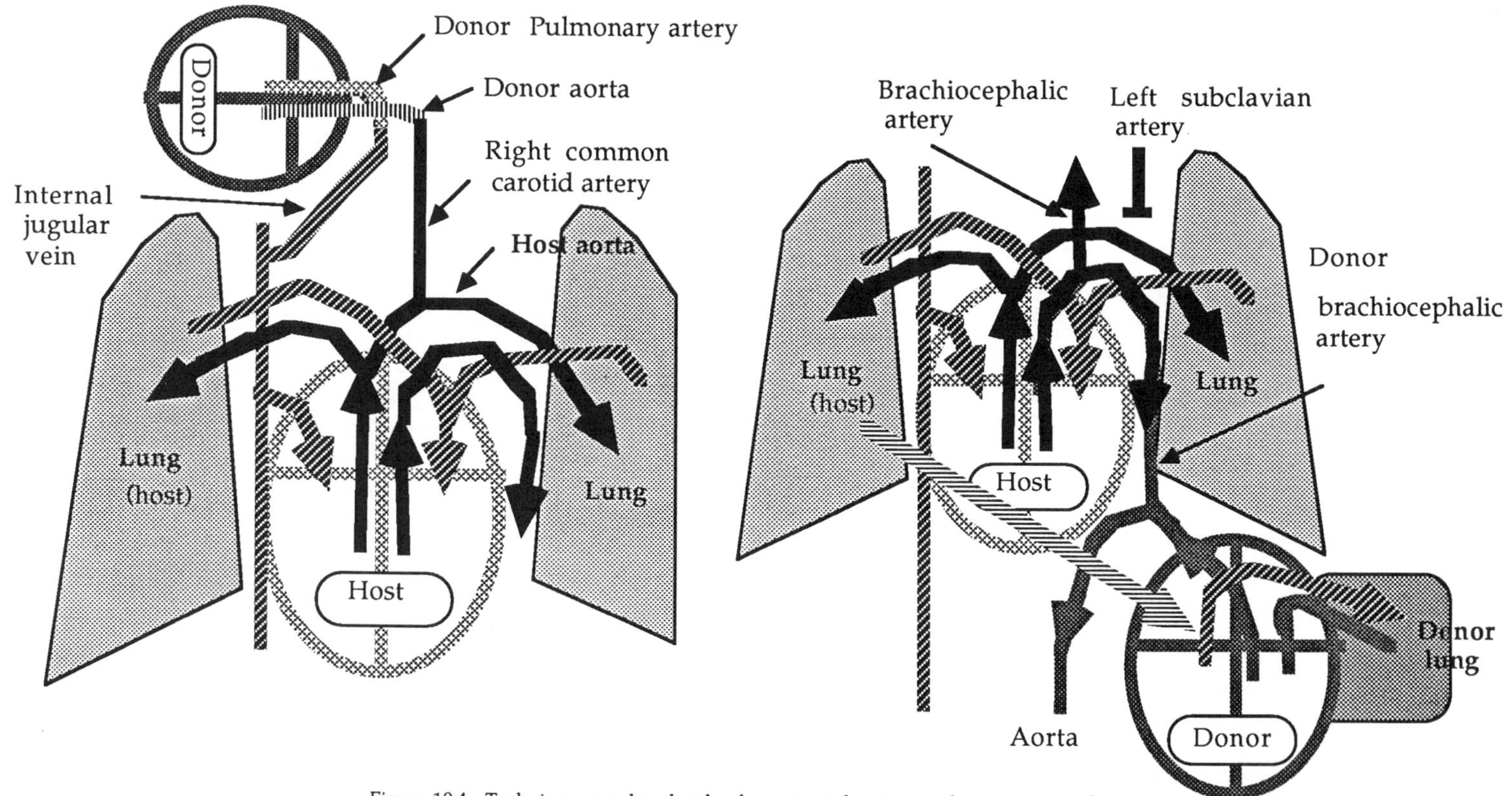

Figure 10.4. Techniques used to develop heterotopic heart transplantation procedures. Diagram illustrating models used to develop heart transplantation procedures including the external coupling through the neck (left) and the internal parallel coupling of a donor heart (right).

(Cooper, 1984).

Prospective candidates for a heart transplant must not be greater than 45 years of age or have an infection, insulin dependent diabetes, malignant tumor, pulmonary infarction, active peptic ulcer, peripheral vascular disease, irreversible dysfunction of any organ or exhibit drug addiction or psychological instability (Commerford, 1984).

Storage of donor hearts in cold saline provides protection against ischemic damage to the myocardium for up to 3-4 hr (Roe *et al.*, 1977). For longer storage periods (up to 48 hr) perfusion methods have been developed that pump a phosphate buffered salt solution containing a variety of nutrients (Cooper *et al.*, 1983) through the heart to be transplanted.

Transplantation in general requires careful matching of donor and recipient to avoid rejection of the transplant. Important tests and procedures in tissue matching include (1) donor-recipient blood antigen (ABO blood groups) compatibility, (2) donor-recipient human leukocyte antigen (HLA) matching, (3) lymphocytotoxicity in mixed leukocyte cultures of donor and recipient cells, and (4) pretransplant perfusion of recipient's blood through donor heart.

Blood-group matching between donor and recipient is necessary to prevent rejection of the heart due to antibodies directed against incompatible A or B red cell antigens present on vascular endothelium of the graft (Starzl *et al.*, 1964; Wilbrandt *et al.*, 1969). In addition, expression of the HLA antigens A, B, and C (class I HLA antigens) on the surface of nucleated cells requires that antibodies present in the recipient's blood not react with antigens present in the donor heart. There are 20 known subclasses or types of A antigens, 40 types of B antigens, and 8 of C antigens. The donor antigenic determinant subclasses are matched to those of the recipient in order to minimize the possibility of transplant rejection. Normally the donor heart will be transplanted into a recipient with the greatest number of matched subclasses of the A, B, and C antigens.

Other antigens present on B-lymphocytes (DR) are tested for using serum from the host while another class of antigen termed D is tested for by mixing cultures of lymphocytes from recipient and donor. Since the mixed lymphocyte culture test requires up to 6 days incubation it is usually not considered as a criterion for donor-recipient matching. Transplants from donors sharing two HLA—DR antigens with the recipient appear to fare better than those with only one or no shared antigens (Goeken *et al.*, 1982).

Transplant rejection involves a cellular response involving T-lymphocytes, killer cells, and macrophages that occurs within hours of the surgical procedure and an antibody dependent response that takes approximately 2 weeks to become effective. T-lymphocytes involved in the rejection process include cytotoxic, helper, and suppressor T-cells. Lysis of transplanted cells in capillaries is the basis of transplant rejection and can be induced by T-lymphocytes, natural killer cells, macrophages, and complement. Macrophages and killer cells require antibody for killing and therefore their effect is delayed until antibody has been produced (Herberman, 1983). Cytotoxic T-cells are activated by direct contact with an antigen on the cell surface (Herberman, 1983) and can kill the transplant cell directly. Antibodies produced by B-lymphocytes in the presence of complement can cause transplant cell death (Muller-Eberhard,1983).

T-cell cytotoxicity is thought to be the primary mechanism by which transplant cells are destroyed (Thomas and Thomas,1980). HLA-A, B, or C antigens are thought to stimulate T-cell activation. Killer-cell cytotoxicity as

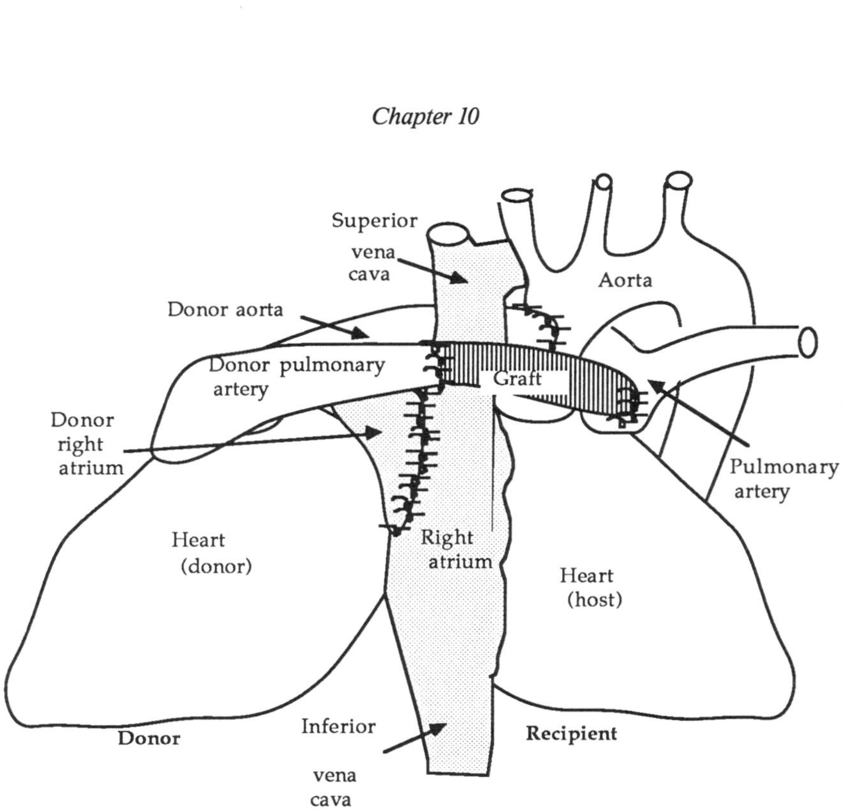

Figure 10.5. Heterotopic heart transplantation. Diagram illustrating perfected surgical procedure for heterotrophic heart transplantation.

well as activated macrophages and complement dependent cell lysis also contribute to rejection (Thomas *et al* 1984; Hall *et al.*, 1978; Descamps *et al.*, 1979).

Two standard operations are used for heart transplantation. They include direct replacement of the recipient's heart with a donor heart in the correct anatomical position (orthotopic procedure) and a second procedure by which the donor heart is placed in series with the recipient's heart (heterotopic procedure, Figure 10.5). Using the second procedure the donor heart is placed in the right chest alongside the recipient organ and blood flows through either or both hearts (Novitzky and Cooper, 1984). Patients that receive transplants using either technique are postoperatively placed on antibiotics and immunosuppressive drugs.

Although the extent of progress of the rejection process cannot be judged based on the gross appearance of the heart, increased heart mass is associated with transplants that are undergoing rejection (Rose *et al.*, 1979; Nowygrod *et al.*, 1983). Short-term (acute) signs of rejection are characterized by mononuclear cells including lymphocytes, monocytes, and neutrophils within the capillary and arterial lumen that pass through the walls of small blood vessels of the transplant. Cardiac macrophages appear activated within the endocardium and endothelial cells appear to be surrounded by fluid (edema) and are even detached (Rose and Uys, 1984). If untreated, additional fluid containing fibrin and activated lymphocytes accumulates between cardiac muscle cells. As the rejection process proceeds major arteries such as the coronary show deposition of fibrin and the presence of blood clots that occlude the vessel thereby decreasing or stopping blood flow to the heart.

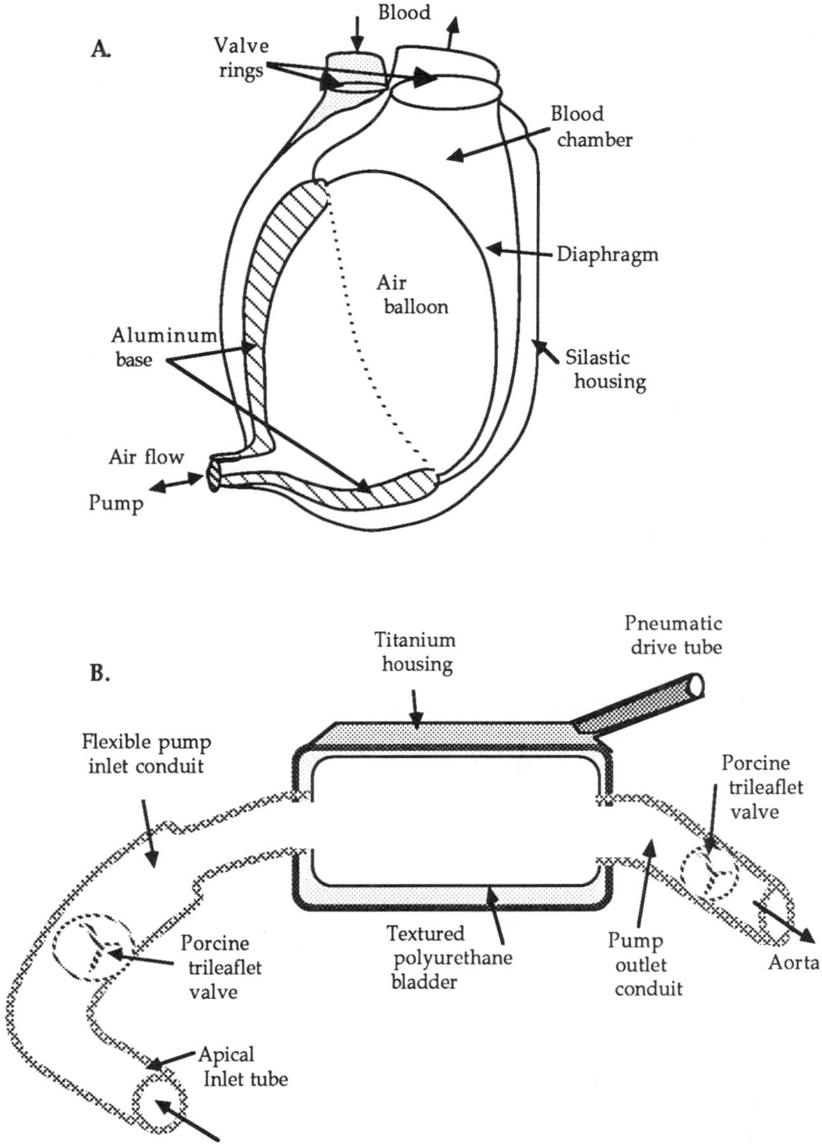

Figure 10.6. Assist and total replacement devices. Diagram illustrating the Jarvik design of an artificial heart (A) and a left ventricular assist device (B).

Administration of an immunosuppressive agent after a period of 1-2 weeks can reverse the signs of acute rejection. Infection is the major cause of death (Baumgartner *et al.*, 1979; Cooper *et al.*, 1983) during the first few months after heart transplantation for patients who are optimally immunosuppressed. At long term the problem becomes that of chronic rejection.

Chronic rejection is characterized by thickened epicardial branches of the coronary arteries with a marked reduction of the lumen for blood flow by blood clots (Rose and Ulys , 1984). Intimal lining cell proliferation leads to progressive obliteration of the lumina of the epicardial branches of the main coronary arteries during chronic rejection. In addition to vascular changes, accumulation of lymphoid cells is also observed.

Patients with depressed heart performance either before or after cardiac surgery can be managed using drugs or mechanical heart assist devices. Patients that fail to respond to pharmacotherapy require left ventricular assist devices (Figure 10.6).

Temporary mechanical-assist devices have been reported to be successfully used in the clinic to increase cardiac output (Bregman *et al.*, 1977). The use of an intraaortic balloon positioned in the thoracic aorta that is inflated during diastole at the closing of the aortic valve dates back to the 1960s (Moulopoulos *et al.*, 1962; Kantrowitz *et al.*, 1969). Failure of intraaortic pumping devices is primarily a consequence of the insertion of the device (Isner *et al.*, 1980) through the femoral artery into the iliac artery and into the aortic arch.

The materials aspect of assist devices have been carefully reviewed in designing blood pumps for assist devices and designs for totally implantable artificial hearts.

Biomer, a polyether polyurethane polymer, has been used to line the pump in the Jarvik-7 artificial heart (Chiang *et al.*,1984). Pumps lined with Biomer have test pumped for more than 4.5 years without failure (Chiang *et al.*, 1984). Reports suggest that this material is inert (Murray and Olsen, 1985), however, some evidence of protein, lipid deposition, and calcification has been reported (Owen and Gone, 1981). Calcification leads to loss of pump lining flexibility and premature failure and has limited many artificial heart designs. Examination of the human recipients of Jarvik-7 total artificial heart (Figure 10.6) showed no evidence of gross calcification (Murray and Olsen,1985).

10.8. Summary

The field of cardiovascular implants has flourished since the 1950s when the first polymeric implants were used to replace damaged areas of the aorta. Development of vascular implants to replace heart valves and small- and large-diameter vessels has been a result of intensive research efforts. Synthetic polymeric tubes of Dacron are the prosthesis of choice for aortic replacement while biological vessel autografts are preferred for coronary bypass procedures. Total heart replacement or heart assist devices continue to be a biomaterials challenge since all synthetic materials currently used in these applications eventually fail due to thrombosis. Design of a total heart replacement is still hampered by complications associated with blood compatibility and therefore scientists must learn more about blood-material interactions to solve this problem.

References

Awad, J.A. and Guidoin, R.G. In: Biocompatibility of Tissue Analogs, edited by D.F. Williams, CRC Press, Boca Raton, Florida, p. 103.

Barnard, F.C., Priestley, J.T., Markowitz, J. and Yater, W.M. Arch. Surg. 26,219,1933.

Barnard, C.N. S. Afr. Med. J. 41,1271,1967.

Baumgartner, W.A., Reitz, B.A., Oyer, P.E., Stinson, E.B. and Shumway, N.E. Curr. Probl. Surg. 16,3,1979.

Bennett, J.G., Trono, R., Norman, J.C. and Cooley, D.A. Cardiovascular Diseases, Bulletin of the Texas Heart Institute, 4,18,1977.

Black, M.M., Daniel, C.L., Lawford, P.V., Stoves, J.L. and Trowbridge, E.A. In: Biocompatibility of Tissue Analogs, edited by D.F. William, CRC Press, Boca Raton, Florida, p. 153.

Blakemore, A.H. and Voorhees, A.B. Jr. Ann. Surg. 140,324,1954.

Boretos, J.W. and Eden, M. Contemporary Biomaterials, Noyes Publication, New Jersey, p. 388,1984.

Boyce, B. In: Biologic and Synthetic Vascular Prostheses, edited by J.C. Stanley, Gune and Stratton, New York, p. 553,1982.

Bregman, D., Bailin, M., Bowman, F.O., Jr., Parodi, E.N., Haubert, S.M., Edie, R.N., Spotnitz, H.M., Reemtsma, K. and Malm, J.R. Ann. Thorac. Surg. 24,574,1977.

Buja, L.M. In: Basic Pathology edited by S.L. Robbins and V. Kumar, W.B. Saunders Co., Philadelphia, p. 312,1987.

Burkel, W.E., Ford, J.W., Venter, D.W., Kahn, R.H., Graham, L.M. and Stanley, J.C. Trans. Amer. Soc. Artif. Intern. Org. 28,178,1982.

Burkel, W.E., Vinter, D.W., Ford, J.W., Kahn, R.H., Graham, L.M. and Stanley, J.C. J. Surg. Res. 30,305,1981.

Burnham, S.J., Flanigan, P., Goodreau, J.J., Yao, J.S.T. and Bergan, J.J. Surgery 8,417,1978.

Campbell, C.D., Brooks, D.H., Webster, M.W., Diamond, D.L., Peel, R.L. and Bahnson, H.T. Surgery 85,177,1979.

Chiang, B.—Y., Cohen, S.R., Virmani, R., Lawanson, W. and Roberts, W.C. Amer. J. Cardiol. 45,260,1980.

Chiang, B.—Y., Olsen, D.B., Dries, D., Hamanaka, Y., Murray, K.D., Iiyia, E. and Kolff, W.J. Trans. Amer. Soc. Artif. Intern. Org. 30,514,1984.

Chignier, E., Monties, J.R., Butazzoni, B., Dureau, G. and Eloy, R. Biomaterials 8,18,1984.

Chin, H.P., Harrison, E.C., Blankenhorn, D.H. and Moacanin, J. Circulation 43/44 (suppl. I), p. 51,1971.

Christensen, J. and Eklof, B. Scand. J. Thorac. Cardiovasc. Surg. 11,43,1977.

Cohn, L.H. Chest 85,387,1984.

Commerford, P.J. In: Heart Transplantation, edited by D.K.C. Cooper and R.P. Lanza, MTP Press Ltd, Lancaster, Pennsylvania, p. 15,1984.

Cooper, D.K.C., Lanza, R.P., Oliver, S., Forder, A.A., Rose, A.G., Ulys, C.J., Novitzky, D. and Barnard, C.N. Thorax 38,822,1983.

Cooper, D.K.C. In: Heart Transplantation, edited by D.K.C. Cooper and R.P. Lanza, MTP Press Ltd, Lancaster, Pennsylvania, p. 1,1984.

Cooper, D.K.C., Wicomb, W.N., Boyd, S.T., Lanza, R.P., Novitsky, D., Hassoulas, J. and Barnard, C.N. Transplant Proc. 15,1232,1983.

Crawford, E.S., DeBakey, M.E., and Cooley, D.A. Arch. Surg. 76,261,1958.

Dardik, D. In: Vascular Graft Update: Safety and Performance, ASTM STP—898, edited by H.E. Kambic, A. Kantrowitz, and P. Sung, American Society for Testing and Materials, Philadelphia, Pennsylvania, p. 50,1986.

Dardik, I. and Dardik, H. Surg, Gyn. Obstet.140,567,1975.

Dardik, H., Ibrahim, I.M. and Dardik, I. Surg. For. 26, 286,1975.

Dardik, H., Ibrahim, I.M., Sussman, B.C., Kahn, M. and Dadik, I. In: Biological and Synthetic Vascular Prostheses, edited by J.C. Stanley, Gune and Stratton, New York, p. 445,1982.

DeBakey, M.E., Cooley, D.A., Crawford, E.S. and Morris, G.C.J. Am. Surg, 24,862,1958.

Descamps, B., Gagnon, R., Van der Gaag, R., Feuillet, M.—N. and Crosnier, J. J. Clin. Lab. Immunol. 2,303,1979.

Edwards, W.S. Surgery 45,298,1959.

Edwards, W.S. and Tapp, J.S. Surgery 38,61,1955.

Ehrenfeld, W.K., Stoney, R.J. and Wylie, E.J. In: Biologic and Synthetic Vascular Prostheses, edited by J.C. Stanley, Gune and Stratton, New York, p. 291,1982.

Goeken, N.E., Thompson, J.S. and Corry, R.J. Transplantation 32,522,1982.

Graham, L.M., Vinter, D.W., Ford, J.W., Kahn, R.H., Burkel, W.E. and Stanley, J.C. Arch. Surg. 15,929,1980.

Graham, L.M., Burkel, W.E., Ford, J.W., Vinter, D.W., Kahn, R.H. and Stanely, J.C. Surgery 91,550,1982.

Graham, L.M. and Bergan, J.J. In: Biologic and Synthetic Vascular Prostheses, edited by J.C. Stanley, Gune and Stratton, New York, p. 563,1982.

Gupta, S.K. and Veith, F.J. Amer J. Surg. 104, 214,1980.

Haimov, M., Giron, F. and Jacobson, J.H. Arch. Surg. 114,673,1979

Halpert, B., DeBakey, M.E. and Jordan, G.L. Surg. Gyn. Obstet.111,659,1960.

Hall, B.M., Dorch, S. and Roser, B. J. Exp. Med. 148, 878,1978.

Hall, G.W., Liotta, D., Ghidoni, J.J., DeBakey, M.E. and Dressler, D.P. J. Biomed. Mat. Res. 1,179,1967.

Hearn, A.R. and Charlesworth, D. Surgery 85,607,1979.

Herberman, R.B. In: The Biology of Immunologic Disease, edited by F.J. Dixon and D.W. Fisher, Sinauer Associates, Inc., Sunderland, Massachusetts, p. 75,1983.

Henry, R., Harvey, R.A. and Greco, R.S. J. Thorac. Surg. 82, 272,1981.

Herring, M., Gardner, A. and Glover, J. Surgery 84,498,1978.

Hufnagel, C.A. Surgery 37,165,1955.

Hylen, J.C., Kloster, F.E., Starr, A. and Oriswold, H.E. Ann. Intern. Med. 72,1,1970.

Hylen, J.C., Hodam, R.P. and Kloster, F.E. Ann. Thorac. Surg. 13,324,1972.

Isner, J.M., Cohen, S.R., Virmani, R., Lawrinson, W. and Roberts, W.C. Amer. J. Cardiol. 45,260,1980.

Kambic, H.E., Kantrowitz, A. and Sung, P. (eds). Vascular graft update. Safety and performance American Society for Testing and Materials, Philadelphia, Pennsylvania,1986.

Kantrowitz, A., Tjonneland, S., Freed, P.S., Phillips, S.J., Butner, A.N. and Sherman, J.R. J. A. M. A. 203,135,1969.

Kantrowitz, A. In: Vascular Graft Update: Safety and Performance, ASTM STP 898, edited by H.E. Kambic, A. Kantowitz, and P. Sung, American Society for Testing and Materials, Philadelphia, Pennsylvania, p. 44,1986.

Kenney, D.A., Sauvage, L.R., Wood, S.J., Berger, K., Davis, C.C., Smith, J.C., Ritenhouse, E.A., Hall, D.G., and Mansfield, P.B. Surgery 92,931-946,1982.

Leather, R.P. and Karmody, A.M. In: Biologic and Synthetic Vascular Protheses, edited by J.C. Stanley, Gune and Stratton, New York, p. 351,1982.

Matsumoto, H., Hasegawa, T. and Fuse, K. Surgery 74,518,1973.

Mann, F.C., Priestley, J.T., Markowitz, J. and Yater, W.M. Arch. Surg. 26,219,1933.

Meade, J.W., Linton, R.R. and Darling, R.C. Arch. Surg. 93,392,1966.

Moulopoulos, S.D., Topaz, S. and Kolff, W.F. Amer. Heart J. 63,669,1962.

Muller-Eberhard, H.J. In: The Biology of Immunologic Disease, edited by F.J. Dixon and D.W. Fisher, Sinauer Associates, Inc., Sunderland, Massachusetts, p. 128,1983

Murray, K.D. and Olsen, D.B. In: New Developments in Cardiac Assist Devices, edited by S. Attar, Praeger Scientific, New York, p. 54,1985.

Noon, G.P. and Debakey, M.E. In: Vascular Grafts, edited by P.N. Sawyer and M.J. Kaplitt, Appleton—Century—Crofts, New York, p. 177,1977.

Novitzky, D. and Cooper, D.K.C. In: Heart Transplantation, edited by D.K.C. Cooper and R.P. Lanza, MTP Press, Ltd, Lancaster, Pennsylvania, p. 103,1984.

Nowygrod, R., Spotnitz, H.M., Dubroff, J.M., Hardy, M.A. and Reemsta, K. Transplant. Proc. 15,1225,1983.

Owen, D.R. and Gone, R.M. Trans. Am. Soc. Artif. Intern. Org. 27,528,1981.

Perloff, L.J. In: Biologic and Synthetic Vascular Prostheses, edited by J.C. Stanley, Gune and Stratton, New York, p. 365,1986.

Rahimtoola, S.H. J. Amer. Coll. Cardiol. 1,199,1983.

Robbins, S.L. and Kumar, V. Basic Pathology, W.B. Saunders Co., New York, p. 345,1987.

Roberts, W.C., Fishbein, M.C. and Golden, A. Amer. J. Cardiol. 35,740,1975.

Roe, B.B., Hutchinson, J.C., Fishman, N.H., Ullyot, D.J. and Smith, D.L. J. Thorac. Cardiovasc. Surg. 73,366,1977.

Rose, A.G., Uys, C.J., Losman, J.G. and Barnard, C.N. S. Afr. Med. J. 56,880,1979.

Rose, A.G. and Uys, C.J. In: Heart Transplantation, edited by D.K.C. Cooper and R.P. Lanza, MTP Press Ltd, Lancaster, Pennsylvania, p. 157,1984.

Sauvage, L.R., Berger, K., Davis, C.C., Smith, J.C., Rittenhouse, E.A., Hall, D.G. and Mansfield, P.B. In: Biologic and Synthetic Vascular Prostheses, edited by J.C. Stanley, W.E. Burkel, S.L. Lindernauer, R.H. Bartlett, and J.C. Turcotte. Gune and Stratton, New York, p. 533,1982.

Sauvage, L.R., Smith, J.C., Davis, C.C., Rittenhouse, E.A., Hall, D.G. and Mansfield, P.B. In: Vascular Graft Update: Safety and Performance, edited by H.E. Kambic, A. Kantrowitz and P. Sung, American Society for Testing and Materials, Philadelphia, Pennsylvania, p. 16,1986.

Sauvage, L.R., Berger, K., Wood, S.J., Nakagawa, Y. and Mansfield, P.B. Surgery 70,940,1971.

Sawyer, P.N., Stanczewski, B. and Mistry, F.D. In: Biologic and Synthetic Vascular Prostheses, edited by J.C. Stanley, Gune and Stratton, New York, p. 476,1986.

Schoen, F.J., Titus, J.L. and Lawrie, G.M. J. Biomed. Mat. Res.16,559,1982.

Schoen, F.J., Titus, J.L. and Lawrie, G.M. J. Biomed. Mat. Res. 16,559,1982.

Schoen, F.J., Titus, J.L. and Lawrie, G. M. J. A. M. A. 249,899,1983.

Schoen, F.J., Goodenough, S.H., Ionescu, M.I. and Braunwald, N.S. J. Thorac. Cardiovasc. Surg. 88,208,1984.

Schoen, F.J. and Levy, R.J. Cardiol. Clin. 2,717,1984.

Schoen, F.J. and Hobson, C.E. Hum. Pathol. 16,549,1985.

Schoen, F.J. J. Biomed. Mat. Res. 21,91,1987a.

Schoen, F.J. J. Card. Surg. 2,65,1987b.

Shah, A., Dolgen, A.M., Tice, D.A. and Trehan, N. Amer. Heart J. 96,407,1978.

Silver, M.D., Hoppenhoefer, H., Heggtveit, H. and Reif, T.H. Artif. Org. 9,270,1985.

Silver, M.D. and Wilson, G.J. Circulation 56,617,1977.

Silver, M.D. J. Thorac. Cardiovasc. Surg. 79,693,1980.

Stanely, J.C., Graham, L.M., Whitehouse, W.M., Linderhauer, S.M., Zelenock, G.B. and Cronenwett, J.L. In: Biologic and Synthetic Vascular Prostheses, edited by J.C. Stanley, Gune and Stratton, New York, p. 333,1982.

Stanley, J.C., Burkel, W.E., Lindenauer, S.M., Bartlett, R.H. and Turcotte, J.G. (eds). Biologic and Synthetic Vascular Prostheses. Gune and Stratton, Inc., New York, 1982.

Stanley, J.C., Graham, L.M. and Burkel, W.E. In: Vascular Graft Update: Safety and Performance, ASTM STP 898, edited by H.E. Kambic, A. Kantrowitz, and P. Sung. American Society for Testing and Materials, Philadelphia, Pennsylvania, p. 33,1986.

Stansel, H.C., Fenn, J.E., Tilson, M.D. and Laks, H. Arch. Surg. 114,1291,1979.

Starr, A. Ann. Thorac. Surg. 42,124,1986.

Starzl, T.E., Marchioro, T.L., Holmes, J.H., Hermann, G., Brittain, R.S., Stonington, O.H. and Talmadge, D.W. Surgery, 55,195,1964.

Synder, R.W. and Botzko, K.M. In: Biologic and Synthetic Vascular Prostheses, edited by J.C. Stanley, Gune and Stratton, New York, p. 485,1982.

Thomas, F. and Thomas, J. In: Renal Transplantation—A Multidisciplinary Approach, edited by S.N. Chatterjee, Raven Press, New York, 1980.

Thomas, J., Thomas, F., Kaplan, A. and Lee, H.M. Transplantation 22,940,1976.
Ulys, C.J. and Rose, A.G. In: Heart Transplantation, edited by D.K.C. Cooper and R.P. Lanza, MTP Press, Ltd., Lancaster, Pennsylvania., p. 243,1984.

Veith, F.J., Moss, C.M., Fell, S.C., Montefusco, C., Rhodes, B.A. and Haimovici, H. Surg. Gyn. Obstet,147,749,1978.

Veith, F.J., Ascer, E. and Gupta, S.K. In: Vascular Graft Update: Safety and Performance, ASTM STP 898, edited by H.E. Kambic, A. Kantrowitz, and P. Sung, American Society for Testing and Materials, Philadelphia, Pennsylvania., p. 25,1986.

Veith, F.J., Moss, C.M., Sprayregen, S. and Montefusco, C. Surgery 85,253,1979.

Voorhees, A.B., Jaretski, A. and Blakemore, A.H. Ann. Surg.135,332,1952.

Wilbrandt, R., Tung, K.S.H., Deodhar, S.R. and Waddell, W.R Amer. J. Clin. Pathol. 51,15,1969.

Wildeviur, Ch.R.H., van der Lei, B. and Schakenraad, J.M. Biomaterials 8,418,1987.

Wright, C.B., White, R.A., Hiratzka, L.F., Morin, R.P. and Mitts, D.L. In: Vascular Graft Update: Safety and Performance, ASTM STP 898, edited by H.E. Kambic, A. Kantrowitz, and P. Sung, American Society for Testing and Materials, Philadelphia, Pennsylvania, p. 60,1986.

Chapter 11

Biomaterials For Drug Delivery

11.1. Introduction

Drugs are most often administered orally by the ingestion of tablets or capsules or via subcutaneous, intramuscular, or intravenous injection. Although these methods of drug administration are well documented they have several drawbacks. These drawbacks include the requirement for multiple dosage regimens, variations in plasma and tissue concentrations, and drug distribution to non-target tissues. Controlled release technology refers to sustained or prolonged release of drugs at constant (zero-order) release rates. Zero-order release kinetics require that the amount of drug released is independent of the drug concentration remaining in the device. Ideally, the drug is delivered only to the affected tissue. Examples of controlled release systems have been reviewed in the literature (Langer and Peppas, 1981; Heller, 1980; Kopecek, 1984; Kligman, 1984; Graham and Wood, 1984; Bruck, 1983a; Chien, 1987) for the interested reader. The purpose of this chapter is to give an overview of the polymeric materials that have been used in controlled release devices and the drugs that have been studied.

11.2. Types of Controlled Delivery Systems

Controlled drug release may be achieved by mechanical pumps, osmotic pumps, diffusion-controlled systems containing reservoirs or matrix systems, chemically controlled systems composed of biodegradable or nonbiodegradable polymers, swelling-controlled systems, and magnetically controlled systems (Table 11.1 and Figure 11.1).

Mechanical pumps have been used for delivery of anticoagulants, insulin, chemotherapeutic and other agents (Bruck, 1983a; Blackshear, 1979). The pump can be either implanted directly into the desired tissue or it can be connected intravenously. Totally implantable pumps require the expense and trauma of

Table 11.1. Control Drug Delivery Devices

Type	Reference
Mechanical pumps	Bruck, 1983b; Blackshear, 1979
Osmotic pumps	Bruck, 1983b; Eckenhoff and Yum, 1981; Theeuwes and Eckenhoff, 1980
Diffusion-controlled systems	
Reservior systems	Langer and Peppas, 1981; Graham and Wood, 1984;
Matrix systems	Bruck 1983b; Khanna and Speiser, 1970; Goodman and Banker, 1970
Chemically controled systems	Langer and Peppas, 1981; Graham and Wood, 1984;
Biodegradable systems	Bruck 1983b; Sidman *et al.*, 1983; Heller *et al.*, 1983;
Drug-polymer systems	Boylan and Banker, 1973; Yolles *et al.*, 1975b; Weadock *et al.*, 1986; Kopecek, 1984
Swelling controlled systems	Langer and Peppas, 1981
Magnetically controlled systems	Langer and Peppas, 1981; Hsieh *et al.*, 1981; Langer *et al.*,1980

pumps is complicated by the incidence of infection with prolonged use.

Osmotic pumps (Bruck, 1983b) consist of a reservoir of water soluble drug in a water-permeable membrane through which the diffusion constant of the drug is very small. Drug is forced through a small channel as a result of water diffusion into the device through the polymer membrane. Release follows first-order kinetics if the solution inside of the pump is saturated with drug (Theeuwes and Eckenhoff, 1980; Eckenhoff and Yum, 1981).

In chemically controlled systems, drug is released either by biodegradation of a polymer in which the drug is physically trapped or by biodegradation of chemical bonds between a polymer and the drug (Langer and Peppas, 1981; Kopecek, 1984; Graham and Wood, 1984; Sidman *et al.*, 1983; Heller *et al.*, 1983; Boylan and Banker, 1973; Yolles *et al.*, 1975a). Systems which involve

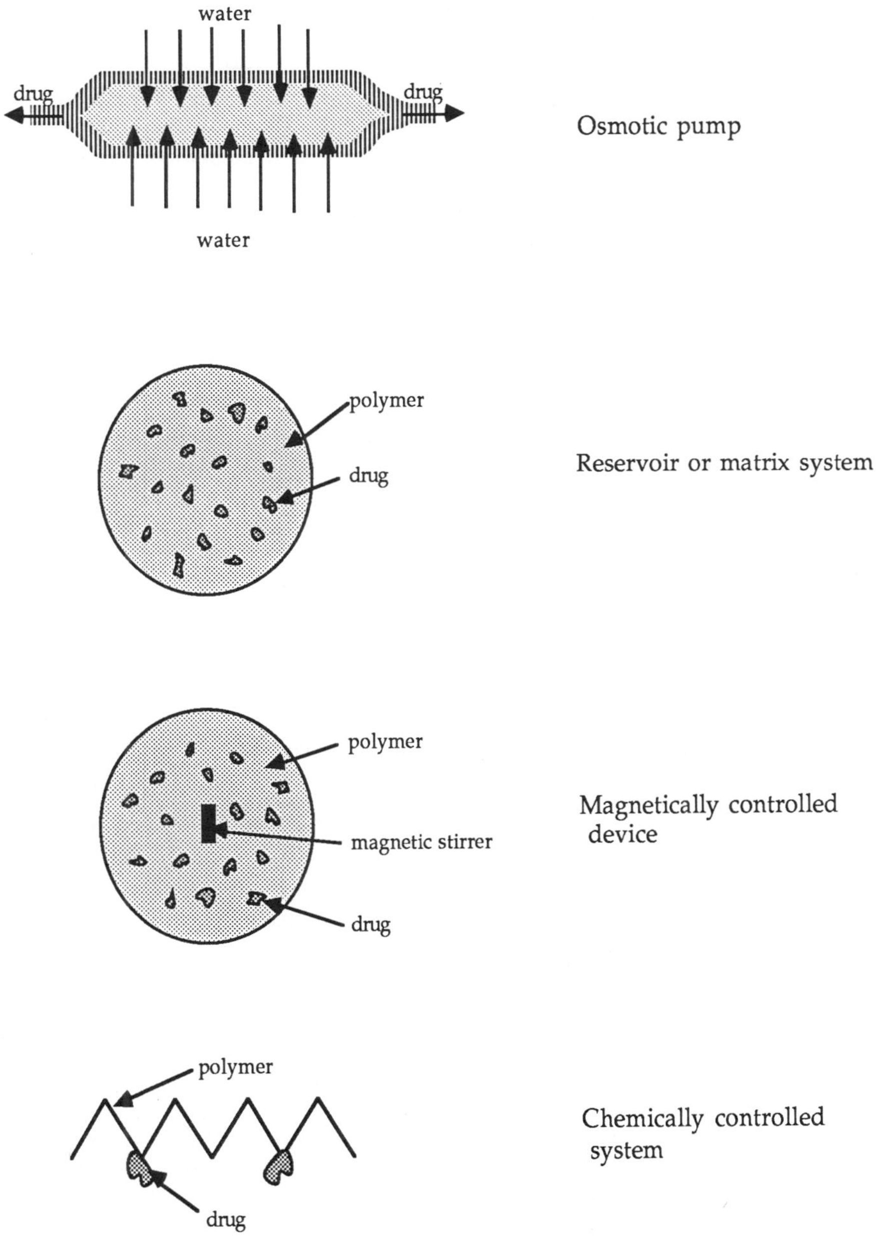

Figure 11.1. Sustained drug release sytems. Sustained drug release is achieved using osmotic pumps, reservoir or matrix systems, magnetically controlled devices, and systems that chemically couple drug to a polymer.

physically trapped drug are formed by coprecipitation of the polymer-drug complex. The polymer can then be crosslinked by conventional means such as exposure to bifunctional reagents or via a free radical process. The activity of the drug must be evaluated after coprecipitation to ensure that it is maintained. Biodegradation of the polymer via enzyme-catalyzed hydrolysis in tissue fluids, releases the drug. Polymer degradation products must be evaluated for short and long term biocompatibility.

Chemically controlled systems consisting of a drug-modified polymer depend on biodegradation of an unstable bond between the drug and polymer for release of the drug. The unstable bond is hydrolyzed in the presence of water.

The physical form of the polymer in chemically controlled systems can vary. For instance, the polymer can be in the form of a bulk matrix or in the form of microscopic beads of polymer-encapsulated drug. The advantage of microporous beads is that they can be injected beneath the skin and do not require surgery prior to their use.

Table 11.2. Drug Release Via Transdermal Applications[a]

Drug released	Pathological condition
Scopolamine	Motion Sickness
Clonidine	Hypertension
Nitroglycerin	Angina Pectoris
Isosorbide dinitrate	Angina Pectoris
Estradiol	Postmenstral Syndrome
Bupranolol	β-adrenergic Blocker

[a] Based on Chien, 1987 and Weiss *et al.*, 1987

Diffusion-controlled systems involve either a uniform distribution of the drug within matrix or reservoir-type devices that use a polymeric membrane to separate the core of drug from the surrounding environment. Matrix devices are composed of solid polymer in which the drug is dissolved or dispersed. Release occurs by diffusion from the matrix. Release from reservoir devices is controlled by diffusion through the polymeric membrane according to Fick's law. If the polymeric membrane swells in an aqueous environment, diffusion may also occur through liquid-filled pores depending on the drug polarity.

Transport of drug molecules in diffusional systems is dependent on the value of the translational diffusion coefficient, D. The value of D depends on the size and shape of the drug molecule as well as the polarity. Details of the mathematics of diffusion are summarized elsewhere (Chien, 1982).

Swelling controlled systems involve dispersion of a drug in a polymer matrix and evaporation of the solvent. Initially there is very little drug diffusion through the polymer since the polymer chains are rigid; however, after the polymer chains begin to swell and become flexible the rate of diffusion increases. Magnetically controlled systems contain magnetic beads that are uniformly dispersed within the polymer matrix. The device initially operates in a similar manner to a diffusion-controlled system; however, when an external magnetic field is applied the rate of drug release is increased (Langer and Peppas, 1981; Hsieh *et al.*, 1981; Langer *et al.*, 1980).

Controlled drug delivery can be achieved by surgical implantation of a system, injection of bead systems beneath the skin and via diffusion through the skin, (transdermal controlled medication).

Transdermal-controlled medication involves placing or adhering the release system on the surface of the skin. The release system is designed such that drug diffuses through the stratum corneum of the skin into the dermis where uptake into the blood and transport via the circulation occurs. Transdermal delivery has many advantages including that it does not require surgical intervention. Table 11.2 gives a list of drugs that have been released via the transdermal approach. Transdermal drug delivery is effective for drugs that have molecular weights of 500 or below. Chemicals that enhance the permeation of high-molecular-weight drugs through the skin are a subject of intensive investigation.

11.3. Polymers Used in Drug Delivery Systems

A variety of polymers are used in drug delivery systems. These include silicone polymers, polyethylene, polypropylene, polyvinyl chloride, polyvinyl fluoride, polyvinylidene fluoride, polymethyl methacrylate, polyurethane, hydrogels and collagen (Table 11.3). Biocompatibility criteria vary depending on whether the device will only contact the epidermis such as for transdermal drug delivery or if it is used internally.

11.4. Design of Sustained Release Systems: Basic Concepts

Administration of drugs to combat pathological as well as normal physiological processes is rather complex because of the need to transport the drug to the tissue site, prevent enzymatic or other modification of the drug, and absorption of the drug into target cells. Therefore it is necessary to understand how drugs enter the blood stream and find their way to target cells without exerting toxic or undesired effects.

A measure of the risk-to-benefit ratio of a drug is assessed by the therapeutic index. This index is a ratio of the minimum drug concentration required to result in a high degree of toxicity to the maximum drug concentration that is required to result in a high degree of response and a low probability of toxicity. Drugs with therapeutic indexes below 3 are difficult to effectively administer without a high probability that toxicity will result.

The toxicity of a drug and resulting therapeutic index is complicated by the fact that the concentration of a drug required to result in toxicity is dependent on the cell type. Rapidly dividing cells are typically more sensitive to drug toxicity than are dormant cells. Therefore, the therapeutic index of a drug

Table 11.3. Studies of Polymers Used in Drug Release Devices

Polymer	References
Silicone	Lucas and Moore, 1979; Rudolf and Abraham, 1980; Folkman and Long, 1964
Polyethylene	Williams, 1979; Bradley *et al.*, 1973; Desai *et al.*, 1965
Polypropylene	Apple *et al.*, 1984; Mook *et al.*, 1983
Polyvinyl chloride	Taylor *et al.*, 1973; Sharma and Vasudevan, 1982; Monroe *et al.* 1980
Polyvinylidene fluoride	Aoshima *et al.*, 1982
Polymethacrylates	Gourlay *et al.*, 1978; Cowsar *et al.*, 1976; Ebert *et al.*, 1980; Tuttle *et al.*, 1980
Polyurethane	Lyman *et al.*, 1972
Hydrogels	Yale and Greer, 1982; Dickinson *et al.*, 1981; Langer *et al.*, 1981; Marchant *et al.*, 1983
Collagen (gelatin)	Heller, 1980; Chvapil *et al.*, 1973; Doillon *et al.*, 1984; Klein, 1983; Shakespeare and Griffiths, 1980
Ethylene-vinyl acetate (copolymer)	Langer and Folkman, 1976; Langer and Peppas, 1981; Stewart and Novak, 1978; Pharris *et al.*, 1976; Zador *et al.*, 1976; Creque *et al.*, 1980; Preis and Langer, 1979
Polylactic acid, polyglycolic acid, and copolymers	Nilsson *et al.*, 1975; Abrahams and Ronel, 1975; Schwope *et al.*, 1975; Kulkarni *et al.*, 1966; Kulkarni *et al.*, 1971; Frazza and Schmitt, 1971; Yolles *et al.*, 1970; Woodland *et al.*, 1973; Yolles *et al.*, 1975b; Mason *et al.*, 1976; Jackanicz *et al.*, 1973; Anderson *et al.*, 1976

varies from tissue to tissue.

It is an oversimplification to state that all sustained drug delivery systems should release a constant amount of drug per unit area and per unit time interval to ensure that blood levels of the drug are time independent. However, because of the complexity of drug metabolism, the zero-order design approach is a good approximation for the design of a device. For instance, a release system may be desired to initially release a high concentration of drug followed by a constant concentration. This would allow for an initial saturation of drug receptors and/or binding sites thereby resulting in a maximally rapid therapeutic response. One might also design a system that releases increasing amounts of drug over increasing time intervals to overcome drug deactivation or binding to tissues. A variety of possible requirements can be conceived for the time rate of release of drugs from delivery systems. These requirements are dictated by the type of drug, mechanism of action of the drug, and the route of administration.

Most of us are familiar with oral administration of drugs. It is convenient to introduce a tablet, capsule, or caplet into the mouth and swallow it along with water. However, once we swallow a drug in the form of a pill that contains binders or a capsule containing drug, a series of events are triggered that determine what fraction of the drug actually is absorbed into our circulatory system.

The initial event involves disintegration and disaggregation of the binder (tablet) or capsule coating. If these processes do not occur in the gastrointestinal (GI) tract then the drug will not be efficiently absorbed into the blood. After the drug is released from its container, it is dissolved in GI fluids and absorbed through the GI wall into the circulation and is immediately processed by the liver. The efficiency of drug absorption is dependent on the fraction of drug released and enzymatically degraded in the GI tract. Unabsorbed drug is lost in the stool.

The drug may be transformed in the liver, activating or inactivating it, and some fraction of the drug and its metabolites are excreted into the urine. The rate of removal of the drug by the liver is dependent on several factors including binding affinity to red blood cells and blood proteins.

Drugs can be administered by means other than orally to eliminate first-pass metabolism by the liver. Direct diffusion of drugs into the circulation is achieved in the oral cavity (sublingual and buccal) and nasal cavity (nasal). Injection below the skin (subcutaneous) and into muscle (intramuscular) are other routes for direct administration. For some drugs direct injection into the cardiovascular system is possible. The absorption rate for oral, nasal, subcutaneous, and intramuscular routes is dependent on the location, degree of vascularity, and solubility of drug. These alternate routes of administration produce more rapid and complete drug absorption compared to oral drug administration.

The concentration of drug in the blood is dependent on the diffusion rate of the drug through vessel walls into tissue, the extent of binding to tissue components, and the diffusion rate through the cell membrane. Drug is lost due to excretion into bile, by deactivation by the liver, and by excretion in the urine.

There are a wide variety of factors that affect the concentration of drug in cells. All these factors must be considered in designing and administering a constant concentration of a drug to cells. Below we discuss the types of conditions and the drugs of choice that may be of interest to the biomaterials scientist.

11.5. Drugs of Choice

The biomaterials scientist must understand medical pharmacology and the drugs of choice in treating a condition before he or she can effectively design a sustained release system. Therefore, it is necessary to review some of the types of conditions that may be ameliorated or cured by sustained release of a drug. In addition, the drugs that have been proven to be effective for specific clinical manifestations will be discussed.

11.5.1. General Analgesics

Reduction of pain is an everyday problem in medicine associated with chronic conditions and with damage to internal body structures. Drugs of choice to treat chronic pain include opioids and nonopioid analgesics.

Opioids resemble morphine in their ability to bind to specific cellular opioid receptors that are associated with synapses of the brain. The opioids may trigger release of natural substances that mediate the sense of euphoria.

Morphines are prescribed for chronic pain associated with cancer, heart attacks, pregnancy, and fluid build up in the lung. Morphine sulfate (Table 11.4 and Figure 11.2) is the prototype for this type of drug. It is administered orally, intramuscularly, intravenously, and through absorption in the oral cavity. A dose of approximately 20-70 mg of morphine sulfate is required when the drug is taken orally every eight hours.

Nonopioid analgesics do not bind to cell receptors but ameliorate pain by blocking peripheral pain impulses via inhibition of prostaglandin synthesis. In comparison to opioids which are addictive, these drugs are not addictive. Salicylic acid derivatives (aspirin) as well as acetaminophen (for patients who cannot tolerate the gastrointestinal irritation of aspirin) are the drugs of choice. The recommended dosage of these drugs when taken orally is between 0.5 and 1.0 g every 3 hs (Table 11.4).

Sustained Release of Analgesics

Nonopioid analgesics are taken orally every 3 hours to combat chronic pain. Bakan and Powell (1983) reported the microencapsulation of acetaminophen and aspirin (Figure 11.1) to decrease the rate of absorption of these drugs when taken orally. The rate of analgesic absorption is slowed down and the time period over which the drug concentration in the blood is above the threshold required for activity is prolonged.

Microencapsulation techniques involve coacervation and phase separation of the drug and polymer coating into particles or droplets 0.1-5000 μm in diameter. The drug is mixed in a solvent containing a dissolved polymer at high temperature. The polymer comes out of solution as the temperature is lowered and coats the drug particles. Bakan and Powell (1983) studied microencapsulation and long-term stability of ethylcellulose coated microparticles containing acetaminophen or aspirin (Table 11.5).

11.5.2. Drugs Used to Treat Glaucoma

Glaucoma is a disease that is associated with diabetes. This disease is characterized by increased intraocular pressure resulting in damage to the optic nerve and loss of visual field. The eye is composed of a protective transparent layer, the cornea, which covers the anterior chamber containing aqueous humor and is separated from the posterior chamber by the iris. The iris adjusts the amount of light that is focused by the lens on the retina which stimulates the optic nerve and initiates signal transmission to the brain. Aqueous humor is produced by a layer of cells (ciliary epithelium) that lines the posterior chamber

Table 11.4.Types of Drugs and Their Use in Clinical Medicine

Drug classification	Disease or condition	Prototype drug	Method of administration	Amount of drug	Dosing interval
Analgesic-opioid	Chronic pain myocardial infarction Pulmonary Idema	Morphine Sulfate	Oral Intramuscular / subcutaneous Intravenous	20-70 mg 10 mg 2.5 t0 15 mg	2 - 3 times per day
Anagesic-nonOpioid	Internal pain	Salicylic acid	Oral	1g	Every 3 hours
		Acetominophen	Oral	0.5g	Every 3 hours
Mitotics	Glaucoma	Pilocarpine	Ocular	1 drop of 1% or 2% solution	Every 4 - 8 hours
		Epinephrine	Ocular	1.1% solution	Once or twice daily
Vasodilators	Angina	Nitroglycerin	Oral	10 - 40 mg	3 - 4 times per day
			Sublingual	0.15 - 0.6 mg	3 times per 15 minutes
			Ointment	15 mg / inch	Every 4 to 8 hours
		Propanol Hydrochloride	Oral	10 - 20 mg up to 400 mg	3 - 4 times Daily
Anticoagulants	Thromboembolism prophylaxis	Heparin	Subcutaneous	5,000 units	2 hours before surgery and every 12 hours
			Intravenous	5,000 units	Every 4- 6 hours
		Dicumarol	Oral	200 - 300 mg	Daily
Hormones	Premenstrual syndrome	Progesterone/ Megestrol Acetate	Intramuscular Oral	100 - 200 mg 40 mg	4 times daily
	Menopause	Estrogen/ Estradiol	Intramuscular	0.5 - 1.5 mg	2 to 3 times per week
	Osteoporosis	Estrogen/ Estradiol	Intramuscular	0.5 - 1.5 mg	2 - 3 times per week
Contraceptive	Fertility control	Estrogen and progesterone	Oral	0.125 - 10 ng 30 - 100 mg	Once per day
		Progesterone alone	Oral	0.075 - 0.35 mg	Once per day

and is transported into the posterior chamber and then through the pupil into the anterior chamber of the eye.

Glaucoma results from an elevated intraocular pressure which is clinically treated by decreasing resistance to aqueous humor outflow, decreasing aqueous humor production, or reducing the volume of intraocular fluids.

Table 11.4 (cont.) Types of Drugs and Their Use in Clinical Medicine

Drug Classification	Disease or Condition	Prototype Drug	Method of Administration	Amount of Drug	Dosing Interval
Blood Glucose Regulation	Diabetes	Insulin	Subcutaneous	5 to 10 units	Before morning or evening meal
			Intravenous	6 to 10 units	Per hour
			Intramuscular	5 to 10 units	Hourly
		Acetohexamide	Oral	0.5 - 1.5 g	Daily
Infection	Anti-Infectious Agents	Silver Sulfadiazine	Topical	1 to 3 mm thick	once or twice daily
		Mafenide	Topical	1 mm thick	2 or 3 times daily
		Povidone - Iodine	Topical	Varies	Varies
			Oral	1.6 - $3.2X10^6$units	Daily
			Intramuscular		
			Intravenous	1.2 - $24X10^6$	Daily
		Cefazolin sodium	Intramuscular		
			Intravenous	1 - 6 g/day	In 3 - 4 doses
		Erthyromycin	Oral	0.25 - 0.5 g	4 times per day
			Intravenous	1 to 4 g	Daily
		Tetracycline	Oral	0.25 - 0.5 g	4 times per day
			Intravenous	0.25 - 0.5 g	2 times per day
		Chloro-amphenicol	Oral	0.05 mg/kg per day	4 equal doses per day
			Intravenous	0.05 mg/kg per day	4 equal doses per day
		Gentamicin sulfate	Intramuscular	3 to 5 mg/kg	3 equal doses per day
		Sulfadiazine	Oral	4 to 6 g	Daily
Anti-Neoplastic	Cancer	Cyclophos-phamide (alkylating agent)	Intravenous	0.5 -1.5 g/ M^2	2 -4 week intervals
			Oral	0.06 - 0.12 g/M^2	2-4 week intervals
		Carmustine (BCNU) (nitrosurea)	Intravenous	0.200 g/M^2	Every 6 weeks
			Topical	0.5 - 3 mg/ml	Every 6 weeks
		Methotrexate sodium (folic acid analogue)	Oral	2.5 - 5 mg	Daily
			Intramuscular	0.025 g/M^2	Once or twice weekly
			Intraarterial	0.8 g/M^2	14-21 days
		Fluorouracil (pyrimidine analogue)	Topical	5% cream	Twice daily
		Bleomycin sulfate (antibiotic)	Intramuscular	10 - 20 mg/M^2	Once or twice weekly in each case.
			Intravenous	10 - 20 mg/M^2	
			Subcutaneous	10 - 20 mg/M^2	

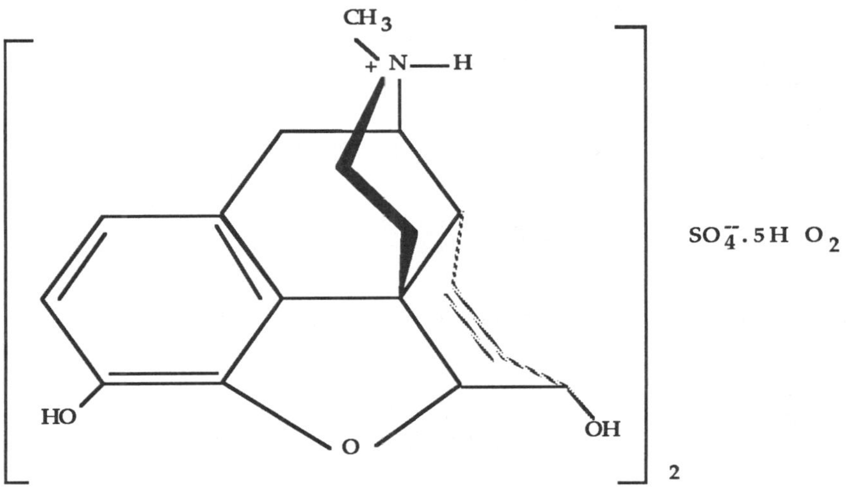

Morphine sulfate - an opioid analgesic

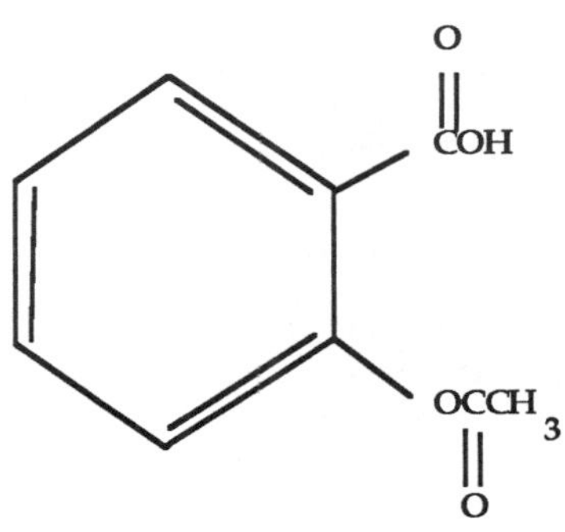

Aspirin- a nonopioid analgesic

Figure 11.2. Chemical structures of opioid and nonopioid drugs. Analgesia is achieved through the use of opioid and nonopioid drugs. An example of each of these classes of drug are aspirin (nonopioid) and morphine sulfate (opioid).

Primary open-angle glaucoma occurs through increased resistance to outflow of aqueous humor. Drug treatment with miotics such as pilocarpine stimulate parasympathetic effector cells that constrict the pupil and ciliary muscle decreasing resistance to outflow and intraocular pressure. Epinephrine, a beta-blocker is also used to decrease resistance to outflow of aqueous humor (Table 11.4).

Table 11.5. Sustained Delivery of Various Drugs

Drug type	Drug delivery system	References
Analgesic		
Aspirin Acetominophen	Ethylcellulose microparticles	Bakan and Powell, 1983
Mitotic		
Pilocarpine	Trilaminar unit containing 2 ethyl-vinyl acetate membrane	Worthen *et al.*, 1974; Stewart and Novak, 1978; Bensinger *et al*, 1978
Antiangina		
Nitroglycerin	Polymeric unit containg drug with adhesive backing	Rezakovic, 1987
Anticoagulant		
Heparin	Polyurethane, Silicone, Polypropylene, Polyester, Teflon	Grode *et al.*, 1968; Yen and Rembaum, 1971; Mijama *et al.*, 1977
Hormone Progestin	Silastic Tubing. Ethylene/vinyl acetate	Diaz *et al.*, 1977; Pharris *et al.*, 1976
Contraceptive		
Estrogen and Progestin Norethisterone	Silicone rubber ring and collagen band, Poly(l-lactic acid)	Anderson *et al.*, 1976; Beck *et al.*,1979; Beck *et al.*, 1981
Blood Glucose Regulation		
Insulin	Ethylene-vinyl acetate hemisphere, Magnetically controlled	Creque *et al.*, 1980; Hsieh *et al.*, 1981; Hsieh and Langer, 1983
Anti-Infective		
Chloramphenicol Penicillin Sulfadiazine Tetracycline Tetracycline	Glycosaminoglycans, Liposomes Poly(lactic acid) Methacrylate copolymer trilaminate Ethylene-vinyl acetate fibers	Sparer *et al.*, 1983; Gregoriadis, 1979; Wise *et al.*, 1979; Olanoff and Anderson 1979; Goodson *et al.*, 1983
Anti-cancer		
Hydrolzed polymer Flourinated nucleotides Anthracycline 5-fluorouracil L-asparaginase 5-flourouracil Mitomycin C Adrianmycin and Daunomycin Actinomyocin D Cytosine arabinoside Methotrexate	Poly(carboxylate)-pyran Polynucleotides DNA from sperm whale Acrylic polymer Lectin Albumin Gelatin Antibodies Liposomes Liposomes liposomes	Ghosh and Sulumar, 1985; Baldwin *et al*, 1985; Alderfer *et al.*, 1985; Trouet *et al.*, 1979; Hartsough and Gebelein,1985; Shier, 1979; Morimoto *et al.*, 1980; Kato *et al.*, 1981; O'Neill, 1979; Ryman, 1983; Gregoriadis, 1979; Rustum *et al.*, 1979; Kimelberg, 1980
Growth Factors		
EGF EGF	Synthetic steroid Silver sulfadiazne	Brightwell *et al.*, 1985; Brown *et al.*, 1986

Sustained Delivery Of Pilocarpine

Alza Corporation has commercially developed a system to continuously deliver pilocarpine to the eye. The Ocusert pilocarpine system consists of two outer membranes (ethylene-vinyl acetate) with a central reservoir of drug. When placed in the corner of the eye, pilocarpine diffuses across the two outer polymeric membranes that control release (Worthen *et al.*, 1974; Stewart and Novak, 1978; Bensinger *et al.*, 1978). Two units are available, Pilo-20 and Pilo-40. These units release pilocarpine at the rate of 20 μg/hr and 40 μg/hr, respectively for 7 days (Table 11.5).

11.5.3. Drugs to Treat Angina

Angina pectoris is a symptom of increased work demand on the heart characterized by attacks of chest pain caused by increased oxygen demand of the heart. Pathologically, the oxygen demand of the heart falls short of inducing heart muscle-cell death (infarction) and the injury induced is thought to be reversible (Robbins and Kumar, 1987). Increased oxygen demand of the heart can sometimes lead to angina in individuals who have narrowed coronary arteries. Effort-induced angina is classically treated with nitrates and calcium-blocking agents.

Nitrates act as vasodilators improving flow to the heart muscle. Nitroglycerin is the drug of choice in treating angina because of its rapid action and low cost. This drug can be delivered orally or through use of a rapidly absorbed aerosol. Intravenously administered nitroglycerin is used for unstable angina in hospitalized patients.

Beta-blocking agents are also administered to block beta receptors and sympathetic nerve impulses which reduces heart rate, cardiac output, and blood pressure. Propanolol hydrochloride is a beta blocker used in long-term management of patients with angina. Its use reduces the frequency of anginal attacks and decreases nitroglycerin requirements in many patients (Table 11.4).

Transdermal Release Of Nitroglycerin

Delivery of nitroglycerin through the skin has been brought to commercial fruition (Table 11.5). Systems have been developed that release this drug from an ointment or a polymeric film (Chien, 1987). The nitroglycerin-impregnated polymer is bonded via adhesive to the skin.

Transdermal nitroglycerin patches are reported to be less effective than other long-acting nitrates (Drug Evaluations, 1986) since absorption through the skin is variable and adequate plasma levels are difficult to achieve (Crean *et al.*, 1985; Parker and Fung, 1984).

11.5.4. Anticoagulant Therapy

Blood clotting can result in the production of an embolism (thromboembolism) that may get trapped in vessels in the lungs and brain. It is a complication of diseases such as cancer, diabetes, heart failure, and surgery.

Damage to blood vessels initiates coagulation of blood proteins and clotting factors involved in the intrinsic and extrinsic coagulation system by exposure to extravascular tissues. The extrinsic pathway is activated by tissue thromboplastin (cellular components of fibroblasts and endothelial cells) which results in platelet activation and aggregation. Factor XII of the intrinsic clotting cascade is activated by contact with exposed collagen fibers in the damaged area and thrombin is generated. Fibrinogen is converted to fibrin in the presence of thrombin which stabilizes platelet aggregates at the site of injury (see Section 4.2.1.1).

Thrombus formation resembles the process of hemostasis that prevents loss of blood when an injury occurs; however, thrombus formation occurs intravascularly. Venous thrombi occur in the absence of damage to the vessel wall in regions of flow that are slow or where a flow disturbance is present. Arterial thrombi have a greater content of platelets and develop at sites of vascular narrowing or flow irregularity. Arterial thrombi do not cause occlusion as readily as venous thrombi.

Heparin, a natural sulfated polysaccharide, reduces fibrin formation by acting to inhibit thrombin activity. Clinically, it is used to help maintain blood fluidity in extracorporeal circulation during open heart surgery and renal hemodialysis. It is also given to patients who exhibit deep venous thrombosis or pulmonary infarction due to thromboembolism. Pulmonary thromboembolism is reduced after abdominal-thoracic surgery when a prophylactic heparin injection is given.

Dicumarol is a coumarin derivative and is an anticoagulant than can be taken orally. It is long acting and is completely absorbed from the gastrointestinal tract (Table 11.4).

Sustained Release of Heparin

Polymeric release systems containing heparin were initially developed as potential nonthrombogenic surfaces for use as a small vessel prosthesis and for the pump lining in the artificial heart. However, it was soon discovered that although heparin inhibited blood coagulation it did not prevent adhesion and aggregation of platelets.

Heparin is a highly polyionic polysaccharide due to the presence of free sulfate as well as free carboxylic groups. Therefore binding of heparin to charged polymer surfaces is easily accomplished. Heparin is bound to nonpolar polymer surfaces via ammonium salts attached to hydrocarbon chains such as triodecylmethyl ammonium chloride (TDMAC). Grode *et al.* (1968) used TDMAC to bind heparin to polyurethane, silicone rubber, polypropylene, polyester, teflon, and a copolymer of polycarbonate and silicone rubber. Other investigators (Table 11.5) used ionic bonding to immobilize heparin to polyvinyl chloride, copolymers (Yen and Rembaum, 1971), polyether polyurethane (Mijama *et al.*, 1977) and polyvinyl chloride homopolymer (Tanzawa *et al.*, 1973).

11.5.5. Treatment of Gynecological Conditions

A great deal of effort has been directed at the use of steroids and other drugs in treating gynecological conditions. Before these pathological conditions are discussed we will give an overview of estrogen and progesterone release and their relationship to the menstruation cycle.

Since we are dealing with a cycle (Figure 11.3) it is possible to follow the cycle starting at any point. Menstruation and loss of the mucosal surface of the uterus stimulates the hypothalamus to act on the pituitary to release follicle stimulating hormone releasing factor which results in follicle development in the ovary and subsequent release of an egg cell down the fallopian tubing. Release of the egg cell is associated with release of the hormone estrogen which causes the follicle to break up. After follicle destruction occurs, the estrogen level decreases, which stimulates the hypothalamus to cause the pituitary to release luteinizing hormone causing the development of the corpus luteum, the site of release of the egg cell in the ovary. Progesterone is released from the corpus luteum. If the egg cell is fertilized, progesterone is continually released from the corpus luteum, maintaining the proliferation of uterine mucosal cells. If the egg is not fertilized, the uterine mucosa is sloughed off resulting in stimulation of the hypothalamus and continuation of the cycle.

Estrogens present in premenopausal women include estradiol 17-B, the major estrogen, and estrone a less potent form of estradiol. Both of these estrogens are secreted by the ovary. Estriol, a much weaker estrogen, is formed by peripheral metabolism of ovarian estrogens.

Estrone is also produced by peripheral conversion of adrenal and ovarian androstenedione, which is converted to testosterone and then into estrogen. In premenopausal women this conversion accounts for 25% of the estrone produced and the balance is secreted by the ovary. In postmenopausal women, peripheral conversion of androstenedione to estrone is the principal source of estrogen. The conversion rate of adrenal and rostenedione is increased in postmenopausal woman.

Natural estrogen or progesterone generally are not useful orally because of rapid deactivation by the liver. Progestins are synthetic compounds possessing progestational activity and are useful for oral and topical administration. The addition of a chemical group to estradiol (17 α-ethinyl) enhances oral activity by inhibiting liver degradation.

Progesterone has been used to reduce the irritability, depression anxiety, and headache associated with premenstrual syndrome (Table 11.4 and Figure 11.4). Oral estrogens are used to treat menopausal symptoms. Osteoporosis is a condition characterized by loss of bone density and occurs in 35-40% of white women over 60 years of age. Treatment of osteoporosis with estrogen (Horsman *et al.*, 1983) and progestin (Upton, 1982) prevents bone loss.

Sustained Release of Estrogens and Progesterones

A variety of studies have been reported concerning release of estrogens and progesterones from delivery systems. Diaz *et al.* (1977) reported the use of silicone tubing for release of progestin, R-2323. Tubes of medical-grade Silastic® tubing were filled with progestin and implanted subcutaneously using an 11-gauge

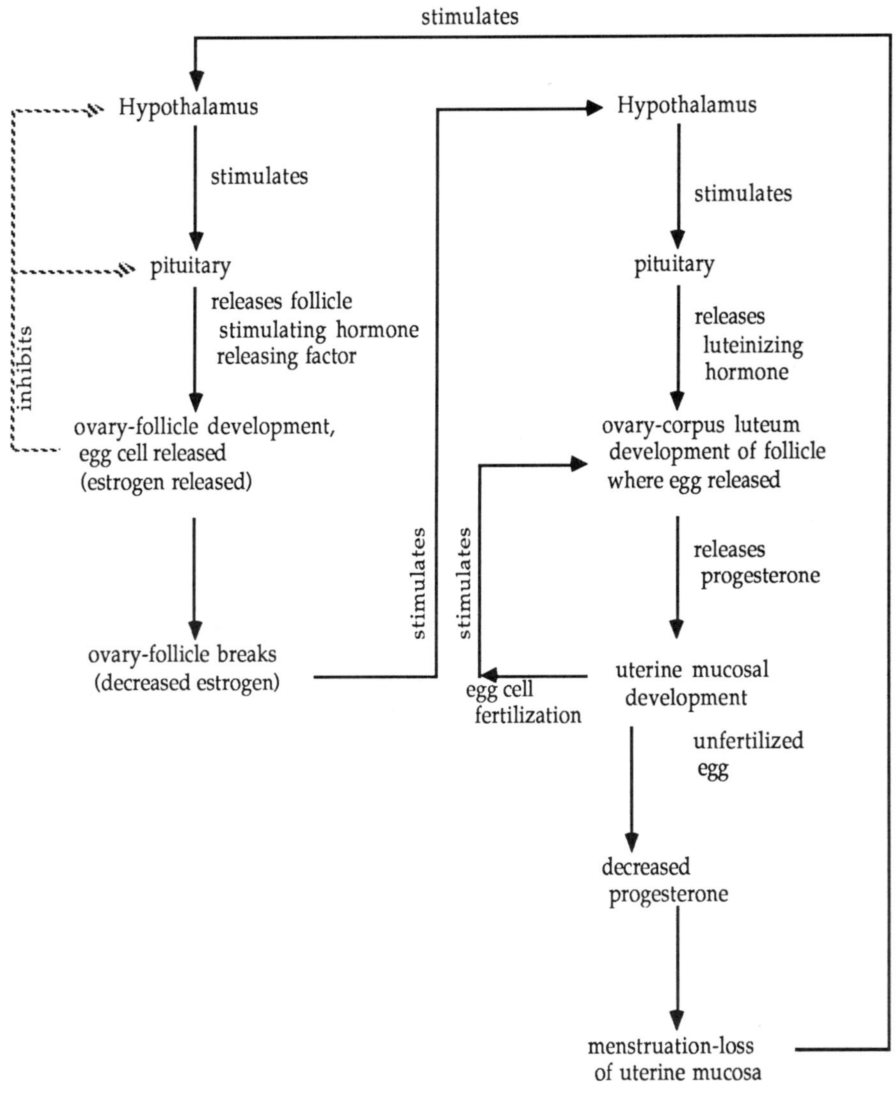

Figure 11.3. Menstration cycle. Diagram illustrating the menstruation cycle and the release of estrogen by the ovary that is followed by stimulation of the hypothalamus. Release of luteinizing hormone by the ovary results in release of progesterone and uterine mucosal wall development. If the egg cell is unfertilized then progesterone present stimulates the hypothalamus to continue the cycle by release of estrogen.

thin- wall trocar. Unfortunately the silicone tubing had to be removed surgically under local anesthesia after the drug was released. Pharriss and coworkers (1976) report release of progesterone by diffusion from an ethylene/vinyl acetate copolymer (Table 11.5).

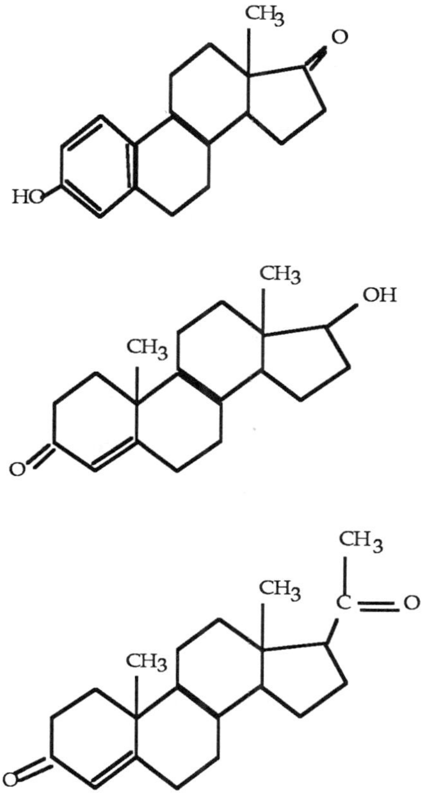

Estrogen

Testosterone

Progesterone

Figure 11.4. Structure of hormones. Diagram of the structure of the hormones progesterone, estrogen, and testosterone that are involved in the female reproductive cycle. Testosterone is converted to estrogen via normal biochemical pathways.

11.5.6. Contraceptive Therapy

Contraception can be prevented at any point in the cycle shown in Figure 11.3. Combinations of orally administered contraceptives or intramuscular release of progestins prevent ovulation.

The most common oral contraceptives include synthetic estrogen (ethinyl estradiol or mestranol) and a progestin (norethindrone and norgestrel). Other preparations contain progestin alone.

A variety of oral contraceptives are available in the United States that are taken daily for 21 days in 28 day cycles. These include combinations of a progestin (0.125-10 mg) and an estrogen (30-100 mg). In general, preparations containing the lowest amounts of steroid consistent with efficacy and tolerable side effects are preferred. A product containing less than 50 mg of estrogen is recommended for a first-time user.

The progestin-only pill causes menstrual side effects and is used only by patients who should avoid estrogen, i.e., women with a history of headaches or hypertension (Table 11.4).

Sustained Release of Contraceptive Agents

Fertility control using sustained release of contraceptive agents has received considerable attention. Oral contraceptives must be taken daily for 21 days in a 28 day cycle to prevent pregnancy effectively. Fertility control would be more effective if daily compliance were not necessary. Therefore the possibility of delivering contraceptive drugs for periods of weeks to months has been explored.

Croxatto *et al.* (1969) studied release of progesterone from silicone rubber capsules. These workers added 30 mg of megestrol acetate to a 22-mm length of tubing that was implanted subcutaneously. Other workers (Victor *et al.*, 1977; Nash *et al.*, 1978) used silastic to deliver a variety of contraceptive hormones (Kincl *et al.*, 1968; Coutinho *et al.*, 1970; Frick *et al.*, 1981).

Victor *et al.* (1977) prepared vaginal rings containing a combination of an estrogen (estradiol benzoate) and a progestin (d-norgestrel). Cylindrical rings of silicone rubber (diameter 60 mm) 9 mm thick were fabricated with a polysiloxane ring and a collagen band containing the contraceptive hormones. They report that the collagen band tended to loosen and separate from the silicone rubber ring.

Anderson *et al.* (1976) extruded rods of poly (*l*, lactic acid) containing 20% by weight norethisterone and then produced drug-loaded polymeric particles 90- 180 µm in diameter by cryogenic grinding followed by sieving. The particles were then suspended in 1% (w/v) methyl cellulose and injected subcutaneously through a hypodermic needle.

Beck *et al.* (1979, 1981) produced microcapsules of poly (*d,l*-lactic acid) containing norethisterone that were injected intramuscularly (Beck *et al.*, 1979). They later evaluated this release system clinically for contraceptive effectiveness (Beck *et al.*, 1981). Other systems involving poly (epsilon-caprolactone) and poly (*d, l*-lactic acid) and their copolymers have been studied (Table 11.5).

11.5.7. Regulation Of Blood Glucose

Diabetes mellitus, more commonly referred to as diabetes, is characterized by a deficiency of insulin activity associated with elevated levels of glucose in the blood and altered lipid metabolism. Associated with this disease is accelerated atherosclerosis affecting the circulation to the kidney and eye as well as peripheral and autonomic pathologies of nervous tissue.

Type I or insulin-dependent diabetes usually is first diagnosed during the adolescent years and is believed to be a result of autoimmune processes that result in destruction or abnormal development of the insulin-producing cells (islet cells) in the pancreas. This form of diabetes is treated by administration of insulin derived from humans or animals. Type II diabetes occurs in overweight middle-aged individuals and is usually treated by careful dietary regulation.

Insulins from porcine, bovine, and human tissues are given either intravenously, intramuscularly, or subcutaneously to type I diabetic patients. The kinetics of absorption and degradation can be modified to give rapid, intermediate, and long-lasting insulins. Crystalline zinc insulin when injected intravenously shows rapid onset of activity and short duration. Long acting insulins are conjugated with large protein molecules.

Sulfonylureas are chemicals that stimulate insulin secretion from beta cells and can be administered orally. The half life of these drugs is 1-36 hr; however, therapeutic effects are only possible if functional pancreatic islet tissue is present. These drugs stimulate release of insulin from pancreatic islet tissue by increasing the sensitivity of beta cells to glucose as well as by increasing insulin activity peripherally (Table 11.4).

Sustained Release Of Insulin

The delivery of insulin on a time-released basis is more complicated than the release of other drugs because the level of insulin in the blood must mirror the level of glucose present. Therefore, the insulin level delivered before meals should be higher than the threshold level present in the blood at other times. A variety of prototype designs have been developed that include a glucose sensor, insulin reservoir and pump, or are simple release devices without a glucose sensor (Blackshear and Rohde, 1983).

Hsieh and coworkers (Hsieh *et al.*, 1981; Hsieh and Langer, 1983) described a hemispherically shaped device made of ethylene-vinyl acetate copolymer into which insulin and a magnet were incorporated. The hemispherical device is implanted subcutaneously and a burst of insulin is achieved by external magnetic stimulation.

Another approach is the incorporation of insulin into a polymeric implant and its release by passive diffusion. Creque *et al.* (1980) reported the sustained release for one month of insulin from an vinyl acetate-ethylene copolymer. Bovine pancreatic insulin was released from this implant after subcutaneous implantation in the abdomen (Table 11.5).

11.5.8. Antiinfectious Agents

Infection occurs through contact with infectious organisms in the environment during day-to-day activities as well as from internal contact arising from surgery or through open wounds. Infectious agents include bacteria, viruses, fungi, protozoa, metazoa (worms), chlamydiae, and rickettsae. The exact mechanism of pathogenesis of these organisms is beyond the scope of this book. We will limit ourselves to a discussion of the commonly used antiinfectious drugs.

Infections associated with burns or other chronic wounds are due to a variety of gram-negative bacteria including *E. coli, Serratia marcescens,* and *Pseudomonas aeruginosa.* These pathogens are normally hospital-acquired and maybe resistant to conventional antibiotics (Drug Evaluations, 1986). Topical application of silver sulfadiazine or mafenide acetate are used. In addition, other agents such as povidone-iodine, silver nitrate, and gentamicin are used for prevention and treatment of infected burns.

Many surgical patients receive an antimicrobial injection to prevent development of a symptomatic infection. Chemoprophylaxis is most often successful when directed against a specific pathogen. However, broad spectrum antibiotics are normally used. Chemoprophylaxis for dental procedures includes penicillin G, erythromycin (for patients allergic to penicillin), ampicillin, cefazolin, and gentamicin.

Penicillins (Figure 11.5) are a group of antibiotics that contain 6-aminopenicillanic acid which contains an intact β–lactam ring required for activity. Penicillin G is extracted from cultures of *Penicillium chrysogenum*. Natural penicillins are also chemically modified by adding side chains to modify the antimicrobial spectrum. Ampicillin is an amino derivative of penicillin G. Penicillins act by interfering with normal development of bacterial cell walls. They have a wide variety of uses for infections involving the central nervous system, ear, lower respiratory tract, pneumonias and bone and joint and septic arthritis (Table 11.4).

Cephalosporins (Cefazolin, see Table 11.4) contain a β-lactam ring similar to the penicillins; however, cephalosporins are derivatives of 7-aminocephalosporanic acid. This compound is a derivative of Cephalosporin C which is a fermentation product of *Cephalosporium acremonium*. This group of antibiotics inhibit the final stage of bacterial cell-wall formation. The broad spectrum bactericidal activity and the high therapeutic/toxicity ratio makes this class of drugs useful in postoperative chemoprophylaxis.

Macrolide antibiotics contain a macrocyclic lactone ring to which sugars are attached. Erythromycin (Table 11.4) is a macrolide that is derived from bacterial strains *Streptomyces erythreus* and *S. antibioticus*. This antibiotic inhibits bacterial protein synthesis but does not affect mammalian cells. It is frequently used as an alternative to penicillin G for a number of infections in penicillin-allergic individuals.

Tetracycline antibiotics (Table 11.4 and Figure 11.5) were discovered by screening soil samples for antibiotic-producing organisms. Tetracycline is a naturally derived product of various species of *Streptomyces* bacteria and contains a hydronaphthacine group consisting of four fused rings. The mechanism of activity involves interfering with protein synthesis. Bacteria take up this drug in greater amounts than do mammalian cells. Clinically, this drug is used as a broad-spectrum antibiotic for a number of infectious diseases.

Chloramphenicol (Table 11.4) is a broad-spectrum antibiotic that is prepared synthetically but derives from *Streptomyces venequlae*. This drug is used to treat typhoid fever caused by *Salmonella* and meningitis caused by ampicillin-resistant organisms. However, because of its side effects on red blood cells it has limited usefulness. Its antibiotic activity occurs by inhibiting bacterial protein synthesis.

Aminoglycosides are naturally produced products derived from soil bacteria. This group includes Streptomycin (Figure 11.5), neomycin, and gentamicin and contains a six-membered aminocyclitol ring. The activity of these drugs is related to inhibition of bacterial protein synthesis under aerobic conditions.

Sulfonamides are synthetically derived molecules similar to para-aminobenzene that inhibit synthesis of bacterial precursors of DNA. These drugs were once widely used in the treatment of infectious diseases but their importance has diminished with the onset of bacterial resistance. They remain the drugs of choice for acute urinary tract infections caused by such organisms as *E coli*.

Sustained Release of Antiinfectious Agents

Delayed release of antibiotics is gaining interest in the research community

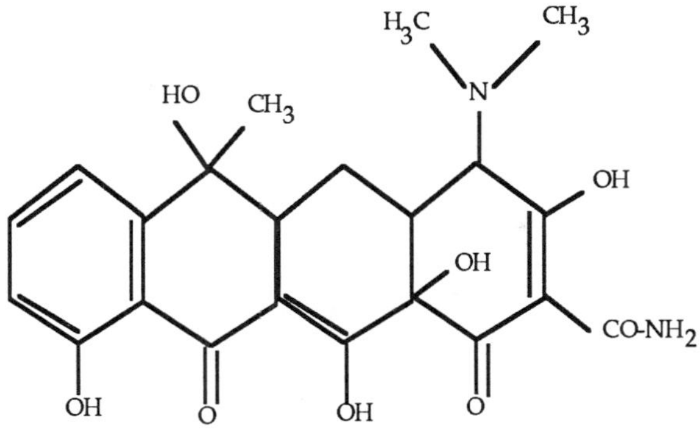

Tetracycline

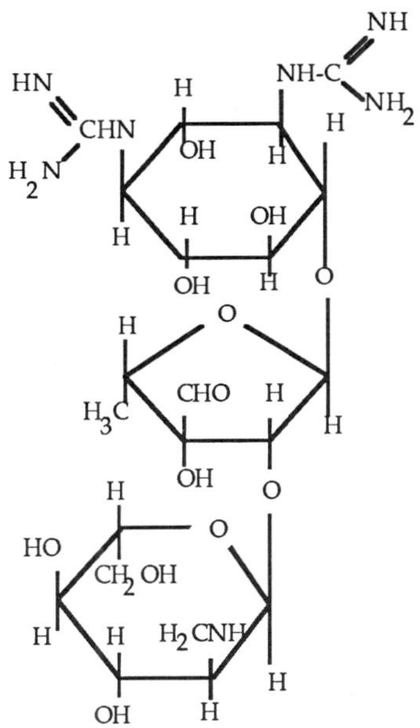

Streptomycin

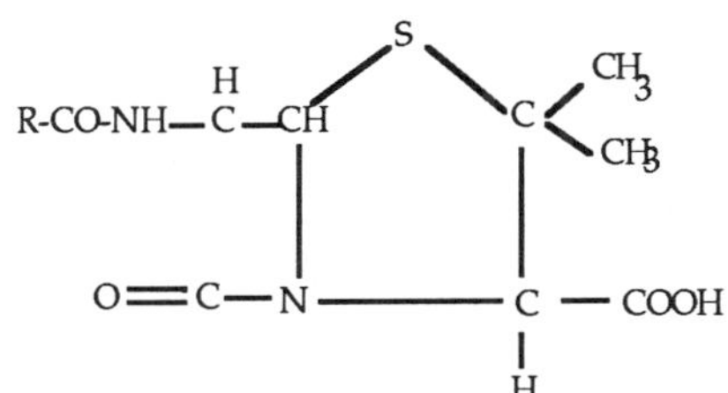

Penicillin

Figure 11.5. Structure of antibiotics. Diagram of the structures of several common antibiotics including streptomycin, tetracycline, and penicillin. Note the different ring structures that are common to different antibiotics.

due to the risk of infection associated with surgical procedures, implantation of medical devices, and the control of chronic open wounds. In general, the approach used involves coupling a broad spectrum antibiotic to a polymer or a complex of molecules, i.e., liposome.

Sparer *et al.* (1983) linked chloramphenicol to glycosaminoglycans and studied the release *in vitro* and *in vivo*. Gregoriadis (1979) described the use of liposomes that form self-contained spherical microspheres composed of lipids for the delayed release of penicillin. Liposomes containing penicillin were found to prolong circulation of the drug in the blood and alter its tissue distribution. Wise *et al.* (1979) reported the release of sulfadiazine in mice from a poly (lactic acid) bead matrix. Thirty-three percent by weight of sulfadiazine was mixed with poly (lactic acid) (molecular weight 150,000) and formed into beads that were 1.5 mm in diameter. Release of sulfadiazine from this system gave effective protection against repetitive weekly bacterial challenges for three months (Table 11.5).

Olanoff and Anderson (1979) developed a trilaminate disc system, similar to the Ocusert system containing pilocarpine, for releasing tetracycline. The membranes used were a copolymer of 2-hydroxyethyl methacrylate and methyl methacrylate and the device was implanted intraperitoneally in rats. Goodson *et al.* (1983) reported the release of tetracycline from a delivery system made of extruded ethylene-vinyl acetate fibers loaded with 25% drug. The fibers were placed and maintained in periodontal pockets for 10 days establishing a drug concentration of approximately 0.06%. Treated sites improved clinically as evidenced by a gain in periodontal attachment and a decrease in pocket depth.

11.5.9. Drugs Used in Cancer Chemotherapy

Cancer ranks second to heart disease in overall mortality rate in the United States. All cancers are masses of cells (tumors) that have unlimited potential to undergo cell division and to migrate via the cardiovascular and lymphatic systems throughout the body.

In the past, chemotherapy was used to treat tumors unresponsive to surgery and radiation therapy. However, it has been determined that chemotherapy combined with surgery and/or irradiation has increased survival rates for a number of solid tumors. The goal of chemotherapy is to achieve selective toxicity against malignant tumor cells and to leave normal cells unaffected. However, antineoplastic agents suppress proliferating cells. Rapidly growing cancers are more susceptible to antineoplastic agents than slower growing cancers. The doubling time of a tumor increases as the tumor volume increases and a tumor must have a diameter of 1 cm to be diagnosed.

Drugs used in chemotherapy specifically interfere with the cell cycle and cell division in cells that are rapidly dividing. Drugs including cytarabine and hydroxyurea exert their effects during DNA replication while vincristine and vinblastine prevent normal separation of chromosomes during cell division. Nitrogen mustards, cyclophosphamide, and nitrosoureas are alkylating agents (Table 11.4) that react with DNA leading to damage or miscoding of genetic material. Methotrexate is an analog of folic acid (Table 11.4) and inhibits DNA, RNA, and protein synthesis while flurouracil is a pyrimidene analog (a derivative of uracil) that prevents normal RNA synthesis. Antibiotics such as bleomycin sulfate interfere with the normal cell cycle (Table 11.4).

Sustained Release of Antineoplastic Drugs

Research into sustained delivery of antineoplastic drugs has been extensive because of the large numbers of patients who are treated with chemotherapeutic agents for prolonged periods. It is anticipated that sustained delivery of these drugs may be more effective in killing target cells with less systemic toxicity.

Polymeric materials and their hydrolysis products have been shown to possess antitumor activity. Review of the literature indicates that a poly (carboxylate) -pyran copolymer in a hydrolyzed form was found active against several cancerous tumors (Ghosh and Sulumar, 1985; Baldwin *et al.*, 1985). A minimum molecular weight of 30,000 as well as an ordered, dense sequence of negative charges are required. Natural and synthetic polymers of nucleic acid may act as anticancer agents by induction of interferon. Fluorinated polynucleotides have been shown to be cytotoxic to leukemia cells (O'Neill, 1979). Anthracycline derivatives have been complexed with DNA extracted from herring sperm and studied for treatment of a wide variety of malignancies (Trouet *et al*, 1979) (Table 11.5).

Another approach involves the copolymerization of an antineoplastic agent with a monomer. Hartsough and Gebelein (1985) report the copolymerization of 5-fluorouracil into an acrylic polymer backbone. Hydrolysis of these copolymers showed zero-order release of 5-fluorouracil for periods of up to 240 hr *in vitro*.

A variety of natural systems have been used to release chemicals, antibiotics, and enzymes that have antineoplastic properties. Lectins are proteins derived from plants, bacteria, or mammals that are well known for their ability to cause cells to clump together. *L*-asparaginase is an enzyme that has been shown to cause regression of tumors involving the lymph nodes (lymphomas) and white blood cells (leukemias). Shier (1979) reviewed the conjugation of *L*-asparaginase to lectins.

Gelatin, the nontriple helical form of collagen, as well as albumin, a protein found in the blood, have been used to form microspheres for delivery of antineoplastic agents. Kato (1983) reviewed the use of albumin microspheres to release 5-fluorouracil. Sugibayashi *et al.* (1977) reported on the tissue distribution of albumin microspheres containing this drug and the antitumor properties (Morimoto *et al.*, 1980). Gelatin sponge has been used to encapsulate mitromycin C and deliver it using a catheter into the blood for treatment of renal cell carcinoma (Kato *et al.*, 1979). Using a similar system the release of mitromycin C was magnetically controlled (Kato *et al.*, 1981) (Table 11.5).

Antibodies are blood-borne proteins that react with soluble foreign molecules or receptors on cell membranes. In theory, they can be specifically targeted to an individual cell line and have been covalently combined with adrinamycin and daunomycin, two drugs with anticancer properties (O'Neill, 1979).

Liposomes are small phospholipid vesicles which are formed into bilayers similar to a cell membrane. Entrapment of actinomycin D, an anticancer drug, in liposomes does not change the therapeutic index, however, it also does not prevent the development of resistance to this drug (Ryman, 1983). Other workers used liposomes to release cytosine arabinoside (Rustum *et al.*, 1979), actinomycin D (Gregoriadis, 1979) and methotrexate (Kimelberg, 1980).

11.6. Growth Factors

A number of factors that cause division of cells found in connective tissue have been reported. These factors cause mitosis of specific cell types as well as lead to biosynthesis and deposition of connective tissue macromolecules

(Gospodarowicz, 1981; Sporn and Roberts, 1986). These factors include epidermal growth factor (EGF), platelet derived growth factor (PDGF), fibroblast growth factor (FGF), transforming growth factors α and β (TGFα ,TFGβ), and cartilage-derived growth factor (CDGF) and are listed in Table 11.6.

Epidermal growth factor stimulates proliferation of a variety of cell lines including the basal cell layer from skin epidermis, corneal endothelial cells, vascular smooth muscle, chondrocytes, and fibroblasts (Gospodarowicz, 1981). EGF is a peptide containing 53 amino acids that is secreted by a variety of cells.

Platelet-derived growth factor is a cationic protein with a molecular weight of 33,000 (Helden *et al.*, 1981). On the molecular level it is a dimer of two chains A and B in the form of A:A, B:B, or A:B. It is chemotactic to fibroblasts, smooth muscle cells, and inflammatory cells (VanBrunt and Klauser, 1988). PDGF is released from the α-granules in platelets as well as from activated macrophages (Sporn and Roberts, 1986).

Fibroblast growth factor causes division of vascular endothelial cells (Sporn and Roberts, 1986) and results in new blood vessel formation (VanBrunt and Klauser, 1988). In addition, it causes fibroblasts to migrate and divide. FGF has been isolated in acidic and basic forms.

Transforming growth factor-α is a structural analog of EGF and binds to the same cell-surface receptor. It is a separate gene product from EGF and is found in a 50-amino-acid form after processing (Sporn and Roberts, 1986). It is found in platelets as well as activated macrophages and stimulates endothelial cells and collagen synthesis (VanBrunt and Klauser, 1988).

Transforming growth factor-β is a peptide with a molecular weight of 25,000 composed of two chains containing 112 amino acids each, that causes proliferation of certain types of cells including fibroblasts (Sporn *et al.*, 1986); however, it also inhibits proliferation of hepatocytes, embryonic fibroblasts, T and B lymphocytes, and keratinocytes. It is found in serum and in α-granules of platelets, in macrophages and activated T-lymphocytes and blocks activity of PDGF, TGF-α, and FGF on specific cell types (Sporn and Roberts, 1986). Increased production of collagen and fibronectin by fibroblasts and osteoblasts occurs in a dos-dose-dependent fashion by increasing the RNA levels (Varga *et al.*, 1987).

Cartilage-derived growth factor is a cationic polypeptide of molecular weight 18,000 derived from articular, scapular, and costal cartilage (Davidson *et al.*, 1985). It stimulates proliferation and migration of cultured fibroblasts, chondrocytes, endothelial, and synovial cells. It regulates the type and amount of proteoglycans that are synthesized by chondrocytes (Hammerman *et al.*, 1986).

Sustained Release of Growth Factors

Recent interest in sustained release of growth factors has developed from the potential of these natural substances to enhance the healing of surgical and chronic wounds (Table 11.5). For instance TGF-β stimulates collagen deposition and has been suggested in the therapy for osteoporosis, a disease where bone density is decreased (Sporn and Roberts, 1986).

Animal and clinical studies have been completed to evaluate the use of EGF in healing corneal defects. Brightwell *et al.* (1985) demonstrated that EGF in combination with a synthetic steroid accelerates the rate of corneal epithelial regeneration. Brown *et al.* (1986) combined EGF with a water-miscible base

Table 11.6. Growth Factors and Their Biological Effects

Factor	Cellular effect
Cartilage-derived growth factor (CDGF)	Proliferation and migration of cultured fibroblasts, chondrocytes, and endothelial cells
Epidermal growth factor (EGF)	Proliferation of skin epidermal cells, corneal endothelial cells, vascular smooth muscle, chondrocytes, fibroblasts
Fibroblast growth factor (FGF)	Proliferation of vascular endothelium; migration and proliferation of fibroblasts
Platelet-derived growth factor (PDGF)	Attracts fibroblasts, smooth muscle cells, and inflammatory cells
Transforming growth factor- α (TGF-α)	Analog of EGF
Transforming growth factor-β (TGF-β)	Proliferation of specific fibroblast cell lines; inhibits proliferation of hepatocytes, embryonic fibroblasts, T and B lymphocytes, and keratinocytes; blocks action of PDGF, TGF-α, and FGF

containing 1% silver sulfadiazine for topical treatment to enhance epithelial regeneration of split-thickness wounds.

11.7. Biocompatibility of Polymers Used in Drug Delivery Systems

A variety of synthetic and natural polymers have been used for sustained delivery. These polymers are listed in Table 11.3 and discussed in Sections 11.4 through 11.5.1. In this section results of standard biocompatibility assays will be presented.

Skin irritation is normally measured using a standard skin patch test in which a known amount of polymer in film form is placed in contact with a region of abraded skin (McCreesh and Steinberg, 1987). This test is sometimes referred to as the "Draize" test and was developed by Draize and coworkers in 1944 (Draize *et al.*, 1944).

According to the 1980 United States Code of Federal Regulations (Code of Federal Regulations, 1980) primary irritation to skin is to be measured by a patch test on abraded intact skin of the albino rabbit, clipped free of hair. One half

Table 11.7. Grading Values for Skin Reactions in Albino Rabbits following Topical Application of Potential Primary Irritants[a]

Skin reaction	Value[b]
Erythema and eschar formation	
No erythema	0
Very slight erythema (barely perceptible)	1
Well-defined erythema	2
Moderate to severe erythema	3
Severe erythema (beet redness) to slight eschar formations (injuries in depth)	4
Edema formation	
No edema	0
Very slight edema (barely perceptible)	1
Slight edema (edges of area well defined by definite raising)	2
Modetate edema (raised approximately 1 mm)	3
Severe edema (raised more than 1 mm and extending beyond the area of exposure)	4

[a] Code of Federal Regulations, 1980.

[b] The value recorded for each reading is the average value for the six or more animals subject to the test.

gram of the material to be tested is placed on the abraded skin and covered with a 1" x 1" patch of surgical gauze and secured by placing adhesive tape over the edges of the gauze. After 24 hr the gauze is removed and the degree of irritation is graded by comparison with Table 11.7. The patch is covered and readings are made again at 72 hr.

The manner in which the federal regulation is written leaves open some aspects of the testing procedure. Abrasion is normally produced by rubbing the tip of a needle over the skin of a rabbit using a syringe needle or plastic scraper. The degree of abrasion is difficult to standardize and may vary from animal to animal.

Irritation produced using the "Draize" test on rabbits tends to underpredict the degree of human skin irritation (Nixon *et al.*, 1975; Philips *et al.*, 1972). The severity of the test can be increased by placing an occlusive film over the gauze dressing to increase the degree of skin penetration and by repeatedly applying the test material for a period of as long as 21 days.

Leachates of the material are tested in a similar fashion in the skin patch test. The leachate can be dissolved in polyethylene glycol or aqueous solvent and is applied to the skin. Acute hypersensitivity to the extract is tested by repeated inoculation of the leachate subcutaneously followed by patch testing.

11.7.1. Skin Patch Testing of Polymeric Implants

Polymeric materials used in transdermal drug delivery should not cause well-defined redness (erythemia) or moderate swelling (edema). The presence of redness in a skin patch test suggests that the polymer or a leachable additive is an irritant. In contrast, a polymeric material or leachate that induces swelling clinically results in contact dermatitis. In general, high-molecular-weight polymeric materials do not in themselves act as irritants or result in contact dermatitis (Fisher, 1986). Irritation and sensitization that is induced by contact with polymeric materials is a result of leachable low-molecular-weight molecules including curing agents, stabilizers, plasticizers, catalysts, and antioxidants.

Medical devices are manufactured using polymers that are prepared containing few, if any, additives therefore medical grade polymers are not highly reactive on skin contact. Data presented below are reported for both medical and nonmedical grade polymers.

Vinyl polymers including polyvinyl chloride, polyvinyl acetate, and polyvinyl alcohol do not sensitize or irritate skin; however, vinyl chloride monomer and other additives may cause dermatitis as well as toxicity to a variety of other tissues and organs (Fisher, 1986; Osbourn, 1964; Sculsinger and Mollgaard, 1980; Walker, 1975). Additives or monomers known to cause dermatological changes on contact are listed in Table 11.8.

Epoxy polymers are polymerized from bisphenol-A and epichlorohydrin in the presence of base. Low-molecular-weight liquid polymers can penetrate the skin and can act as a sensitizer (Fisher, 1986; Fregert and Thorgeirsson, 1977). After crosslinking it is believed that epoxy resins are unreactive; however, epoxy resin dermatitis is observed and believed to be due to release of free bisphenol-A and curing agents (Table 11.8).

Formaldehyde-based polymers or the monomers may induce skin reactions. Phenol-formaldehyde polymers such as Bakelite and the monomers including Bisphenol-A, partertiary phenyl phenol, or para tertiary butyl-phenol are skin irritants (Fregert, 1980; Adams, 1975; Foussereau *et al.*, 1976; Mobacken and Hersle, 1976). This group of polymers and monomers is in general not good for skin contact. Urea-formaldehyde based fibers produce textile dermatitis by breaking down and releasing formaldehyde.

Acrylic polymers containing free methyl methacrylate monomers or free-radical-generating crosslinking compounds including benzyl peroxide can cause skin irritation (Nethercott, 1979; Bjorkner *et al.*, 1980; Pegum and Medhurst, 1971; Fries *et al.*, 1975).

Polyesters such as polyethylene terephthalate (Dacron ®) when completely polymerized are not dermatologic hazards (Fisher, 1986). However, components used in the manufacture of polyesters including phthalic anhydride, polyethylene glycol, styrene, peroxide catalysts, oils, and surfactants do cause skin irritation (Key and Discher, 1966; Bourne and Milner, 1963; Sjoberg *et al.*, 1982). Methyl chloride and styrene monomer released from polystyrene can produce irritation to a variety of tissues either directly or in combination with other chemicals.

Polyurethanes are polymerization products of a polyol (compound containing more than one hydroxyl group) and an isocyanate. Isocyanates are, in general, very reactive chemicals which are mild skin sensitizers but strong pulmonary

Table 11.8. Irritants and Sensitizers Contained in Polymers

Polymer	Chemical agent	Effect on skin
Poly (vinyl acetate)	Dibutyl tin maleate Dibutyl sebacate Glyoxal	Dermatitis Dermatitis Allergic reaction
Poly (vinyl chloride)	Vinyl chloride	Liver tumor, bone lesion, scleroderma*
Epoxy resin	Bisphenol A amines, amides, anhydrides and low -molecular-weight polymer	Dermatitis Dermatitis
Phenol-formaldehyde	polymer, Bisphenol A, para-tertiary butyl-phenol, para-tertiary phenyl phenol, formaldehyde	Dermatitis
Acrylic	Methyl methacrylate, benzyl peroxide cyanoacrylic polymers	Dermatitis, neurologic effects
Polyester	Polyethylene glycol styrene, peroxides, oils, surfactants methyl chloride	dermatitis
Polyurethanes	Toluene-diisocyanate, 4-cyclohexyl-isocyanate, amine catalysts	Dermatitis, lung irritants

[a] scleroderma, thickening and stiffening of skin

allergens (Mowe, 1980; Emmett, 1976; Malten, 1977). Amines used to catalyze the polymerization reaction also contribute to skin irritation (Lachapelle and Ketclear, 1979) (Table 11.8).

11.7.2. Biocompatibility of Polymers Used In Subcutaneous Delivery Systems

Biocompatibility assessment is more detailed for delivery systems that are implanted subcutaneously beneath the skin. In this case, screening initially involves cell culture studies to evaluate cell-polymer interactions followed by *in vivo* implantation to study similar interactions in animals.

Cell culture studies usually evaluate the effect of a polymer used in a delivery system on isolated cells. Studies using fibroblasts are aimed at evaluating the cytotoxic effects of a polymer or leachate as well as the effects on protein synthesis and cell replication. Addition of polymers and their leachate to cultures of inflammatory cells such as monocytes are studied for signs of inflammatory cell activation and proliferation. Cell culture experiments are used to predict the outcome of interactions *in vivo*; however, they sometimes do not reflect synergistic effects of different cells or of different cellular products.

In vivo tests are directed at evaluating the response of fibroblasts and endothelial and inflammatory cells to the implant. The response to a polymer can include a foreign-body response in which the polymer or device is surrounded by a connective tissue coating, a cytotoxic response in which fibroblasts and other cells around the implant are dying, or an inflammatory response in which the implant is surrounded by numerous neutrophils, monocytes, and other white blood cells.

11.8. Summary

Polymeric release systems have been developed for a number of medical applications. Transdermal or ocular sustained release can be accomplished via devices containing synthetic polymeric membranes. These membranes do not need to be biodegradable since they can easily be removed from the skin or eye. Delivery systems for drugs to treat glaucoma and angina have already been developed and are quite successful.

Release of drugs to internal organs or external wounds requires the use of biodegradable polymer delivery systems since the system cannot be easily removed. A variety of applications exist for the delivery of chemotherapeutic drugs, antibiotics, and growth factors. The development of these systems requires the synthesis of new biodegradable polymers and the careful exploration of well-characterized biodegradable polymers.

References

Abrahams, R.A. and Ronel, S.M. J. Biomed. Mat. Res. 9,355,1975.

Adams, R.M. Contact Dermatitis 1,321,1975.

Alderfer, J.A., Loomis, R.E., Soni, S.D., Sharma, M., Bernacki, R. and Hughes, R., Jr. In: Polymeric Materials In Medication, edited by C.G. Gebelein and C.E. Carraher, Jr., Plenum Press, New York, p. 125,1985.

Anderson, L.C., Wise, D.L. and Howes, J.F. Contraception 13,375,1976.

Aoshima, R., Kanda, Y., Takada, A. and Yamashita, A. J. Biomed. Mat. Res. 16,289,1982.

Apple, D.J., Mamalis, N., Brady, S.E., Loftfield, K., Kavka-Van Norman, D. and Olson, R.J. Am. Intra—Ocular Implant Soc. J. 10,53,1984.

Bakan, J.A. and Powell, T.C. In: Controlled Release Systems, edited by T.J. Roseman and S.Z. Mansdorf, Marcel Dekker, Inc. New York, p. 177,1983.

Baldwin, J.R., Carrano, R.A., Imondi, A.R., Iulucci, J.D. and Hagerman, L.M. In: Polymeric Materials In Medication, edited by C.G. Gebelein and C.E. Carraher, Jr., Plenum Press, New York, p. 139,1985.

Beck, L.R., Cowsar, D.R., Lewis, D.H., Gibson, J.W. and Flowers, C.E. Amer. J. Obstet. Gyn. 135,419,1979.

Beck, L.R., Ramos, R.A., Flowers, C.E., Lopez, G.Z., Lewis, D.H. and Cowsar, D.R. Amer. J. Obstet. Gynecol. 140,799,1981.

Bensinger, R., Shin, D.H., Kass, M.A., Podos, S.M., and Becker, B. Invest. Ophthalmol.15,1008,1978.

Bjorkner, B., Dahlquist, I. and Fregert, S. Contact Dermatitis 6,405,1980.

Blackshear, P.J. Sci. Amer. 241,66,1979.

Blackshear, P.J. and Rohde, T.D. In: Controlled Drug Delivery, Vol. II, edited by Stephen D. Bruck, CRC Press, Boca Raton, Florida, Chapter 5,1983.

Bourne, L.B. and Milner, F.J. Brit. J. Ind. Med. 20,100,1963.

Boylan, J.C. and Banker, G.S. J. Pharmaceut. Sci. 62,1177,1973.

Bradley, S.A., Engler, P. and Carr, S.H. Appl. Polym. Symp. 22,269,1973.

Brightwell, J.R., Riddle, S.L., Eiferman, R.A., Valenzuela, P., Barr, P.J., Merryweather, J.P. and Schultz, G.S. Invest. Ophthamol. Vis. Sci. 26,105,1985.

Brown, G.L., Curtsinger, L.,III, Brightwell, J.R., Ackerman, D.M., Tobin, G.R.,

Polk, H.C., Jr., George—Nascimento, C., Valnzuela, P. and Schultz, G.S. J. Exp. Med. 163,1319,1986.

Bruck, S.D. (editor) Controlled Drug Delivery, Vols. I and II, CRC Press, Boca Raton, Florida, 1983a.

Bruck, S.D. In: Controlled Drug Delivery, edited by S.D. Bruck, CRC Press, Boca Raton, Florida, Chapter 1,1983b.

Chien, Y.W. (ed) Transdermal Controlled Systemic Medications, Marcel Dekker, Inc., N.Y.,1987.

Chien, Y.C. Novel Drug Delivery Systems Fundamentals, Developmental Concepts, Biomedical Assessments, Marcel Dekker, New York.,1982.

Chvapil, M., Kronenthal, R.L., and van Winkle, W. Int. Rev. Connect. Tiss. Res. 6,1,1973.

Code of Federal Regulations. Title 16, part 1500.41, 1980.

Coutinho, E.H., Ferreira, D.A.M., Prates, H. and Kind, F. J. Reprod. Fert. 23,345,1970.

Cowsar, D.R., Tarwater, O.R. and Tanquary, A.C. In: Hydrogels for Medical and Related Applications, edited by J.D. Andrade, ACS, Washington. p. 180,1976.

Crean, P.A., Crea, F., Dearfield, J., Sharom, M., Davies, G. and Maseri, A. Amer. J. Cardiol. 55,900,1985.

Creque, H.M., Langer, R. and Folkman, J. Diabetes 29,37,1980.

Croxatto, H., Diaz, S., Vera, R., Etchart, M. and Atria, P. Amer. J. Obstet. Gyn. 105,1135,1969.

Davidson, J.M., Klagsburn, M., Hill, K.E., Buckley, A., Sullivan, R., Brewer, P.S. and Woodward, S.C. J. Cell Biol. 100,1219,1985.

Desai, S.J., Simonelli, A.P. and Higuchi, W.I. J. Pharm. Sci. 54,1459,1965.

Diaz, S., Parez, M., Qunteros, E., Robertson, D.N. and Croxatto, H.B. Contraception 16,155,1977.

Dickinson, H.R., Hiltner, A., Gibbons, D.F. and Anderson, J.M. J. Biomed. Mat. Res. 15,577,1981.

Doillon, C.J., Whyne, C., Berg, R.A., Olson, R.M. and Silver, F.H. Scan. Elect. Microsc. 11,1313,1984.

Draize, J.H., Woodard, G. and Calvery, H.O. J. Pharmacol. Exp. Ther. 82,377,1944.

Drug Evaluations, American Medical Association, Chicago, Illinois, 6th edition, 1986.

Ebert, C.D., McRhea, J.C. and Kim, S.W. In: Controlled Release of Bioactive Materials, edited by R.W. Baker, Academic Press, New York, p.107,1980.

Eckenhoff, B. and Yum, S.I. Biomaterials 2,89,1981.

Emmett, E.A. J. Occup. Med. 18,802,1976.

Fisher, A.A. Contact Dermatitis, Lea and Febiger, Philadelphia, Pennsylvania, p. 546,1986.

Folkman, J. and Long, D.M. J. Surg. Res. 4,139,1964.

Foussereau, J., Cavelier, C. and Selig, D. Contact Dermatitis 2,254,1976.

Frazza, E.J. and Schmitt, E.E. J. Biomed. Mat. Res. Symp. 1,43,1971.

Fregert, S. and Thorgeirsson, A. Contact Dermatitis 3,301,1977.

Fregert, S. Contact Dermatitis 6,493,1980.

Frick, J., Bende, T., Aulitzsky, H. and Schmidt, F. Eur. Urol. p. 73,1981.

Fries, J.B., Fisher, A.A., and Salvati, E.A. J. Bone Jt. Surg. 57—A,547,1975.

Ghosh, M. and Sulumar, M. In: Polymeric Materials in Medication, edited by C.G. Gebelein and C.E. Carraher, Jr., Plenum Press, New York, p. 103,1985.

Goodman, H. and Banker, G.S. J. Pharmaceut. Sci. 59,1131,1970.

Goodson, J.M., Holborrow, D., Dunn, R.L. and Dunham, S. J. Periodontal 54,575,1983.

Gospodarowicz, D. Ann. Rev. Physiol. 43,251,1981.

Gourlay, S.J., Rice, R.M., Hegyeli, A.F., Wade, C.W.R., Dillon, J.G., Jaffe, H. and Kulkarni, R.K. J. Biomed. Mat. Res. 12,219,1978.

Graham, N.B. and Wood, D.A. In: Macromolecular Biomaterials, edited by G.W. Hastings and P. Ducheyne, CRC Press, Boca Raton, Florida, Chapter 8, 1984.

Graham, N.B. and McNeill, M.E. Biomaterials 5,27,1984.

Gregoriadis, G. In: Drug Carriers in Biology And Medicine, edited by G. Gregoriadis, Academic Press, New York, p. 324,1979a.

Gregoriadis, G. In: Drug Carriers in Biology and Medicine, edited by G. Gregoriadis, Academic Press, New York. p. 287,1979b.

Grode, A.S., Anderson, S.J., Grotta, H.M., and Falk, R.D. Trans. Amer. Soc. Artif. Int. Organs, XV,1,1968.

Hammerman, D., Sasse, J. and Klagsburn, H. J. Cellular Physiology 127,317,1986.

Hartsough, R.R. and Gebelein, C.G. In: Polymeric Materials in Medication, edited by C.G. Gebelein and G.E. Carraher, Jr., Plenum Press, New York, p. 115,1985.

Helden, C.—H., Westmark, B. and Wasteson, A. Biochem. J. 193,907,1981.

Heller, J. Biomaterials 1,51,1980.

Heller, J., Helwing, R.F., Baker, R.W. and Tuttle, M.E. Biomaterials 4,262,1983.

Horsman, A., Jones, M., Francis, R. and Nordin, C. N. Engl. J. Med. 309,1405,1983.

Hsieh, D.S., Langer, R. and Folkman, J. Proc. Natl. Acad. Sci., USA, 78,1863,1981.

Hsieh, D.S.T. and Langer, R. In: Controlled Release Delivery Systems, edited by T.J. Roseman and S.Z. Mansdorf, Marcel Dekker, New York, p. 121,1983.

Jackanicz, T.M., Nash, H.A., Wise, D.L. and Gregory, J.B. Contraception 8,227,1973

Kato, T., Nemoto, R., Mori, H., Unno, K. Goto, A., Harada, M. and Homma, M. Proc. Jpn. Acad. 558,470,1979.

Kato, T., Nemoto, R., Mori, H. Abe, R., Unno, K., Goto, A., Murota, H., Harada, M., Kowamura, K. and Homma, M. J. Jpn. Cancer Ther. 16,1351,1981.

Kato, T. In: Controlled Drug Delivery, edited by S.D. Bruck, CRC Press, Boca Raton, Florida, p. 189,1983.

Key, M.M. and Discher, D.P. Cutis 2,1,1966.

Khanna, S.C. and Speiser, P. J. Pharmaceut. Sci. 59,1398,1970.

Kimelberg, H.K. In: Liposomes in Biological Systems, edited by G. Gregoriadis and A.C. Allison, p. 219,1980.

Kincl, F.A., Benagiano, G. and Angee, I. Steroids 11,673,1968.

Klein, A.W. J. Amer. Acad. Dermatol. 9,224,1983.

Kligman, A.M. American Heart Journal 108,200,1984.

Kopecek, J. Biomaterials 5,19,1984.

Kulkarni, R.K., Pani, K.C., Newman, C. and Leonard, F. Arch. Surg. 93,839,1966.

Kulkarni, R.K., Moore, E.G., Hegyeli, A.F. and Leonard, F. J. Biomed. Mat. Res. 5,169,1971.

Lachapelle, J.M. and Ketclear, M.J. Cont. Dermatitis 5,55,1979.

Langer, R. and Folkman, J. Nature 263,797,1976.

Langer, R., Rhine, W., Hsieh, D.S.T. and Folkman, J. J. Membr. Sci. 7,333,1980.

Langer, R.S. and Peppas, N.A. Biomaterials 2,201,1981.

Langer, R., Brem, H. and Tapper, D. J. Biomed. Mat. Res. 15,267,1981.

Lucas, M.S. and Moore, D.J. J. Prosthet. Dent., 42,447,1979.

Lyman, D.J., Hill, D.W., Stirk, R.K., Adamson, C. and Mooney, B.R. Trans. Am. Soc. Artif. Intern. Organs, 18,19,1972.

Malten, K.E. Cont. Dermatitis 3,344,1977.

Marchant, R., Hiltner, A., Hamlin, C., Rabinovitch, A., Slobodkin, R. and Anderson, J.M. J. Biomed. Mat. Res. 17,301,1983.

Mason, N., Thies, C. and Cicero, T.J. J. Pharm. Sci., 65,847,1976.

McCreesh, A.H. and Steinberg, M. In: Dermato—toxicology edited by F.N. Marzulli and H.I. Maibach, Hemisphere Publishing Co., New York, New York, p. 153,1987.

Mijama, H., Harumiya, N., Mori, Y. and Tanzawa, H. J. Biomed. Mater. Res. 11,251,1977.

Mobacken, H. and Hersle, K. Cont. Dermatitis 2,59,1976.

Monroe, J.M., Ligana, R.C. and Williams, M.C. Biomat. Med. Dev. Artif. Organs 8,103,1980.

Mook, P.H., Wong, P., Wildevuuer, C.R., Mayes, P.J. and Gaylor, J.D. Trans. Am. Soc. Artif. Intern. Organs 29,215,1983.

Morimoto, Y., Kimoto, M., Sugibayashi, K., Nadai, T. and Kato, Y. Chemc. Pharm. Bull. 28,3087,1980.

Mowe, G. Cont. Dermatitis 6,44,1980.

Nash, H.A., Robertson, D.N., Young, A.J.M. and Atkinson, L.F. Contraception 18,367,1978.

Nethercott, J.R. Br. J. Dermatol. 98,541,1979.

Nilsson, C.G., Johansson, E.D., Jackanicz, T.M. and Luukkainen, T. Am. J. Obstet. Gyn. 122,90,1975.

Nixon, G.A., Tyson, C.A. and Wertz, W.C. Toxicol. Appl. Pharmacol. 31,481,1975.

Olanoff, L. and Anderson, J.M. J. Pharmaceut. Sc. 68,1151,1979.

O'Neill, G.J. In: Drug Carriers in Biology and Medicine, edited by G. Gregoriades, Academic Press, New York, p. 27,1979.

Osbourn, R.A. J. A. M. A 188,1159,1964.

Parker, J.O. and Fung, H—L. Amer. J. Cardiol. 54,471,1984.

Pegum, J.S. and Medhurst, F.A. Br. Med. J. 2,141,1971.

Pharris, B.B., Place, V.A., Sendelbeck, L. and Schmitt, E. J. Reprod. Med. 17,91,1976.

Philips, L., Steinberg, M., Maibach, H.I. and Akers, W.A. Toxicol. Appl. Pharmacol. 21,369,1972.

Preis, I. and Langer, R. J. Immunol. Meth. 28,193,1979.

Rezakovic, D.E. In: Transdermal Controlled Systemic Medications, edited by Y.W. Chien, Marcel Dekker, New York, p. 245,1987.

Robbins, S.L. and Kumar, V. Basic Pathology, W.B. Saunders Co., Philadelphia, Pennsylvania p. 319,1987.

Rudolf, R. and Abraham, J. Ann. Plast. Surg. 4,14,1980.

Rustum, Y.M., Dave, C., Mayhew, E. and Papahajopoulas, D. Cancer Res. 39,1390,1979.

Ryman, B.E. In: Controlled Release Systems, edited by T.J. Roseman and S.Z. Mansdorf, Marcel Dekker, New York, p. 27,1983.

Salthouse, T.N. and Matlaga, B.F. Surg. Gyn. Obstet. 142,544,1976.

Schulsinger, C. and Mollgaard, K. Cont. Dermatitis, 6,477,1980.

Schwope, A.D., Wise, D.L. and Howes, J.F. Life Sci. 17,1877,1975.

Sharma, R. and Vasudevan, P. Biomaterials 3,109,1982.

Shakespeare, P.G. and Griffiths, R.W. Lancet 1,795,1980.

Shier, W.T. In: Drug Carrier in Biology and Medicine, edited by G. Gregoriadis, Academic Press, New York, p. 60,1979.

Sidman, K.R., Steber, W.D., Schwope, A.D. and Schnaper, G.R. Biopolymers 22,547,1983.

Sjoberg, S., Dahlquist, I., Fregert, S. and Trulson, L. Cont. Dermatitis 8,207,1982.

Sparer, R.V., Ekwuribe, N. and Walton, A.G. In: Controlled Release Delivery Systems, edited by T.J. Roseman and S.Z. Mansdorf, Marcel Dekker, New York, p. 107,1983.

Sporn, M.B. and Roberts, A.B. J. Clin. Invest. 78,329,1986.

Sporn, M.B., Roberts, A.B., Wakefield, L.M. and Assolan, R.K. Science 233,532,1986.

Stewart, R.H. and Novak, S. Ann. Ophthalmol. 10,325,1978.

Stewart, R.H. and Novak, S. Ann. Ophthalmol. 85,794,1978.

Sugibayashi, K., Morjmoto, Y., Nadai, T. and Kato, Y. Chem. Pharm. Bull. 25,3433,1977.

Tanzawa, H., Mori, Y., Harumiya, N., Miyama, H., Hori, M., Ohshima, N. and Idezuki, Y. Trans. Amer. Soc. Artif. Intern. Organs XIX,188,1973.

Taylor, J.A., Abodeely, R.A. and Fuson, R.L. Trans. Amer. Soc. Artif. Intern. Organs 19,175,1973.

Theeuwes, F. and Eckenhoff, B. In: Controlled Release of Bioactive Materials, edited by R. Baker, Academic Press, New York, 1980.

Trouet, A., Camprez—De Campeneere, D., Baurain, R., Huybrechts, M. and Zenebergh, A. In: Drug Carriers in Biology and Medicine, edited by G. Gregoriadis, Academic Press, New York, p. 87,1979.

Tuttle, M.E., Baker, R.W. and Laufe, L.W. J. Membr. Sci. 7,351,1980.

Upton, G.V. J. Reprod. Med. 27,1,1982.

VanBrunt, J. and Klauser, A. Biotechnology 6,25,1988.

Varga, J., Rosenbloom, J. and Jimenez, S.A. Biochem. J. 247,597,1987.

Victor, A., Nash, H.A., Jackamiez, T.M. and Johansson, E.D.B. Contraception 16,125,1977.

Walker, A.E. Br. J. Dermatol. 93,22,1975.

Weadock, K.S., Ph.D. Thesis in Biomedical Engineering, Rutgers University, May, 1986.

Weiss, I., Wolff, H.-I, Cordes, G. and Cawello, W. In: Transdermal Controlled Systemic Medications, edited by Y.W. Chien, Marcel Dekker, New York, Chapter 12,1987.

Williams, D.F. Proceedings of the Conference on Plastics Medicine and Surgery, Plastics and Rubber Institute, London, p. 5,1979.

Wise, D.L., McCormick, G.J., Willet, G.P., Anderson, L.C. and Howes, J.F. J. Pharm. Pharmacol. 30,686,1979.

Woodland, J.H.R., Yolles, S., Blade, D.A., Helrich, M. and Meyer, F.J. J. Med. Chem. 1,897,1973.

Worthen, D.M., Zimmerman, T.J. and Wind, C.A. Invest. Ophthalmol. 13,296,1974.

Yale, B.H. and Greer, R.T. J. Biomed. Mat. Res. 16,171,1982.

Yen, S.P.S. and Rembaum, A. J. Biomed. Mat. Res. 1,83,1971.

Yolles, S., Eldridge, J.E., and Woodland, J.H.R. Polymer News 1,9,1970.

Yolles, S., Leafe, T.D. and Meyer, F.J. J. Pharm. Sci. 64,115,1975a.

Yolles, S., Leafe, T.D., Woodland, J.H.R. and Meyer, F.J. J. Pharm. Sci. 64,348,1975b.

Zador, G., Nilsson, B.A., Nilsson, B., Sjobert, N.D., Westrom, L. and Eiese, J. Contraception 13,559,1976.

Chapter 12

Metals, Ceramics, Composite Materials

12.1. Introduction

Volume 1 of this series is devoted to polymers, long chain molecules, which have numerous applications in replacement of soft biological tissues. Use of this type of material has led to successful replacement of tissues such as skin, tendon, ligament, blood vessels, nerve, and cartilage. However, because of the inherent low tensile strength and modulus and high elongation at failure, polymers by themselves are insufficient for the replacement of hard tissue such as bone and tooth. For this reason Volume 2 of this series is devoted to metals, ceramics, and composites, materials that have tensile strengths and moduli significantly higher than those found for polymers.

In addition to evaluation of the physical structures and properties of metals, ceramics, and composites, Volume 2 will present results of studies on biocompatibility and additional end-use applications of all materials. This will give the biomaterials scientist a better understanding of the breadth of the field.

12.1.1. Metals

Metals are crystalline materials composed of atoms that share outer valence electrons. Positive metal ions are arranged in different packing patterns (lattices also termed unit cells) depending on their size and whether one or more types of metal ions are present. Valence electrons hold the lattices together via what are termed metallic bonds with electrons that are free to move from metal ion to metal ion.

When two or more metal ions are melted and cooled they form mixtures termed alloys. If two atoms are roughly the same size and have similar bonding tendencies, they form similar crystal structures as formed by the individual pure atoms. However, if the atoms are of a different size they form crystalline structures that are not identical to those formed from the constituent atoms. In the case of iron, smaller carbon atoms are added to the crystalline structure to increase the resistance to deformation of individual rows of atoms. This results in increased stiffness and tensile strength.

Metals used as biomaterials include titanium and its alloys, cobalt-chromium alloys, stainless steel and iron alloys, tantalum, gold, and platinum.

12.1.2. Ceramics

Ceramics are materials that contain atoms that are held together by ionic and/or covalent bonds. They form crystal lattices similar to those formed by metallic atoms, but because of ionic charges present, they are difficult to deform via the mechanism where rows of atoms slide by their nearest neighbors. Due to the difficulty in deforming ceramics, they have higher strengths than polymers and metals, but fail at lower elongations and therefore are brittle.

Ceramic materials such as alumina and synthetic hydroxyapatite are used for a variety of hard-tissue applications. In these applications porous ceramics are desired so that cellular ingrowth can anchor the implant.

12.1.3. Composite Materials

Composites are, as their names implies, materials that contain one or more class of materials such as polymers, ceramics, and metals. The virtue of this class of material lies in the ability to design devices with physical properties that are not attainable using any single material alone.

Normally, composites consist of high-strength fibers (polymer, metallic, or ceramic) in a ductile matrix (usually polymeric but sometimes ceramic). The fibrous component contributes strength to the composite while the more ductile matrix prevents premature failure of the composite especially when it is loaded cyclically.

Composites used to replace tissues include carbon fiber reinforced materials such as poly (ethylene) and poly (tetrafluoroethylene) as well as composites of collagen and other naturally occurring polymers.

12.2. Summary

A variety of materials are used to replace damaged soft and hard tissues. These include polymers, metals, ceramics, and composites. Polymers are used to replace soft tissues but are not rigid enough to be used as a replacement for hard tissue. However, metals, ceramics, and composites have higher stiffnesses and tensile strengths, which makes them more suitable for hard tissue substitutes. Volume 2 of this series presents the chemistry, physics, biology, and engineering of the latter classes of materials.

INDEX